Research Methodology

Peter Pruzan

Research Methodology

The Aims, Practices and Ethics of Science

 Springer

Professor Emeritus Peter Pruzan, Sc.D., Ph.D.
Department of Management, Politics
  and Philosophy
Copenhagen Business School
Frederiksberg
Denmark

ISBN 978-3-319-80084-4   ISBN 978-3-319-27167-5   (eBook)
DOI 10.1007/978-3-319-27167-5

Printed on acid-free paper

This Springer imprint is published by SpringerNature
The registered company is Springer International Publishing AG Switzerland

Foreword

To perform "good research" in the natural sciences, the practitioner must draw upon an inquisitive mind, an appreciation of the methods, aims and limitations of science, and, of course, skill in applying the "tools of the trade."

This book provides an extremely well-written, lucid, and, quite often, thought-provoking approach to these prerequisites of "good research." It is distinctive in the diverse sets of topics it covers and in the various examples it draws upon from the different areas of the natural sciences. Another distinguishing characteristic is its appeal to the reader's reflection, not just on the "nuts and bolts" of science and questions of "how to," but also on more fundamental matters as to "why," even in sections dealing with such down-to-earth, practical matters as measurement, data collection, design of experiments, and testing of hypotheses. In addition, it provides valuable advice on a matter of importance for all researchers but particularly, graduate students in the natural sciences—how to write up one's findings in a form suitable for publication.

There are few, if any, books that address these needs in one place. This book succeeds on all these fronts.

I can illustrate the book's balancing the reflective with the practical by referring to the treatment of uncertainty in science. The motivation for choosing to work on a particular problem depends critically on a very good familiarity with already existing literature. The work may be done to test the predictions of a particular hypothesis, model or theory, or it may even end up in a discovery for which a new or an extended version of an existing model or theory may be required. In each case, it is necessary to know how compelling the evidence obtained is. The chapter dealing with probability and statistics is excellent from this point of view. Not only does it consider uncertainties in measurements, but also how to determine whether one's data can be said to confirm or reject the investigation's hypotheses. The so-called Frequentist and Bayesian approaches to probability and statistics are very clearly detailed, a rather unique feature of this book. Cautionary remarks about the pitfalls of blindly using various software for statistical analyses of data is a valuable addition.

Once a researcher has obtained new, reliable results, the next task is to present the results of the work in the form of a thesis or in the form of an article to be sent for publication in a peer-reviewed journal. In this endeavor, the Internet and other repositories of relevant information can be of great help. The author takes great pains to guide the researcher on how to acknowledge the sources one has drawn upon and how to avoid infringement of copyrights, plagiarism, etc. He even provides clear, detailed checklists to assist in the planning and writing of research proposals and dissertations, as well as in the oral defense of a thesis.

To pave the way for reflection on such more practical aspects of research methodology, the introductory sections of the book present and exemplify important topics in the philosophy of science. These include the distinction between science and non-science, the aims and claims of science, and the role of mathematics in science. The author also explains the basic concepts such as realism, reductionism, epistemology, and ontology.

These introductory sections provide a far broader and more inclusive perspective than one ordinarily meets in publications on research methodology; most books on the subject focus on a particular area of specialization, implicitly assuming as given, the efficacy and validity of the methods presented, and thereby essentially ignoring the role of the aspiring scientist's own thinking, qualifications, and motivations.

The book's closing chapter on ethics and responsibility in scientific research builds upon the preceding chapters. It combines reflection on ethics and *un*ethics in science as well as on one's personal responsibility as a scientist with practical, down-to-earth guidelines for ethical practice in research.

All in all, I can highly recommend the book, in particular to Ph.D. students in the natural sciences as well as to those interested in the philosophy of science. It describes in some detail what "good research" in natural sciences is all about. It is a unique, provocative, and illuminating read!

Dr. Mosur K. Sundaresan
Former Distinguished Research Professor
Theoretical Physics
Carleton University
Ottawa, Canada

Acknowledgements

I acknowledge the contributions provided by the scientists and philosophers of science I have drawn upon, both their written words and my interactions with a number of them. In particular, I acknowledge the support and inspiration provide by the author of the book's Foreword, former Distinguished Research Professor in Theoretical Physics, Dr. Mosur K. Sundaresan, Carleton University, Canada.

I note in this connection that it is one thing to receive inputs via reading and interactions, and it is quite a different matter to interpret and select amongst the enormous amount of information received. In this connection, the focus and the power to discriminate was a result of the demands to write so that my students could understand and be enriched by what they read. So I gratefully acknowledge the feedback provided by my students in several different fields and spanning a period of more than 50 years; they have taught me much of what I had to learn!

Finally, I most humbly and gratefully acknowledge the inspiration and guidance provided by Bhagavan Sri Sathya Sai Baba (1926–2011), Founder Chancellor of Sri Sathya Sai Institute of Higher Learning, India.

Contents

Chapter 1
Introduction

The enormous influence of modern science and technology on our abilities to work effectively, to communicate, to travel, to perform ordinary household duties, to investigate the depths of outer space as well as the mysteries of the foundations of matter and of life itself, is accompanied by a concomitant growing interest in the role of science on the human condition. For large segments of the world's population, scientific research, the systematic investigation of reality to advance knowledge, is regarded as having a privileged access to truth—to knowledge of the physical world and its inhabitants as well as of the universe. Indirectly, science also exerts a powerful influence on our fundamental perspectives on the purpose and quality of life. Therefore, reflection, in particular by scientists and students of the natural sciences, on the aims, powers and limitations of science and research is vital for the well-functioning of societies and of all of us.

This is especially relevant in our time due to the significant growth in specialization in science as reflected in the increase in the number of scientific disciplines/sub-disciplines, the number of scientists, and in the number of science journals and journal articles (Ware and Mabe 2009; 18, 23). A result is that younger scientists are learning more and more about less and less. The metaphor of not being able to see the wood for the trees is increasingly applicable. Our universities and research programs are "producing" scientists who lack a broader and more inclusive/holistic appreciation of science and its methodology—and of their own potentials for contributing to the development of science and to the common good.

It is the modest aim of this book on the methodological aspects of scientific research to stimulate and contribute to such reflection, especially by students of science and of the philosophy of science, but also by those who already are members of "the scientific community".

When I agreed in 2004 to develop a course in research methodology for M.Phil. and Ph.D. students in the natural sciences at Sri Sathya Sai Institute of Higher Learning in India, I anticipated that this would be an easy and straightforward task. I had spent most of my professional life working in academia, where I had seriously reflected on science and its epistemology—about science as a field of professional practice and about how scientists learn and know things about the world. I was mistaken; the task of developing such a course turned out to be

© Springer International Publishing Switzerland 2016
P. Pruzan, *Research Methodology*,
DOI 10.1007/978-3-319-27167-5_1

extremely demanding. While there exists an abundance of literature dealing with topics such as the philosophy of science, the history of science, and the sociology of science, and of course with the content and methods of the specific fields of science, there is very little literature dealing specifically with research methodology in the natural sciences. What I was looking for, and had difficulty finding, was a book that spanned the reflective, more fundamental aspects of research—a book that primarily focused on the methodology—the "why's" of one's methods —rather than on the more practical aspects—the specific methods, the "how to's". I note that such "why questions" are not only significant for the methodology of science; according to the renowned theoretical physicist and cosmologist Stephen Hawking, they are fundamental to all of science: "Up until now, most scientists have been too occupied with the development of new theories that describe *what* the universe is to ask the question *why*" (Hawking 1988; 174).

This was the background for the development of teaching notes that have been fine-tuned, based on more than 10 years of classroom experience with graduate students in the natural sciences. Hopefully, the result in book form will help you to improve the insights and knowledge you acquire via your participation in research —and therefore enable you to answer questions such as:

What do we really mean by science and scientific knowledge? What are the aims, claims and limitations of science and how can one distinguish between science and other forms of intellectual activity?

Does science have a metaphysical[1] basis?

Is there a "standard" approach to performing research that is widely accepted? In other words, are there general principles or "rules of the game" that one should follow when performing research?

If there are such general principles, are they more or less independent of one's major field of science and of one's culture? Or does each branch of science, perhaps even each specialization within a branch (e.g. quantum mechanics in physics, genetics in biology, artificial life in computer science, palaeontology in geology, polymer organic chemistry in chemistry, galactic astronomy in astronomy) have its own research methodology?

How can I carry out and present my research in such a way that more experienced peers will evaluate it as being "good science" and not "poor science", or even "non-science"?

More specifically, how can and should I choose and justify my choice of hypotheses or research questions? My data collection procedures? My experimental designs? My analyses and conclusions?

How can I provide rational arguments, based on my experiments and observations, that will be accepted as having scientific validity?

How can I use probability and statistics to take account of the many uncertainties that characterize research?

---

[1] What is shared by most definitions of the term *metaphysical* is its reference to the principles and fundamental nature of reality which transcend any particular field of science. Etymologically, the word derives from the Greek words "metá" (beyond or after) and "physiká" (physics) and refers to Aristotle's writings on matter that followed after his writings on "physics". Today "metaphysics" is understood as generalizations that characterize the fundamental nature of physical reality, being and truth—of ultimate reality.

How can I complement my specific and specialized research with an appreciation of the importance and demands of multidisciplinary and interdisciplinary research?

How should I plan my research, including writing my thesis, in such a way that I can complete my research project within the allotted time without having to experience fear and stress?

Is there a particular way that I should write my thesis/article/book? Are there internationally accepted guidelines for structuring scientific publications and are such guidelines independent of one's field of study?

Are there ethical issues I must pay attention to in my research and in my writing? To what extent should such issues and human values influence my research and its mediation?

Does being a scientist imply particular responsibilities to science as a field of inquiry, to the scientific community, and to society in general?

It must be emphasized here that this book is primarily oriented towards students and practitioners of the natural sciences, i.e. those dealing with the empirical study of the natural world/nature/the universe—including such broad domains as physics, chemistry, bioscience, geology, palaeontology, astronomy, oceanography and the like that are considered to follow laws of a natural origin. Although there are significant differences between the particular methods and techniques employed in these domains, for example between those that tend to rely on active experimentation (e.g. chemistry, medical science) and those that tend to rely more on more passive forms of observation (e.g. geology, astronomy), they nevertheless share many fundamental perspectives on the nature of science (as distinguished from non-science) and on how it is to be performed.

Thus, the book is not oriented towards the social sciences, including such domains as sociology, anthropology, archaeology, economics, management, law, political science, history, philology and the like. These tend to have more in common with each other as regards methodology than with the natural sciences. To a great extent this is due to the different emphasis the natural sciences and social sciences tend to put on the roles of experimentation and quantification, on the qualitative aspects of research, and on the nature of the subjects of their investigations (where the social sciences directly or indirectly study the activities of humans, other sentient beings, and of organisations, cultures and societies). Nevertheless, the chapters on Science; Scientific Statements, Uncertainty, Probability and Statistics; and Ethics and Responsibility in Scientific Research provide valuable background material for students and practitioners of the social sciences and can be used to complement the teaching materials that are specifically developed for them.

Neither does the book deal at length with the field of mathematics which, together with other formal science fields such as logic, theoretical computer sciences, information theory, microeconomics and systems theory is essentially deductive, whereby its reasoning and content are independent of empirical investigation.[2] I note

---

[2]Crowe (1997) challenges both the popular conception that deduction is the sole method of mathematics (pp. 260–1) and that the methodology of mathematics is radically different from that of natural science (pp. 271–3).

however that since mathematics is vital for many fields of science, Sect. 2.2.8 provides a (hopefully!) provocative discussion of the role of mathematics in the sciences and of the relationship between reality and mathematics.

It was originally tempting to try to develop a "nuts and bolts" course aimed at answering questions in scientific investigation dealing primarily with "how to"— just as a carpenter must learn how to cut wood and build doors. But a person pursuing a career in science, and particularly in scientific research, must be able to provide more than good answers to such "how to" questions; he or she must be able to deal with more fundamental questions dealing with "why" and "what"—and this implies developing what we can refer to as a "scientific mind-set". By this I mean developing a vocabulary and a way of looking at the world that enables one to become a reflective member of the "scientific community". In fact, contributing to the development of such a reflective scientific mind-set, whereby one is consciously aware of, reflects on, and, when appropriate, challenges the (often taken for granted) underlying assumptions and practices of science, is the primary aim, the overall ambition of this book; an alternative title I considered earlier was *Research Methodology—For Reflective Scientists*.

In recent years, almost all people who are recognized as being scientists have obtained a Ph.D.—otherwise they could not have obtained employment as a scientist and would have been compelled to seek another livelihood, no matter how potentially creative and productive they could have been as scientists. Nevertheless, and this may challenge the self-conception of many practitioners of science, it is my observation that most scientists have never gone through a methodological training that prepared them to develop such a reflective scientific mindset. Instead, they spent their early research time learning about a particular scientific domain and then, during their doctoral research, specialized in a specific, narrow and well-defined aspect of that domain. In fact, this is probably a major reason underlying the apparent dearth of good literature on Research Methodology. A result is that during their formative days most scientists follow a learning path comparable to one followed by an apprentice who trains to be a skilled craftsman.

This apprenticeship-like relationship in the natural sciences has typically been characterized by the Ph.D. student working on such a well-defined empirical investigation, where the goals, tools of investigation, its resources, limitations, etc. have been more-or-less defined or at least strongly influenced by his or her advisor. Such an approach to performing a research project tends to omit more fundamental inquiry as to the nature of science, its history, its presuppositions, its aims, its limitations, its relationships to other approaches to generating and evaluating knowledge, its ethical dimensions, and so on. The typical path appears to be primarily one of learning by doing, in relative isolation, under the critical supervision of a research guide.

This may be related to the very strong tendency in most, if not all, branches of science towards increased specialization referred to earlier—a movement that feeds on and contributes to the amazing developments in modern technology. According to one of the few modern and inspiring books I have found with the term "scientific method" in its title (Gauch 2003; xv): "The thesis of this book ... is that there exist general principles of scientific method that are applicable to all of the sciences, but excessive specialization often causes scientists to neglect the study of these general principles, even though they undergird science's rationality and greatly influence science's efficiency and productivity."

To this can be added that increased specialization has implications of a social nature: With an emphasis on specialization the scientist is not inspired to develop awareness as to both the potential effects of his research on others or on the world at large, and as to his responsibility as a scientist.

Therefore, the book in your hands aspires to provide information and reflections on research and its methodology in the natural sciences that can assist you to:

- appreciate the nature, rationale, aims and limitations of the natural sciences
- understand and live up to the demands of the scientific community as to quality research, in particular as regards justification, measurement and experimentation
- develop the qualities and competencies required to undertake such research in your field of study as well as to evaluate research performed by others
- develop an appreciation of differing approaches to uncertainty in research
- develop an awareness as to the ethical and spiritual dimensions of research
- cultivate skills in publishing the results of your research.

The perspectives and reflections provided build upon my own personal experience as well as on the experiences of others as presented in literature that directly or indirectly deals with the broad field of Research Methodology. In particular I have drawn upon the following publications as general background material: Chalmers (1999), Elliot and Stern (1997), Gauch (2003), Godding et al. (1989), Hacking (1983), Klemke et al. (1998), Kuhn (1970), Lee (2000), Penrose (1991), Reichenbach (1951) , as well as various websites; unless otherwise specified, all websites referred to were accessed in 2015. I will *not* specifically refer to such sources each and every time that I have derived inspiration from them; that would make the book extremely boring and difficult to read. Instead, I here broadly acknowledge their influence on and contributions to my writing. Of course, whenever I directly quote a passage from a reference or find it appropriate to refer to a specific page or section as providing inspiration, acknowledgement of the source will be made.

Chapter 2
Science

2.1 A Very Brief, Incomplete and Highly Selective History of Science

My experience indicates that Ph.D. students, and therefore, presumably, younger scientists in general, implicitly consider science to be a systematic, rational, time honoured and stable approach to generating knowledge, and that there is little need to consider its historical development. The lack of awareness among younger researchers of the major developments that have taken place in science is quite understandable. After all, the research fields they work in may be undergoing such rapid change that it appears that almost all of the results that are directly relevant for their research have been achieved and published in recent years. Furthermore, there may be little or no emphasis placed on the history of science in their curricula and their advisors may not raise questions that stimulate them to delve into historical matters.

Thus, although they clearly recognize that major changes have taken place as regards the tools of science as well as its results (theories, technologies, publications), there is but limited realization among younger scientists of the fact that the role of science in society and its fundamental perspectives on reality have undergone significant changes over time—and that in recent years some of its most fundamental characteristics have been seriously challenged.[1]

---

[1] A highly cited example of such challenges is the book *Against Method* (Feyerabend 1975). Based on his studies of the history of science, the noted and controversial Austrian-American philosopher of science Paul Feyerabend (1924–1994) argues that there are no general norms or rules of the game for science—that we cannot distinguish between science and non-science and that the special status of science today is not due to its methodology, but rather to its results. Instead he argues for an extreme form of methodological pluralism—that advocating a particular methodology for science will stifle scientific practice—and that "The only principle that does not inhibit progress is: *anything goes.*" (Ibid.; 23).

© Springer International Publishing Switzerland 2016
P. Pruzan, *Research Methodology*,
DOI 10.1007/978-3-319-27167-5_2

In the sequel, I will at various times discuss some of these challenges as they deal directly with several of the most important concepts in science, including how science progresses and the role of verification in science. Here however I will simply provide a very brief overview of the development of science as an introduction to the next section that deals with the fundamental question: "What is science?" I emphasize that these introductory perspectives on the history of science are *extremely* selective and are limited to reflections as to two epochs of great significance for the development of science as it is practiced in the world today: Ancient Greece and the period in Europe in the 16th and 17th centuries that later became known as the Scientific Revolution. Such a selective and capsulated view most certainly does injustice to the major developments in other parts of the world. I ignore for example the early and significant contributions to the development of science in the ancient Near East (Babylonia, Egypt), India, China and the Islamic world. And I also ignore the developments in science and in the philosophy of science since the Scientific Revolution. The reason is simply that, given the restrictions on time and space here, as well as my own cultural biases and limitations, I consider these two periods as the most important in order to develop an (albeit highly superficial) appreciation of the developments that have lead up to modern science, no matter where it is practiced.

Ancient Greece, where the corner stones were laid for what was later to be modern science, did not distinguish between science and philosophy, between objective and subjective, or between factual descriptive accounts of the world (e.g. in terms of structural properties of things and laws governing them) and an evaluative or even a moral interpretation of the world as embodying an order, pattern, beauty, purpose and even goodness. In the *Republic*, Plato (ca. 428–348 Before the Common Era, BCE) characterized what modern scientists tend to refer to as objective reality in terms of an idea or eternal form of "the good" as the ultimate basis of a unified, ordered whole. Therefore, according to Plato, in order to understand experience we must have knowledge of the purpose which pervades all experience and which unifies it in a coherent and meaningful fashion. From this perspective, empirical observation can at best provide only limited insights into reality. To obtain a fuller understanding of our experiences, the methods of "science" (the concept as we know it today did not exist then) must be complemented by insight into the more basic principles and patterns which govern everything. Therefore the emphasis was not on observation and experimentation but on deduction from first principles that were assumed to be true and were not questioned. In fact geometry with its clarity and proofs based on deduction was a model for Plato's and many other ancient Greek thinkers' vision of methodology. "The old rationalist Plato admired geometry and thought less well of the high quality metallurgy, medicine or astronomy of his day." (Hacking 1983; 4)

In contrast, Hippocrates (somewhere between ca. 450–370 BCE), commonly referred to as "the father of medicine" as a profession, is known for advocating the systematic study of clinical medicine whereby it became separated from superstition. He is perhaps most widely known for the so-called "Hippocratic oath" he established and which to this day serves as the basis for oaths taken by new doctors throughout the world as to their duties and ethical responsibilities. In particular, two of his statements regarding science in contrast to opinion and belief are still often cited to this day: "Science is the father of knowledge, but opinion breeds ignorance." (Warrell et al. 2005; 7) and "There are in effect two things: to know and to believe one knows. To know is science. To believe one knows is ignorance."[2]

Aristotle (384–322 BCE), who was the most famous pupil of Plato at his Academy in Athens, integrated Plato's respect for deduction in his own view of science. However, in contrast to his teacher, he also accepted the reality of the world as it appears to us.[3] He argued that we live in a physical, biological and social world and it is the job of "natural philosophy", what we today would call science, to study that world—to find causes and laws for the changes that take place in nature and society. While Plato focused on ideal ideas, Aristotle focused on classifying the multitude of individual objects and argued that nature is simply the combination of form and matter that gives reality to things. Taken together, he argued, classifications, causal explanations and laws can be said to constitute the principles that determine nature and society—and that knowledge is belief that is justified as being true on the basis of proper reasoning, empirical evidence and a rational method of inquiry. In this manner, he developed an inductive-deductive approach to pursue truth as to physical reality. It is interesting to note that in spite of his approach to determining truth based on observation and reasoning (corresponding to what is today referred to as a "correspondence concept of truth"), he also left a negative mark on the development of science by his lack of interest in experimentation. His view of science concerned nature as it is, not in learning from its manipulation, and his mark on science was so profound that this impeded the development of experimental science for more than 1500 years.

While there was a contrast between Plato's idealistic and rational approach to science and Aristotle's more empirical and common sense approach whereby ideas are general terms that result from induction, they both considered science to be based on first principles that form the basis for all of physical reality, independent of time and place: According to their thinking, the world is an ordered cosmos where the same structures, causes and laws characterize and determine the physical world and society at all times. It is thus the aim of science to find, behind the appearances of phenomena, laws that could provide certain knowledge.

This view of science and reality, in particular the thinking of Aristotle, dominated philosophical and scientific thinking (at least in the West) until the end of the European Renaissance in the 1600s and the development of an empirically and experimentally based natural science. With the advent of classical mechanics with

---

[2]See http://www.goodreads.com/author/quotes/248774.Hippocrates; accessed 27.03.2015.

[3]The school that Aristotle established in 335 BCE. (Lyceum; in Greek, Lykeion) lasted for over 860 years which is longer than any other university. It had its own library and the world's first museum of natural history.

its emphasis on the use of a mathematical language (e.g. formulas, deduction) and quantitative concepts (e.g. mass, force, acceleration), science took a giant step forward.

However, this "giant step" did not take place all at once. Many years went by before most members of the "scientific community" no longer regarded such fields as e.g. alchemy and astrology as worthy of the qualification "science". According to (Thagard 1998; 72) "...a theory can be scientific at one time but pseudo-scientific at another. ... Astrology was not simply a perverse side line of Ptolemy and Kepler, but part of their scientific activity, even if a physicist involved with astrology today should be looked at askance. ... Rationality is not a property of ideas eternally: ideas, like actions, can be rational at one time but irrational at others."

The following example illustrates the immensity of this huge step forward that occurred when, perhaps for the first time in 2000 years, knowledge in the West was *systematically* sought not by referring to first principles or to authority, especially that of Aristotle or of the Bible, but by referring to experimentation, observation and measurement—to what we today refer to as empirical-based research. Based on his observations of falling bodies, Galileo Galilei (1564–1642) challenged the axiom of Aristotle (that had not been challenged—believe it or not—for almost 2000 years) that the speed with which an objects falls is proportional to its weight. According to this axiom, if two objects having the same weight are released at the same height at the same time, they will hit the ground at the same time. Therefore he reasoned that if they are tied together, they will have double the weight and the combined object should therefore fall twice as fast as when its parts were separate. He was convinced by this thought experiment that such reasoning was counter-intuitive and this led him to theorize that it is necessary to remove weight as the determining factor for the speed of fall.

Although it is not documented, there is a widely accepted story that after he had formed his hypothesis, Galileo performed a "test of hypothesis" by simultaneously dropping two balls, one weighing one pound and one weighing 100 pounds, both of which had been placed on the edge of the parapet of the leaning tower of Pisa in northern Italy. This was witnessed by those university professors that had attacked his hypothesis—they all saw that the two balls fell evenly and hit the ground at the same time.[4]

---

[4]Although this gave considerable support to Galileo's hypothesis, a more exact test could not take place until roughly 50 years later when the discovery in 1658 of the air pump permitted the creation of a vacuum. This permitted controlled experiments where the influence of air resistance was removed and, so to speak, a feather and a cannon ball could be shown to follow the same "law of gravity" as expressed by the mathematical formula that relates the distance an object falls to the surface of the earth in a vacuum directly to the square of the time it has fallen: $s = \frac{1}{2} at^2$, where 's' is distance, 'a' is the constant of acceleration, and 't' is time.

Galileo's teachings and writings paved the way for Newton (1642–1727) to develop what we now call the inverse square law of gravity (the force between any two masses is proportional to the product of the masses divided by the square of the distance separating them).[5] Newton's theories as to mechanics emphasized that legitimate knowledge could be obtained via induction based on observation and experimentation.

Since we will focus on the relationship between deductive and inductive reasoning later on, I briefly note here that this was in contrast to the approach of another great thinker, René Descartes (1596–1650), who preceded Newton. Descartes had emphasized reasoning as the surest source of truth (an example is his famous deductive statement: "Cogito ergo sum"—"I think, therefore I am"), rather than what he referred to as uncertain observations and risky induction. According to (Salmon 1967; 2), "… Descartes shows a complete lack of appreciation of the empirical approach. For Descartes, science is not an experimental enterprise in which one attempts to investigate clearly defined phenomena by observation of highly controlled situations. For him the order is reversed. One understands the broadest aspects of nature by deduction from indubitable first principles, the details come at the end rather than the beginning. The first principles are grounded in pure reason."

Newton described his own method as empirical and inductive and he strongly criticized Descartes for his emphasis on preconceived rational principles that should not be based on observations and experiments. I note too that even to this day the relative degree to which emphasis is placed on deduction and induction characterizes debates about research methodology. We will treat this matter far more thoroughly later on, particularly in connection with reflection on the justification of theory via verification and falsification.

Thus a mechanical picture of the world developed that challenged the focus on deduction from first principles (and the authority of Aristotle and the Bible) that had dominated science in the West for so long. First principles, faith and reason were no longer sufficient—ideas had to be tested in the real world. Nature was looked upon as a vast machine (characterized by its matter and motion) that is governed by quantitative laws, by relationships that are, so to speak, written in the language of

---

[5]In his main scientific publication, *Philosophiæ Naturalis Principia Mathematica* from 1687 (better known today as *Principia*), Newton presented the principles of theoretical mechanics (commonly referred to as today as classical or Newtonian mechanics). These included not only the movement of physical bodies on or near the surface of the Earth, but the movement of any physical body *anywhere* in the universe, including the movement of the moon around the earth and the movement of the Earth and the other planets around the sun. Classical mechanics formed *the* basis for physics and astronomy up until the time that Einstein developed his theory of relativity. And of course classical mechanics is still highly accurate for most calculations, including e.g. those concerning the trajectory of rockets, where the velocities of interest are much smaller than the speed of light. Einstein's relativistic equations reduce to the classical equations when $v/c \ll 1$, where v is the velocity of an object and c is the speed of light.

mathematics.[6] Rational methods of inquiry via observation, experiment, and measurement of the world around us became the royal road to meaningful knowledge.[7]

This development took place to a great extent outside of the universities. It reflected the political and ideological conflict between the Roman Catholic Church, that to a great extent dominated the universities, and the new empirically oriented natural sciences that were developing in private academies, often under the protection of kings and noblemen, and that challenged much religious dogma. However, towards the end of the 1700s, when major political, economic and religious changes were taking place in European countries, scientific research began to be based in universities that were now becoming more independent of both church and state. Science could now develop as an area of rational intellectual pursuit that was independent of philosophy, theology and the powers of vested interests. Thus, university teaching began to be based on knowledge obtained via observation and experiment—and this knowledge was now open for criticism! This led in turn to science being accepted as setting new standards for what could count as genuine knowledge. The practitioners of science were referred to then as "natural philosophers"; in fact, the term "scien*tist*" was first coined in 1834 by members of British Association for the Advancement of Science to describe students of nature, by analogy with the existing term "ar*tist*".

From about 1800 more specialized and application-oriented institutions of higher learning developed (dealing with engineering, agriculture, pharmacy, commerce, etc.). For example, today we find the word science connected to many fields of study that are far removed from the natural sciences, including not only those well-established fields within the social sciences (economics, sociology, anthropology, political science, etc.), but also such fields as library science, management science, dairy science, computer science …

These developments, characterized by the independence of science and by disciplinary differentiation, were also characterized by a number of basic beliefs and attitudes regarding science that still dominate our views of science today. These include:

---

[6]This perspective on science and nature was stated as early as 1623 by Galileo Galilei in his scientific-philosophical publication, *The Assayer (Il Saggiatore* in Italian). Here he described a research methodology that built on measurable data, mathematically formulated hypotheses as to law-like relationships between phenomena, and experimental testing of hypotheses. The book contained Galileo's famous statement that mathematics is the language of science. His method also included the use of thought experiments, something he himself used with considerable success in his actual experiments (Lübcke 1995; 150).

[7]According to (Hacking 1983; chap. 9) times have once again changed, at least as regards the way that historians of science treat their domain. The history of the natural sciences is today almost always written as a history of theory, and not of experiment. This perspective is supported by (Gooding et al. 1989; xiii): "Experiment is a respected but neglected activity. …it is all the more surprising that students of science have paid so little attention to how and why this particular activity has become so significant."

- science is, and *should* be (a moral perspective), autonomous, that is, responsible solely to its own norms,
- science leads to the highest form of knowledge as compared to e.g. common sense or belief,
- there are meaningful distinctions between pure/basic science and applied science, and
- science is neutral and value-free in its investigation of an objective, material reality.

I will return to these (more or less tacit) assumptions later on. Here I can simply note that this institutionalization of science, together with its emphasis on providing a privileged access to "truth" via a particular "scientific method", has also resulted in a growing divide between science and other fields of intellectual inquiry (Snow 1993). In any case, today it is considered completely natural that teaching at universities and other institutions of higher learning is scientifically based—with the authority and responsibility this leads to.

This development with increased disciplinary differentiation and specialization can also be said to lead to a growing divide *within* science. We have continually spoken here of "science"—but are their different sciences or is there just one science? And are there different methodologies for each of these different sciences —is it reasonable to speak of one general "scientific method"? This last query is particularly relevant as branches of science have become more specialized and yet have also joined together in investigating increasingly complex systems. Such questions are treated in greater detail in Chap. 7 (Scientific Method and the Design of Research) as well as in Chap. 9, Sect. 9.2 (Multidisciplinary and Interdisciplinary Research).

The "authority" referred to above is said to be provided by the so-called "scientific community". I have already referred a number of times to this term, and I have, for example, spoken of "the demands of the scientific community". What is this community and what characterizes it? In fact, not that much is known about the way that scientists organise themselves and how they determine the norms that they apply to characterize good science and good scientific behaviour. Thomas Kuhn in his provoking and challenging book on the history and sociology of science provides some tentative generalizations as to what is required for membership in such a community (Kuhn 1970; 169). These include:

1. a concern to solve problems about the behaviour of nature
2. the solutions that satisfy the scientist cannot be merely personal but must be accepted by many in the community
3. the group that shares them must be a well-defined group of the scientist's peers
4. the group is recognized for its unique professional competence and as the exclusive arbiter of professional achievement, and therefore
5. by virtue of its members' training and experience, the community is seen as the sole possessor of the norms, standards and rules of the game for passing unequivocal judgements. Kuhn notes (Ibid.; 164): "there are no other

professional communities in which individual creative work is so exclusively addressed to and evaluated by other members of the profession".

We will return to Kuhn's analysis of the history and sociology of science in Sect. 2.2.5 of this chapter where traditional and more commonly accepted perspectives on the history and development of science are challenged.

Before ending this brief overview of the history of science, it is necessary to introduce the theme "positivism" that has for many years exerted a dominating influence on science and its methodology, and, in spite of severe criticisms, to this day is the implicit starting point for most researchers in the natural sciences, at least for those who do not investigate the micro-world of quantum reality.

From the 1920s and until what its critics celebrate as its philosophical death around 1960 (although it still is alive and kicking in the mind-set of most scientists) this school of thought, referred to at times as logical empiricism or logical positivism or just positivism, dominated scientific rationality.[8] The central idea of positivism is that true scientific statements can only be based upon sensory-experience (empirical observations) and logical inferences based on these observations. This perspective on the establishment of true scientific statements had no place for metaphysics and denies that there are presuppositions for science; what is real is measurable. I note that in the literature one will meet various synonyms for what I often refer to as metaphysical assumptions; these include the terms: presuppositions, a priori assumptions, first principles and axioms.

Although there are many different interpretations of the concept of positivism and no single, widely accepted definition, the following underlying assumptions, based on an in-depth study by Lincoln and Guba (1985), appear to characterize most understandings of positivism:

- There is a single, knowable, measurable physical reality that can be reduced to elements that can be studied independently of each other.
- This external reality can in principle be exhaustively described in scientific language, where true propositions are in a one-to-one relation to facts about reality, including facts that are not observable.
- Independent observers can study reality since the knower and the known are independent.
- Observations can be made independent of existing theory and of an observer's values.
- There are real causes that temporally precede or are simultaneous with their effects.
- It is possible to make time- and context-free generalizations.

---

[8]This perspective on science and truth owes much to the philosopher David Hume (1711–1766), whose scepticism led him to argue that science should only aim at describing observations and to avoid philosophical speculation about an external physical reality and about the ability of science to obtain objective truth. See in particular (Hume 1748/1955).

Summing up, the underlying goal of a positivist perspective on scientific investigation was and is the creation of a clear, well-defined version of science based on firm data and free of philosophical speculation.

Yet problems emerged and led to the erosion of positivism's credibility amongst many philosophers of science (but not most scientists!) by 1960. These problems, many of which we will return to later on in the discussions dealing with the justification of theories, included, amongst others, the following:

(a) Theories cannot be tested individually by means of their observational consequences. For example, theories about planetary motion implicitly imply the acceptance of other theories about space and time—and the observations they are based on imply theories about optics, perception and the like. A result is that if a theory's predictions fail when confronted with observational data it can be unclear whether the theory itself is at fault or whether one of its auxiliary theories is wrong (or whether an instrument of observation was faulty). Thus, empirical data can neither falsify nor confirm a theory—truth is elusive.

(b) Positivism did not include considerations of the fact that humans (as observers, analysers, mediators) are fallible. For example, our observations are determined to some extent by the theories we as scientists implicitly accept as well as by our values; inquiry is not and cannot be value-free. In addition, we may implicitly use a multiple of criteria when choosing between alternative theories, each of which can be supported by the same data. Nature does not necessarily constrain theory choice.

(c) A great deal of scientific investigation is not based on direct sensory experience but on our own mental activities. Interaction with the external world is mediated via the use of technologically advanced instruments that have been constructed on the basis of theories which are products of our cognition and not of direct sensation. And the questions that we ask most often arise not from data obtained via the senses but from already existing knowledge. In this connection, the role of an intuitional leap from experience must be noted. According to Einstein, "For the creation of a theory, the mere collection of recorded phenomena never suffices—there must always be added a free invention of the human mind that attacks the heart of the matter"; quoted in (Pagels 1982; 58).

Thus, our modern period is characterized by a diversity of views on science. In a nutshell, a rationalist tradition has emphasized reason and logic, while an empiricist tradition has emphasized sensory experience and empirical evidence. My own experience, including my discussions with many other scientists, leads me to conclude that to be a reflective practitioner of science today means that one must appreciate and balance the roles played by evidence, logic and the presuppositions that underlie science. This implies that there is a need for you as a researcher to master both your field of specialization and the basics of research methodology, as well as to appreciate your field's dependence on and interplay with other fields of science. Perhaps doing this in a manner similar to that of a musician who aspires to become respected for her talent and virtuosity—and perhaps as well for her creativity; s/he must learn how to master her musical instrument (often more than one), must learn to work together with other musicians in a band or orchestra, and must have an understanding and appreciation of music as a field, including its relevant theories and the cultural traditions and expectations of music lovers.

For those interested in easily accessed information on the history of science, there are a number of online resources, including the following:

http://www.ebscohost.com/academic/history-of-science-technology-and-medicine Website of the History of Science Society that describes its database as "The Definitive International Database for the History of Science, Technology and Medicine" and that "reflects the influences of these fields on society and culture from prehistory to the present."

http://web.clas.ufl.edu/users/rhatch/pages/03-Sci-Rev/SCI-REV-Home/05-RSW-Sci-Rev.htm Website dealing with the Scientific Revolution and the role of historians of science.

http://plato.stanford.edu/contents.html Website of the comprehensive *The Stanford Encyclopaedia of Philosophy*. Although primarily oriented towards philosophy, it also includes a large number of brief articles on subjects and individuals that are particularly relevant for the history of science.

http://web.clas.ufl.edu/users/rhatch/pages/10-HisSci/links/index.htm Website providing informative links on the history of science.

2.2 What Is Science?

We have now used the word "science" a number of times without defining it, relying on the reader's familiarity with the word. Such lack of attention to definition is not uncommon, even amongst authors who write about science; an often-cited book (Chalmers 1999) with the title *What is this thing called science?* never really tells us what the author himself means by "science".

Presumably, this lack of definition did not cause you any difficulties up until now. "Science" is an everyday word, used by practicing scientists, the so-called "man-on-the-street", journalists, and university teachers—all of whom assume they know what is meant by "science", at least in a manner that is sufficient for them to communicate meaningfully with others. It is my experience that a scientist will say that he knows from experience and common sense what he is doing, that he does not need a definition to guide him in his work—that a conventional learned wisdom in his field will guide him. While this could appear to be a strange reaction by just those people who are engaged in the demanding, abstract, intellectual, competitive activity we call science, scientists appear to simply carry on the tradition they were brought up in. When they defended their Ph.D. dissertations they were not challenged with more reflective and philosophical questions dealing with methodology.

They might have been asked why they used a particular instrument, or about the reliability of some measurement, or about what other scientists have written on the subject, or about their use of statistics. But they did not face questions dealing with more fundamental topics such as whether, when they describe objects that cannot be seen (e.g. electrons or black holes) they are describing physical reality or whether their descriptions simply provide meaningful models of an unobserved/unobservable reality—or in what sense a theory can be said to be verified by the observation of some favourable instances (or falsified by an anomalous observation). Scientific investigation appears to be considered as a practical, although highly skilled, art by most scientists—who tend to ignore the potential benefit of reflecting on the meaning of such central concepts as science, truth, validity, proof, etc. According to (Ziman 1968 in Klemke et al. 1998; 50) "No scientist really doubts that theories are verified by observation, any more than a common law judge hesitates to rule that hearsay evidence is admissible."

However, the definition of science, or rather the attempt to understand the concept "science", can play an important role in our ability to understand and appreciate what science aims at, what its limits are—and what it tacitly presumes.

An experience I had a number of years ago illumines the importance of focusing on the concept of science as such, and not just on its specific procedures and contents. An internationally known and highly respected physicist who was also the Vice-Chancellor/Rector/President of the University of Copenhagen had written an essay (in Danish) on science and spirituality.[9] It started as follows (my translation):"At a social gathering I was asked by the person sitting beside me what I thought about such phenomena as reincarnation, aura, healing and the like. 'Badly' I replied, 'it's pure nonsense.'" In his rejection of these phenomena as "nonsense", he implied that they did not exist and were not worthy of scientific investigation. I wrote a letter to this highly esteemed person, whom I knew professionally, and asked him the very simple question: "How can science show that something does not exist?" I argued that science—and common sense for that matter—can demonstrate that something does exist. For example, if one wanted to demonstrate that cockroaches exist in the well-known five-star hotel in Copenhagen, Hotel D'Angleterre, it suffices to find just one such insect anywhere in the hotel. But can a scientific investigation, no matter how thorough, demonstrate that they do not exist?[10]

---

[9]The essay by Professor Dr. Scient. Ove Nathan was the first in a series on 'Science and Spirituality' in the major Danish morning newspaper *Berlingske Tidende* in 1993; it appeared on January 1, 1993.

[10]This question is most tricky—see if you can answer it. The same applies to its 'opposite': Can scientific investigation prove universal claims—such as "all ravens are black"? One might note that this claim corresponds to the statement that there are no ravens that are not black—and we are back to the original question of whether, based on evidence, one can prove that something does not exist (here, non-black ravens)! I note that the famous Michelson-Morley experiment (1887), which was designed to demonstrate the existence of the so-called "aether wind" as a medium for the movement of light waves, can be said to have provided the first strong evidence that such an all-pervading aether does *not* exist. The experiment with great accuracy demonstrated the absence

Perhaps I should have presented this anecdote later on when we discuss the limits of science—but I chose to bring it here as it clearly demonstrates that even experienced scientists, no matter how knowledgeable they may be in their field of specialization and how significant a role they play in the development of science, can have a rather limited concept of what science is (and what it is not).

2.2.1 Science as Facts

So let's start by looking at the way my former M.Phil. and Ph.D. students understood the meaning of "science". In 2004, when I first developed and taught the course on research methodology at Sri Sathya Sai Institute of Higher Learning in the state of Andhra Pradesh in India, after some discussion the students arrived at a consensus as to a definition of science:

> Science is the observation, collection, and analysis of facts.

This definition is in line with the thinking of many of my students during the decade I taught the course as well as of many observers and practitioners of science. The underlying thought is that so-called "facts" provide the basis for theory development and verification and for knowledge about the natural world in general. Scientific knowledge is said to have a special status due to its being based on facts, that is, on claims about the physical world that can be established by careful, systematic, unbiased use of our senses and reasoning, rather than by common sense, personal opinion, hearsay, belief, or imagination. In other words, a "fact" is generally understood as something that has actually happened or is "true". If in addition, "the reasoning that takes us from this factual basis to the laws and theories that constitute scientific knowledge is sound, then the resulting knowledge can itself be taken to be securely established and objective." (Chalmers 1999; 1)

Therefore let us temporarily use this definition as a starting point for reflections as to the nature of science.

The very simple definition provided by the students raises a number of questions. For example, consider the question: what are facts in a scientific context?[11] We have mentioned that they can be considered to be observations/statements based on the careful, systematic and unbiased use of our senses and reasoning. In the

---

(Footnote 10 continued)

of an anticipated drift of the aether as the earth moved through it without obstruction. In 1905, Einstein abolished the existence of the medium (Pais 1982; 20–21).

[11]Of course we could also raise other questions regarding the idea that "Science is the observation, collection, and analysis of facts." For example: Are there criteria as to the quality of one's observation, collection and analyses of facts? What do we mean by facts if, at least at the quantum level, the nature of reality is probabilistic? Doesn't science also deal with more than facts—what about "understanding" and "meaning"? Doesn't science also deal with the establishment of verifiable theories/laws? And so on! We will return to such important questions later on.

natural sciences in particular, such facts are considered to be: (a) data that are obtained by careful and well-documented investigation (observation/experiment), and b) that are so accepted by one's peers that it becomes difficult to consider other interpretations of the data. In other words, facts could be measurements that have been carefully performed and recorded, where we know details about the precision of the equipment employed, and where the observation/experiment has been performed accurately and in accordance with generally accepted standards as well as ideals as to objectivity.

But things are not all that simple. For example, according to (Chalmers 1999; xxi): "...the idea that the distinctive feature of scientific knowledge is that it is derived from the facts of experience can only be sanctioned in a carefully and highly qualified form, if it is to be sanctioned at all. ... (There are) reasons for doubting that facts acquired by observation and experiment are as straightforward and secure as has traditionally been assumed. ... a strong case can be made for the claim that scientific knowledge can neither be conclusively proved nor conclusively disproved by reference to the facts, even if the availability of these facts is assumed."

The arguments that Chalmers provides to support such scepticism are primarily based on analyses of the nature of observation and of the history of science, as well as on the nature and limitations of logical reasoning. He argues that there are three implicit, and challengeable, assumptions that underlie a definition of science based on facts. These are:

1. Facts are directly available to careful, unprejudiced observers via the senses
2. Facts are prior to and independent of theory
3. Facts constitute a firm and reliable formulation of scientific knowledge.

Let us briefly consider these assumptions; in particular, we will go into more detail as to the second assumption when we consider what is meant by a "scientific theory". Let us start by considering the role of human perception and focus on eyesight. Consider the following drawings of a "staircase" from (Chalmers 1999; 6) and of a "cube" from (Hanson 1958 in Klemke et al. 1998; 343) (Fig. 2.1).

Start by looking at the drawing of the "cube". Look at it for some time and you may find that what you see changes spontaneously. At one moment it appears to be a cube that you look at from above, only to become a cube that you view from

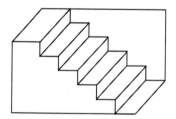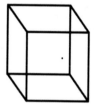

Fig. 2.1 A "staircase" and a "cube"

below—and vice versa. Or perhaps you simply see it as a wire frame for a kite, or perhaps simply as some criss-crossed lines in a plane. If we all see the same thing (the same drawing), why do we see different things? According to (Hanson 1958 in Klemke et al. 1998; 343) a traditional perspective on seeing leads scientists to (mistakenly) argue that the explanation is that we interpret things differently. The traditional argument is that since there does not appear to be a place in the seeing for these differences, the disparities in accounts of the observations must be due to ex-post interpretations. But Hanson argues that this is not so; "one does not first soak up an optical pattern and then clamp an interpretation on it." His argument is that although normal retinas are impressed by the lines on the paper in the same way, some observers will see the cube viewed from above, others from below—and the same observer may see it, at one moment, as viewed from above, and at another moment as viewed from below. No interpretation or thinking is involved.

Let us continue this line of thought. Look now at the drawing of what appears to be a staircase. Once again, if you look at the picture for some time, you may find that what you see changes spontaneously, involuntarily, and without any conscious thought involved. At one moment it appears to be a staircase that you look at from above, only to become a staircase that you view from below—and vice versa. Yet it must be the same object that you view since no one has changed the picture and the retinal images do not change. Whether you see the picture as a staircase view from above seems to depend on something other than the image on your retina.

If what you see changes, even though the object itself does not change what are the "facts" here? Is what you see "firm and reliable"? The answer is neither a clear "no" nor a clear "yes". The above arguments tend to evoke an answer of "no". On the other hand, there is an ambiguity that results from representing three dimensional objects on a two dimensional space like a sheet of paper. However this ambiguity can be overcome, for example by employing a stereoscopic representation by using stereo cameras. This would certainly be done in situations where the depth perspective is crucial to an observation. So since it is possible to overcome the ambiguity mentioned, there are good reasons for answering "yes"—we are able in science to obtain firm and reliable representations of the objects to be studied. One might say that scientific observation is not limited but that ordinary human perception is. The problem remains of course that so much of the way science develops is affected by human perception.

Let us continue with this matter of perception. According to (Chalmers 1999; 6), "the results of experiments on members of African tribes, whose culture does not include the custom of depicting three-dimensional objects by two-dimensional perspective drawings, nor staircases for that matter, indicate that the members of those tribes would not see the drawing as a staircase at all". They could not see the drawing as you do. Since, if we assume that you and the members of an African tribe have reasonably good eyesight and you are both seeing the same thing (with your eyes, neural connections, brain), why do you see different things? It seems that perceptual experiences in the act of seeing are not uniquely determined by the images on one's retina but depend on the experiences and expectations of the observer. So "facts" may not be independent of who you are and what you know in advance.

To elaborate on the question of whether facts are independent of the observer, consider now the situation where a medical student who has not yet had training in working with X-ray photos and an experienced radiologist both look at the X-ray images of a patient being examined for lung cancer. The perceptual experiences of the senior, skilled observer are not identical to those of the untrained novice; he sees things that the student does not see and is able to make a diagnosis that the student cannot. To be a competent observer requires training and experience. Perception is influenced by our background, knowledge, worldviews, and expectations. An infant looking at a book would only see marks and lines where you see letters, words and sentences. Humans do not have direct perceptual contact (like film in an X-ray camera) with the physical world, and this is not the same as saying that different people who see the same object or phenomenon interpret it differently; they simply *see* different things!

According to physicist Amit Goswami who has studied the relationship between human consciousness and physical reality, "...the picture is not the object. The map is not the territory. Is there even a picture out there? All we know for sure is that there is some sort of picture in our brains, a truly theoretical image. In any event of perception it is this theoretical, very private image that we actually see. We assume that the objects we see around us are empirical objects of a common reality—quite objective and public, quite open to empirical scrutiny. Yet in fact, our knowledge about them is always gathered by subjective and private means." (Goswami 1993; 142)

Thus arises the old philosophical question: What is real—the theoretical image that I (whether I am a scientist or a member of an African tribe) actually see but only privately, or the empirical object that I do not seem to see directly but about which I form a consensus (with other scientists or tribe members)? As we shall see later on in Sect. 2.7 of this chapter, this question as to the authenticity of empirical objects that we never experience without the intermediary of a theoretical image, is at the heart of "realism", one of the metaphysical foundations of the natural sciences.

So it appears that facts as observable states of affairs in the physical world can be fallible; different scientists may not see the same thing even though they have good eyesight, are using the same equipment, and are looking at the same object of study. On the other hand, it may also be argued that: (a) use of proper equipment (e.g. stereo cameras to photograph what would otherwise be two dimensional representations), (b) proper education and shared experiences (e.g. whereby African tribesmen who are familiar with staircases can see staircases as we do), and (c) experience (e.g. where a young medical student eventually develops into an experienced radiologist), together with critical application of the methods of science, can provide science with reliable data—with "facts".

Since we have just considered X-rays, we may extend this line of reasoning to ask what the objects of the physical universe would look like if we had X-ray eyes, that is, if we were not able to see in the frequency range that we do as humans ($4 - 8 \times 10^{14}$ Hz), but in the frequency range corresponding to X-rays ($3 \times 10^{16} - 3 \times 10^{19}$ Hz). Clearly our aesthetics would be totally changed—for example, our concept of a "handsome" person would not be based on the person's exterior, but on the person's skeleton. But more important here, all our data as to the physical universe would be different. We would not be able to "see" many things we now see, and we would be able to "see" many things that we cannot see now with our human eyesight. It is reasonable to assume that the same scientific laws would exist, at least potentially, since what we have come to accept as scientific laws are assumed to have universal legitimacy, independent of how they are arrived at. I write "potentially" to indicate that they might not be discovered/developed in a world where vision was limited to the X-ray range, and that the formulation of such laws would of necessity be transformed to fit a world perceived by people with such X-ray vision. In addition, they might discover other laws that we have not yet been able to uncover, since they would have access to different data/facts than we do. It is also most likely that the history of science, and of the world, would be immensely different than we have known them to be.

Consider now the following closely related reflection from (Hawking and Mlodinow 2010; 91) as to how we gather knowledge of the world via our ability to see: "It's probably no accident that the wavelengths we are able to see with the naked eye are those in which the sun radiates most strongly; it's likely that our eyes evolved with the ability to detect electromagnetic radiation in that range precisely because that is the range of radiation most available to them. If we ever run into beings from other planets, they will probably have the ability to 'see' radiation at whatever wavelengths their own sun emits most strongly". It follows that such beings would have had access to different data than we do and would most likely have developed different sciences than we have been able to! And it follows too that 'scientific truth' can be said to be relative; what is a fact/true for humans may not be a fact/true for such beings and vice versa.

Note that in science, the word "fact" does not only refer to an observable state of nature, it also refers to a *statement* that expresses the observable state of nature. Science is based on such statements, and these, unlike the observable states of nature, do not enter the brain by way of the senses. Rather, facts as statement are products of human cognition—and this too is notoriously fallible.

(Chalmers 1999; 15–16) provides a good example. He notes that prior to the realization that the earth spins on its axis and orbits the sun the statement that "the earth is stationary" was a fact, confirmed by observation. We cannot see or feel it move; if we jump in the air, it does not move away from under us. But we now know that what was earlier a "fact"—the observation statement that "the earth is stationary"—is false.

Let us briefly reflect on this historical development. We know that even though we at the surface of the earth are moving in a horizontal direction at more than 100 m per second, there is no reason why our relative position should change when we jump in the air; there are no horizontal forces acting on us that change our speed. So we retain the speed we share with the earth's surface when we jump and land again. To appreciate why this is so requires understanding the great 17th century discovery of inertia by Galileo (which also is the first of Newton's three laws of motion that laid the foundation for classical mechanics). This discovery/law was not available earlier since the "fact" of the earth's stationarity appeared to be obvious to earlier scientists. This demonstrates how judgment as to the truth of an observation statement ("fact") depends on the knowledge one has prior to making the judgment. The Scientific Revolution involved not just a transformation of theory, but also of what were considered to be empirical facts.

The above reflections as to "facts" imply that the definition: "science is the observation, collection, and analysis of facts" is subject to challenge. As is the statement that facts can be firmly established by observation. We have shown that one challenge concerns the extent to which our perceptions are influenced by our background, our culture, and our expectations—so that what appears to be an observable fact for one observer may not be so for another. Another challenge concerns the extent to which what we already know or assume affects our judgments about the truth of observation statements. And yet another challenge arises from the dependence of our observations on the limitations of our senses. This is not to say that we should simply give up trying to establish a factual basis for our theories as to phenomena, relationships etc., but that we must be aware of the fact that facts are fallible—they may change as we and our knowledge of the world change!

So what appeared to be a straightforward, common sense definition of science based on "facts" is not so straightforward indeed. We may therefore ask whether there exist other, more reliable—and less challengeable—ways of defining science. Let us briefly consider another rather common definition.

2.2.2   Science as Generalization—and as Establishing Verifiable and True Theories

Many see generalization as the hallmark of the natural sciences. For the present purposes, we can loosely consider generalization to be the act of inferring from that which has been observed to that which is unobserved. Indeed, the "science" of our ancient forefathers that served as a basis for more modern science resulted from making quite a few generalizations of a comprehensive nature—for example that the laws of astronomy determine the progress of time and that laws of geometry, as developed in particular by Euclid in Greece roughly 300 BCE, hold for all space without exception. It is interesting to note in this connection that, as often has happened in the history of science, the assumption as to the universality of Euclidean geometry was shown to be false in 1764 when Thomas Reid developed the first non-Euclidean geometry; since then other types of geometries, each based on its own set of axioms, have been developed.[12]

Just as was the case with defining science as facts, defining science as a means of establishing reliable generalizations certainly contains important elements of what

---

[12]Euclidian and non-Euclidian geometries are equally consistent; if one of them contains a contradiction, so does the other (Salmon 1967; 36).

most members of the scientific community would consider to be science. But it too raises a number of questions, such as:

- How do/should scientists establish reliable generalizations/theories? Is verification the only way to determine the "goodness" of a theory?
- Does science have a special way of establishing true statements about the physical world—and does it have a special way of distinguishing between true statements about the physical world and statements which are not true?
- Can science "prove" that something is true, and if so, how?
- To what extent does e.g. usefulness in solving problems or developing technology influence the development of theory?

Once again we find that an attempt at providing a simple, straightforward definition of science leads to a number of fundamental questions. Instead of attempting to answer the above questions here, I suggest that no matter how we attempt to provide definitions of "science" we will encounter a series of questions that compel us to reflect a bit deeper as to what we are doing when we perform "science" and what distinguishes science from non-science.

This should not surprise us. We meet the same challenges when we try to define similar fundamental concepts such as e.g. "ethics" (which we will consider in relation to science in Chap. 10). There are a large number of publications that provide different perspectives on what the term ethics means, perspectives that are far more important, nuanced and meaningful for a philosopher than for the so-called man-on-the-street who is satisfied with an intuitive understanding of the term—in particular as to what is *un*ethical! What is surprising is that although students of philosophy cannot avoid considering what they, and in particular what noted philosophers throughout history mean by "ethics", students of science have not usually been motivated or compelled to ask themselves what they mean by "science". Nor have most practitioners of science.

To conclude this digression on "what is science" on a more constructive note, I suggest that science is more than just the organised collection of facts, concepts, and knowledge about the structure, characteristics and behaviour of the physical universe; that it is more than the systematic investigation of reality; and that it is more than the rational establishment of verifiable theories—each of these being common definitions. It is all of these and more, and I therefore propose the following very broad definition:

Natural science is a special way of looking at the universe – a rational approach to discovering, generating, testing, and sharing true and reliable knowledge about physical reality.[13]

---

[13] Although the concepts of "truth" and "reality" have not been dealt with in any depth so far, I feel they should nevertheless be included in the definition of science. I note at this early point that most scientists tend, consciously or unconsciously, to adhere to what has been called the "correspondence theory of truth" within a framework of "realism", both of which will be presented shortly in Sect. 2.4 where we discuss the Aims and Claims of Science.

One might inquire why I have delimited the above definition to "physical reality". The reason is twofold. First of all, to distinguish "natural science" from other branches of science, such as e.g. the social sciences, that do not share all of these aims and that are characterized by subject matters (e.g. human and social activity) that may be investigated with other aims than those listed above regarding "true and reliable knowledge" (for example, how to make better decisions). Secondly, to exclude phenomena that *may* have non-physical characteristics and causes—for example consciousness, a theme we will reflect on several times later in the present chapter on the essence, aims and limitations of science, as well as in the final chapter on science and ethics.

This "special way of looking at the universe" has led to activities we call science, to a profession we call science, and to an institution we call science. In the remainder of this book, we will continually reflect on the nature of science and on how you can develop a worldview and the research methodology that will enable you to become a respected, reflective practitioner of science.

2.2.3 Some Distinctions When Describing Science

In order to further characterize science, briefly consider the following distinctions that are commonly employed when describing science (as an activity, a profession, or an institution):

First, there is an often stated distinction between science as creating knowledge (what is generally referred to as *pure/basic science*) and science as applications of knowledge (*applied science*). As was seen in the discussion of the history of science, this distinction was made very early in the development of science when special institutions of higher learning were established to promote the applied aspects of science (such as schools of engineering, agriculture, medicine) leaving so-called pure science to the disciplines that were developing in the universities of Europe.

Within pure science we find another distinction: between the *formal sciences* such as logic, computer science and mathematics and the *empirical sciences*, such as physics, chemistry, biology, botany, psychology, sociology and economics.

And digging even deeper we find in the empirical sciences (and these are the sciences we are primarily concerned with in this book) a number of commonly applied distinctions between major domains of science: the *natural sciences* (concerned with the nature of matter and energy and that study the physical and natural world, nature's phenomena—physics, chemistry, biology, astronomy, the earth sciences ...), the *life sciences* (concerned with living things, their components and interactions—biology, zoology, pharmacology, botany....), the *behavioural sciences* (concerned with human behaviour—psychology, anthropology, social neuroscience ...), and the *social sciences* (sociology, economics, management, political science ...).

There are several scientific disciplines that cross the boundaries of such categories of science (natural, life, behavioural …). For example, one often sees biology classified as both a natural and a life science, and anthropology may be considered both a life, behavioural and social science. In particular, in an age characterized by huge developments in specialization as well as in multidisciplinary and interdisciplinary research, an increasing number of new "hybrid" scientific disciplines are emerging (e.g. social neuroscience, evolutionary psychology, genetic engineering) that are difficult to classify as belonging to any one category. Of course any such categorization can be considered artificial and arbitrary; "Knowledge … is one. Its division into subjects is a concession to human weakness." (Lee 2000; 3)

This development whereby disciplines differentiate and specialize is a result of and a characteristic of a traditional reductionist and materialist perspective on science that provides focus and amazingly detailed information about both micro (DNA, quarks) and macro phenomena (animals, black holes, an expanding universe). From such a perspective, matter is the fundamental substance in nature such that all that exists is material, and all physical reality is built up of individual units of matter that obey the laws of nature.

At the same time, this perspective on reality also makes it increasingly difficult for a scientist to appreciate a perspective whereby everything that appears to exist does so in relation to everything else—that all of reality is interconnected. Another way of saying this is that the way science is developing increases our tendency to look upon reality, at whatever level, as being ontologically separate, although perhaps functionally interrelated.[14]

Keeping to the empirical sciences, one also meets distinctions between the *observational/non-experimental sciences* (typically associated with the more descriptive approaches to science as in astronomy, zoology, meteorology or geology) and *experimental sciences* (typically associated with a more analytical approach, as in much of physics or medical research). I note that since all natural sciences observe and describe, it is difficult to speak of a non-observational or a non-descriptive science, but that some branches of science are primarily oriented towards describing and answering questions while others are more oriented towards active experimentation and testing hypotheses. For example, scientists may be interested in answering questions such as: What is the surface of Venus like? What species of plants live in the Himalaya Mountains? What patterns of nucleotides make up human genes? Such questions are more representative of a descriptive rather than an analytic/causal approach to scientific inquiry.

To answer such descriptively-oriented questions a study might rely primarily on library research, photos, extracting physical samples and the like, and not on

---

[14]An alternative to such a reductionist perspective is provided by a holistic/systemic perspective whereby all of nature, including all its living and non-living components, is self-regulating and can only be understood via systemic thinking. This later perspective is commonly referred to as the "Gaia Hypothesis", originally developed by the chemist and environmental scientist Dr. James Lovelock in the 1960s and 1970s (Lovelock 1979) in connection with his work at the Jet Propulsion Laboratory in California on methods of detecting life on Mars.

experiments. For example, in some branches of life sciences, like botany or zoology, pictorial descriptions of a species based on photos of members of the species might be sufficient to establish that something new has been found. Some researchers may carry out such work all their lives and add more flora and fauna to their list. The work may remain primarily non-experimental and descriptive and still provide a background for theory development and make a valuable addition to scientific knowledge.

Yet another distinction, closely related to the distinction above between observational and experimental sciences, is the distinction between *law finding sciences* (physics, chemistry ...) that attempt to discover universal laws that apply to nature (always and everywhere), and *fact-finding sciences* (geography, history....) that are primarily descriptive. Note that even the fundamental concept of a 'scientific law' has been challenged by the renowned mathematician and philosopher of science, Alfred North Whitehead (1861–1947): "People make the mistake of talking about 'natural laws'. There are no natural laws. There are only temporary habits of nature." (Whitehead (1954), *Dialogues of Alfred North Whitehead*, Boston, USA: Little Brown; quoted in (Sheldrake 2012; 99).

Finally here, we can refer to a distinction between *historical science* (a longer time-scale characterizes the process involved—e.g. as in geological processes such the formation of the Himalaya Mountains) and *non-historical science* (where a far shorter time-scale is involved—e.g. as in chemical reactions). Even slow chemical reactions are extremely fast when compared to geological processes (Siever 1968).

2.2.4 Science as a Social Activity

A further characterization of science deals with its social aspects. Essentially all modern approaches to science assume that the facts and theories of science must survive critical study and testing. From this perspective, science also deals with obtaining consensus of rational opinion in the scientific community. According to (Ziman 1968 in Klemke et al. 1998; 51–2): "Technology, art and religion are perhaps possible for Robinson Crusoe, but law and science are not ... The young scientist ... learns by imitation and experiences a number of conventions that embody strong social relationships ... He learns to play his *role* in a system by which knowledge is acquired, sifted, and eventually made public property."

The ability to play such a role is developed, slowly but surely, via a process of socialization. The young, budding scientist observes how others at his laboratory/university behave and speak, and he or she receives guidance as to dissertation research, attends conferences at other institutions, reads articles and perhaps interacts with the editor of a journal in connection with a paper submitted for publication, receives feedback on grant proposals, and so on.

These social aspects of science are manifested in its institutionalization via more or less universally accepted norms as to what is "good" scientific behaviour, including:

- lack of bias
- maintaining data and records of one's research
- giving due credit to others who have contributed to one's research
- making the results of one's research available to the scientific community
- judging the scientific work of others solely on its merits.

We will directly or indirectly return to the concept of scientific norms when we consider such topics as: acceptance, peer review and the scientific community (in Chap. 3); the scientific method (Chap. 7); and the role of values and ethics in research (Chap. 10).

As we shall soon see, the next section on scientific revolutions concludes with a rather special and controversial perspective on the social aspects of science, one that turns ordinary more traditional approaches to science on their head. While by far the most common perspective on science is that it is a higher standard for what is true (as compared to e.g. common sense), the position to be considered is that science is not a higher standard which determines both the scientific community's as well as the layman's assent to truth—but vice versa: What makes a statement scientific is that scientists make the statement. Sociology replaces method in this rather radical account of what is scientific.

2.2.5 Scientific Revolutions and Paradigms

Based on the brief presentations earlier of the history of science and of the characteristics of science, it is tempting to accept the traditional assumption that science develops gradually over time and that it provides a continually improving standard for what is true knowledge. This section presents a contentious challenge regarding this traditional understanding of progress in science.

The challenge is to a great extent due to the path-breaking research on the history and sociology of science performed by Thomas Kuhn (1922–1996). In his book *The Structure of Scientific Revolutions* (1962, 1970),[15] Kuhn introduced the now widely referred to concepts of "paradigm" (from the Greek word *paradigma* meaning "pattern") and "paradigm shift" or its synonym, "scientific revolution". The term paradigm is widely used by scientists today to refer to more-or-less universally accepted scientific positions or frameworks that, for a time, provide a community of scientific practitioners with a model for formulating and solving problems. It is not simply a current theory or set of theories, but rather a shared cultural lens—the whole worldview in which theories exist within a field, including the broadly accepted pre-suppositions, theories, practices, terminologies, schools of thought, and even values. A more concise and more popular phrasing would be: a scientific

---

[15]Originally published in 1962, the second and enlarged edition in 1970 includes a postscript where Kuhn replies to criticisms and where he provides sketches of revisions based on the criticisms.

worldview that shapes our perceptions of reality within a discipline. Examples of paradigms include Newtonian mechanics, Darwinian evolution, Maxwellian electromagnetics, Einsteinian relativity, quantum mechanics, and the psychoanalytic model of the unconscious mind. The term paradigm is now a very popular term that is (mis)used in many other realms of human experience than science.

In order to reflect on Kuhn's challenge to the traditional assumption that science develops gradually, continually and cumulatively, we must consider the following concepts he developed: "paradigms", "normal science", "anomalies", "crises" and "revolutions".[16] I introduce these terms by considering the so-called Copernican revolution (Russell 2003; 20–25). For roughly 1500 years, astronomers had interpreted their observations based on the geocentric (earth-centred) model developed by the first century Greek philosopher Claudius Ptolemy (100–178). According to this model, the sun, moon, planets, and stars all revolve around the earth. This was a most reasonable assumption for these astronomers, even though there were problems with the model. For example, although the stars appeared to move along circular orbits, the planets appeared to wander among the stars; their orbits appeared to wobble, their speeds varied, and at times they appeared to reverse direction. These were examples of *anomalies* that the geocentric model could not explain. To get around such problems and to maintain the fundamental geocentric model, astronomers developed the concept of epicycles whereby the planets follow circular paths around larger circular paths (this enabled the paths to reverse directions as seen from the earth). When more accurate data showed that even this model had irregularities, more complex epicycles were developed, with circles rolling around circles rolling around circles, all to maintain the belief in circular motion of the celestial objects around an earth that was not moving and at the centre of the universe.

In the 16th century, Nicolaus Copernicus (1473–1543) showed that the existing geocentric model was incompatible with the radically different heliocentric model he had developed whereby the planets revolve around the sun. According to Copernicus, the apparent motion of the celestial bodies was an illusion (e.g. the rising and setting of the sun) caused by the motion of observers on earth—which was not stationary and was not even at the centre of the universe; it was just another planet revolving about its axis as well as around the sun. This challenged what everyone, including the scientists of his time, believed and that common sense supported; that the earth stands still and the planets, sun and stars all move around it. So radical, so heretical was this model that the clergyman Copernicus dared not publicize it for roughly 30 years, and only did so on his deathbed; the first copy of

---

[16]In his richly documented historical study of creativity Arthur Koestler presents similar reflections on the "jerky, unpredictable" development of science and relates the phases in the evolution of ideas in the individual scientist to that of the branch of science he works in: "The Eureka act proper, the moment of truth experienced by the creative individual, is paralleled on the collective level …The collective advance of science as a whole, and of each of its specialized branches, shows the same alternation between relatively brief eruptions which lead to the conquest of new frontiers, and long periods of consolidation." (Koestler 1989; 225–26)

his little book *On the Revolutions of Celestial Spheres* was placed in his hands on the day he died.

The shift in the direction of a new *paradigm* was given a big shove in 1609 by Galileo Galilei whose observations in his newly invented telescope[17] supported Copernicus's ideas. Amongst other things he had observed that Venus moved through phases, like the moon, indicating it circled the sun, and that moons orbited Jupiter, both of which went against the accepted paradigm which included the supposition that all heavenly bodies circled the earth.[18] It should be noted here that when Galileo published his findings the Pope (the leader of the worldwide Roman Catholic Church) demanded that he retract his heretical ideas, which he did to save his life. Later, in 1632 he once again published his observations and defended Copernican theory—and once again he was forced to retract his statements and was then condemned to house arrest for the remainder of his life. According to Cardinal Bellarmine who participated in the trial of Galileo, "To assert that the earth revolves around the sun is as erroneous as to claim that Jesus was not born of a virgin." (Russell 2003; 24)

The *paradigm shift* received another shove when the German mathematician Johannes Kepler, using data provided by the Danish astronomer Tycho Brahe, showed that even if the planets were orbiting the sun, they were not following circular orbits. Kepler was able to explain all the apparent irregularities in the orbits of the planets if they were assumed to follow elliptical orbits. The reason for such orbits was first provided 70 years later when Newton showed that all heavenly bodies are governed by exactly the same laws as earthly objects and that *any* orbiting body would move in an ellipse. This completed the paradigm shift or *scientific revolution*. It took 150 years and significant breakthroughs by scientists from five countries before the key idea developed by Copernicus—that the sun is at the centre of our solar system—became accepted. However, it was not until 1992

---

[17]Although another person had already developed a similar tool (he was denied a patent), and although "Galileo astonished the world when he turned the telescopic toys, invented by Dutch opticians, to astronomic use." (Koestler 1989; 102), Galileo Galilei's refraction telescope from 1609 was the first well-functioning and patented telescope. Already in March, 1610 he published his book *Sidereus Nuncius* (translated into English as *The Sidereal Messenger* by Albert van Halden in 1989, but often referred to as *The Starry Messenger*) where he described his telescope and his observations of the moon, stars and the moons of Jupiter.

[18]In addition to demonstrating that these heavenly bodies did not orbit the Earth, Galileo also provided a bold conjecture about motion and friction that lent support to the "Copernican Revolution". Since Aristotle, philosophers/scientists had believed that in order for an object to sustain its motion, an external source of energy is required. Therefore, if the earth is revolving about its axis as it orbits the sun, a force must be pushing the Earth, which would result in a huge wind from the East. Galileo Galilei conjectured that without friction to slow it, an object will sustain its speed without any force "pushing" it, whereby it made sense that the Earth (together with the air around it) turns on its axis. It can be added that Newton (1642–1727) built upon Galileo's analyses of motion and inertia; his famous first law of motion essentially states that a force is not needed to keep an object in motion; in fact a force such as friction is required to bring a moving object to rest.

that the Vatican, the central governing body of the Catholic Church, formally apologized for its treatment of Galileo!

Another, more "modern" example of a "scientific revolution", and one that took place in a far shorter period of time, is when Newtonian mechanics gave way to Einstein's relativity. This paradigm shift occurred when experiments piled up data that falsified Newtonian mechanics for velocities approaching that of light, but fitted the theory of relativity.

Based on the study of these, and many other developments in science, particularly in physics, Kuhn developed his thesis that the development of knowledge does not take place in a purely cumulative manner, as commonly assumed, but in a far more discontinuous and revolutionary manner, such that the transition from one paradigm to the next is not smooth. Thus, according to Kuhn, the history of science has alternating episodes of *normal science*, during which scientists refine and apply an accepted paradigm, and episodes of *revolutionary science*, during which scientists switch to a new paradigm, and thereby to new norms for the practice and understanding of normal science.

In a nutshell, according to Kuhn, such a switch occurs when a stable period of *normal science*, characterized by *puzzle solving*, is replaced by a *crisis*, arising from observed and persistent *anomalies*—failures of the current paradigm to take into account observed phenomena. Such crises can lead to periods where bold scientists embark on *revolutionary science*. They explore alternatives to the long held, and what have seemed to be obvious, assumptions characterizing the existing paradigm.

During periods with anomalies, most practicing scientists do not lose faith in the established paradigm and in the normal science they practice, since no credible alternative is available. Were they to lose faith, they would face huge challenges to their role and identity as scientists and as members of the scientific community. Occasionally the explorations lead to the identification of one or more new candidates that challenge the established frame of thought. These will of course be severely countered by the practitioners of normal science—and often such practitioners' confidence in the established frame of reference has been vindicated. But if the challenges due to the anomalies persist, and if credible alternatives become available, according to Kuhn a phase develops where more-or-less incompatible and incomplete theoretical frameworks coexist. If, over time, the challenging paradigm demonstrates its ability not only to solve the problems solved by normal science, but to convincingly deal with the anomalies that have persisted and resisted solution using normal science, a transition period obtains that leads eventually to a scientific revolution—a paradigm shift.

Kuhn speaks of revolutions because he argues that different paradigms, before and after a shift, are incommensurable—not only do scientific beliefs about nature change, but also the standards and criteria of scientific judgement change. Textbooks are rewritten—often the history of science is rewritten. Therefore scientific revolutions are like religious conversions. There is a transfer of allegiance that is not simply brought about by solid logic and objective evidence. The process of such paradigmatic change is a most complex social process, where many beliefs, standards and explanations are replaced—and where the status, self-conceptions, language, and practices of scientists are modified.

This leads to the position that science is not a higher standard which determines the scientific community's assent to truth—but vice versa. Sociological phenomena replace scientific method in this rather radical account of what is scientific, where what makes a belief scientific is not the underlying methodology of investigation, but what scientists say about the belief. Science, according to this perspective, is simply what scientists say it is.

For most scientists, these are radical and upsetting ideas. They share adherence to more-or-less traditional and stable worldviews as to science and its methods. They reply that science certainly does progress in a continual manner and that paradigm shifts, if they occur at all, are not characterized by incompatible world-views. Certainly at the level of practical day-to-day scientific research, our ability to develop technology based on scientific findings demonstrates that the methods of science find much that we can call truth and that we are able to confirm. From this more traditional perspective, progress and truth are intertwined and science is far more than what scientists simply say it is; the strength of its statements lies in its methods and the vigilance of the scientific community in controlling its results.

Having reflected for some time now on what "science" is we should now be prepared to consider now what it is not.

2.3   Science and Non-science/Pseudo-science

Science (as an activity, profession and institution) is highly regarded by most people, even though there have been serious criticisms of science and of scientists for contributing, directly or indirectly, to the development of debatable technologies. These include weapons of mass destruction, products and production processes that pollute the environment, techniques for controlling our movements and invading our privacy, and for engineering our genes. But on the whole, "science" is considered by the great majority of people to be a positive buzzword—it is widely regarded as providing not only advanced technologies that can contribute to our wealth, well-being and health, but also to knowledge that enriches our lives and make them meaningful.

But what distinguishes science from non-science? I note that many philosophers of science prefer to use the derogatory but perhaps more precise term "pseudo-science" when the promoters of say astrology or faith-based theories such as creationism present their methods and results as though they are in accord with traditional scientific standards.

In the seminal work "Science: Conjectures and Refutations", Karl Popper (Popper 1957 in Klemke et al. 1998) attempted to distinguish science from non-science. He referred to:

> ... the autumn of 1919 when I first began to grapple with the problem, '*When should a theory be ranked as scientific?*' or '*Is there a criterion for the scientific character or status of a theory?*' The problem which troubled me at the time was neither, 'When is a theory true?' nor, 'When is a theory acceptable?' My problem was different. I *wished to distinguish between science and pseudo-science*; knowing very well that science often errs, and that pseudo-science may happen to stumble on the truth." (Ibid.; 38)

His interest for these matters developed in Austria after the First World War based on his readings of Einstein's theory of relativity, Marx's theory of history, Freud's psychoanalysis and Adler's so-called "individual psychology".[19] He became increasingly dissatisfied with the last three theories and their claims to scientific status: "... these other three theories, though posing as sciences, had in fact more in common with primitive myths than with science; they resembled astrology rather than astronomy." (Ibid.; 39) The underlying reason was that these theories appeared to be compatible with the most divergent human behaviour; they could be interpreted so they appeared to always lead to correct predictions. What in the eyes of their admirers appeared to be strong arguments supporting the theories, for Popper began to resemble weaknesses.

Things were quite different with Einstein's theory. The British astronomer, mathematician and physicist Arthur Eddington (1882–1944) had just performed observations that confirmed Einstein's gravitational theory which implied that light must be attracted by heavy bodies (such as the sun). He had subjected Einstein's theory to a strong test: If observation had shown that the predicted bending of the light was absent, strong arguments would exist for refuting the theory. Such considerations led Popper to the following conclusions (Ibid.; 40; my synthesis):

1. It is easy to obtain confirmations/verifications for nearly every theory—if we look for them.
2. Confirmations should count only if they are the result of risky predictions.
3. Every good scientific theory is a prohibition—it forbids certain things to happen. The more it forbids, the better the theory.
4. Theory that is not refutable by any conceivable event is non-science.
5. Every genuine test of a theory is an attempt to falsify it, to refute it.
6. Some theories are more testable than others in that they forbid more outcomes, take so to speak a greater risk.
7. Evidence should not count as a confirmation unless it is the result of a genuine test—a serious but unsuccessful attempt to falsify it.
8. Some genuinely testable theories, when found to be false, are still upheld by their admirers—for example by introducing some auxiliary assumptions or reinterpreting the theory—this is at the price of lowering their scientific status.

---

[19]Alfred Adler (1870–1937) was an Austrian medical doctor and psychologist, founder of the school of individual psychology and especially known for introducing the concept of "inferiority complex". Adler had a major effect on the development of psychotherapy in the course of the 20th century.

Summing up the above, Popper concluded (Ibid.; 40): *"The criterion of the scientific status of a theory is its falsifiability, or refutability, or testability."*

I underscore here that Popper was not concerned with the meaningfulness, significance, utility, truth or acceptability of a theory, but only with the problem of drawing a line between statements founded in empirical science and all other statements. This has since been referred to as "the problem of demarcation". We will return to the concept of falsification in more detail later on in Chap. 4 as it is one of the major ingredients in consideration of how theories can be justified. Although many scientists today disagree with some of Popper's conclusions, they are all worthy of serious contemplation and discussion by reflective practitioners of science.

Let us now reflect on this issue of the demarcation of science from non-science/pseudo-science by considering specific several examples provided by (Lee 2000; Chap. 7), who argues that it is better to reject a few correct ideas than to accept many incorrect ones and that well-established theories should not be discarded easily. Therefore pseudo-scientists must provide strong evidence in order to overthrow conventional scientific knowledge; the more outrageous the claim, the stronger the evidence must be (Ibid.; 134).

Astrology: This is the classic case in almost all writings on non-science/pseudo-science. According to Lee, the assessments made are typically vague and this leaves much room for interpretation which makes people believe its predictions were specifically made for them, when in fact they apply to many people. He also provides evidence that when astrology is subjected to careful experiments, it fails.

Note however that (Thagard 1978 in Klemke et al. 1998; 67–69) provides strong arguments that challenge such experimental approaches to declaiming the scientific nature of astrology. In addition, Thagard argues that astrology cannot be condemned as pseudo-scientific on the grounds typically proposed, such as those dealing with its origin, its lack of a physical foundation, or its lack of verifiability. Rather, he argues that astrology is pseudo-science for the following two principle reasons that are said to provide a basis for distinguishing between science and pseudo-science:

1. It has been less progressive than alternative theories over a long period of time, and faces many unsolved problems (progressive here refers to the success of a theory in adding to the set of facts it explains and the problems it solves; my comments);
2. In spite of this, the community of practitioners has made little attempt to develop the theory towards solutions of the problems, shows no concern for attempts to evaluate the theory in relation to others, and is selective in considering confirmations and disconfirmations.

I note in this connection that Thagard also provides strong arguments *against* a statement from 1975 and signed by 192 leading scientists, including 19 Nobel Prize laureates, attacking astrology as pseudo-scientific. He does so not because he disagrees with their conclusion, but because of their arguments, each of which he argues could be applied to what are now firmly established domains of science. (Ibid.; 70–71) There is much food for thought here for reflective scientists!

Lysenkoism: The theories of Lysenko (that rejected much of what was at the time accepted by most biologists, especially Mendelian genetics) strongly affected the agricultural and biological sciences in the Soviet Union from the 1930s to the 1960s —with disastrous consequences. Lysenko's theories concerned, amongst others, the transformation of species. He argued that young plants do not compete with others of the same species in a given environment and that if they are planted in clusters, most will sacrifice themselves for the good of the species and let one or a few survive. This was in line with the collectivist-ideology of the Communist regime and led Stalin to order the planting of vast belts of forests in the grasslands of the southern Soviet Union so as to alter the climate and make the region better for agriculture. The results were disastrous for the grasslands and the farmers. Lysenko also claimed that a species can be transformed into other species quickly (in opposition to most Darwinian-based theories of evolution), and that if a plant lives in an environment where it is poorly suited to survive, within that plant, seeds or buds of a better adapted species will develop. Many scientists supported his work, claiming that they had experimental observations that supported his theories; this is believed to have been a result of the great pressures on them by the government as well as of the chance of improved working conditions if they supported Lysenko's theories. Those scientists who opposed his thinking were removed from their positions, or suffered a much worse fate. There was no real peer review; no open and honest criticism was permitted.

Creation "Science": This "science" is based on the belief that the book of Genesis in the Old Testament of the Bible provides an accurate description of Earth history. According to the interpretations made by the supporters of "Creationist Science" (often also referred to as "Intelligent Design", although this term implies a less strict or fundamentalist interpretation of the Bible and emphasizes instead a rejection of Darwinian evolution), the Earth is 6–10,000 years old. According to this "science", fossils are remnants of the Great Flood from the time of Noah. In particular, the theory of evolution is said to be invalid because according to the Bible all plant and animal species were created in their present form.

Creationism is promoted in particular by certain fundamentalist Christian groups in the USA and is supported by a number of scientists as well as politicians who share its religious foundation.[20] Its method is to start with "theory" (the interpretations of the book of Genesis) rather than with existing evidence and to seek evidence that supports the creationist perspectives and that goes against other theories which are a result of established science. For example, Creationists challenge the technique of radioactive dating that shows the Earth to be roughly 4.55

---

[20]Not only fundamentalist Christians support creationist perspectives. For example, the spiritual organisation, the Brahma Kumaris, based in Mount Abu in Rajasthan, India, which emphasizes social and environmental action, also supports similar perspectives. They do so within a framework of a cyclical cosmology, whereby history repeats itself every roughly 5000 years when a new phase of creation occurs. It is interesting to note that in spite of the fact that the organization ascribes to a cosmology that is rejected by modern science, since 1983 it achieved consultative status with the Economic and Social Council at the United Nations.

billion years old. They use fossil records to argue that the theory of evolution is false because fossils showing transition from one species to another are not found—while biologists say they have such evidence. Likewise, Creationists argue there is no widely supported theory for the origin of life on Earth, and therefore argue that this must have been divinely accomplished.

In a nutshell, while the theory of evolution as well as the theory that the Earth is billions of years old is well supported by evidence, Creation Science is poorly supported by empirical evidence.

With the provocative title: "Believing Where We Cannot Prove," (Kitcher 1982/1998) provides an in-depth treatment of the perspective by Creationists that the theory of evolution is wrong and that it is not part of science because, according to them, science demands proof and evolution cannot be proved. Kitcher's point is that "...science is not a body of demonstrated truths. Virtually all of science is an exercise in believing where we cannot prove. Yet scientific conclusions are not embraced by faith alone". In his concluding remarks, he notes: "Like Newton's physics in 1800, evolutionary theory today rests on a huge record of successes. In both cases we find a unified theory whose problem-solving strategies are applied to illuminate a host of diverse phenomena. Both theories offer problem solutions that can be subjected to rigorous independent checks. Both open up new lines of inquiry and have a history of surmounting apparent obstacles. The virtues of successful science are clearly displayed in both." (Ibid.; 32/78)

UFOs (Unidentified Flying Objects): The evidence provided by believers in UFOs is challenged by many scientists who claim that human perception and memory are not reliable under conditions like those where UFOs are reported to have been spotted. Nevertheless, there are many respected people that have given detailed descriptions of UFOs; see e.g. the website http://www.ufoevidence.org/researchers/index.asp and the website of the international organisation promoting research on UFOs: http://www.ufoevidence.org/.

Among the other domains that Lee (2000; Chap. 7) categorises as pseudo-science are *paranormal/extrasensory phenomena and perception* and *health related pseudo-science* (including essentially the whole field of so-called alternative medicine).

In contrast to Lee's position, which is mainly in the form of common sense and scientific reasoning rather than empirical investigation, in *The Conscious Universe: The Scientific Truth of Psychic Phenomena*, (Radin 1997) provides extensive empirical evidence based on controlled experiments of the existence and effectiveness of the very paranormal/extrasensory phenomena that Lee categorises as pseudo-science. Included in Radin's investigations are mind-to-mind connections (telepathy), perception of future events (precognition), perception of distant objects and events (clairvoyance) and mind-matter interaction (psychokinesis).

In this connection I note that the history of science clearly demonstrates that what at one time was considered to be non-science by the scientific community later on became accepted as powerful pillars of existing science; "Newton's notions of a field of force and of action at a distance and Maxwell's concept of electromagnetic waves were at first decried as 'unthinkable' and 'contrary to intuition'." (Feller 1968; 2)

According to Kuhn: "Closely examined, whether historically or in the contemporary laboratory, that enterprise (normal science; my comment) seems an attempt to force nature into the pre-formed and relatively inflexible box that the paradigm supplies. No part of the aim of normal science is to call forth new sorts of phenomena; indeed those that will not fit the box are often not seen at all. Nor do scientists aim to invent new theories, and they are often intolerant of those invented by others." (Kuhn 1970; 24)

So it appears that demarcating science and pseudo-science is not all that simple and straightforward. Although a number of positions have been presented here regarding the distinction between what is science and what is non-science or pseudo-science, what Popper referred to as the problem of demarcation is not solved once and for all. While you may have an intuitive feeling regarding such a distinction, the following exercise indicates that the intuition of even highly educated people, including those with backgrounds in science, can lead to conclusions and behaviour that are at odds with scientific thinking.

2.3.1 An Exercise as to Science and Pseudo-Science

The following is a "case study" that concludes the section of (Klemke et al. 1998) dealing with "Science and Pseudo-science" (pp. 99–100):

"The following is a letter which was received by the editor of a science journal.
'Dear Sir:
I am taking the liberty of calling upon you to be the judge in a dispute between me and an acquaintance who is no longer a friend. The question at issue is this: Is my creation, umbrellaology, a science? Allow me to explain…For the past 18 years assisted by a few faithful disciples, I have been collecting materials on a subject hitherto almost wholly neglected by scientists, the umbrella. The results of my investigations to date are embodied in the nine volumes which I am sending to you under separate cover. Pending their receipt, let me describe to you briefly the nature of their contents and the method I pursued in compiling them. I began on the Island of Manhattan. Proceeding block by block, house by house, family by family, and individual by individual, I ascertained (1) the number of umbrellas possessed, (2) their size, (3) their weight, (4) their colour. Having covered Manhattan for many years, I eventually extended the survey to other boroughs of the city of New York, and at length completed the entire city. Thus I was ready to carry forward the work to the rest of the state and indeed the rest of the United States and the whole known world.

It was at this point that I approached my erstwhile friend. I am a modest man, but I felt I had the right to be recognized as the creator of a new science. He, on the other hand, claimed that umbrellaology was not a science at all. First, he said, it was silly to investigate umbrellas. Now this argument is false, because science scorns not to deal with any object, however humble and lowly, even to the 'hind leg of a flea.' Then why not umbrellas? Next, he said that umbrellaology could not be recognized as a science because it was of no use or benefit to mankind. But is not truth the most precious thing in life? Are not my nine volumes filled with the truth about my subject? Every word in them is true. Every sentence contains a hard, cold fact. When he asked me what was the object of umbrellaology I was proud to say. "To seek and discover the truth is object enough for me." I am a pure scientist; I have no ulterior motives. Hence it follows that I am satisfied with truth alone. Next, he said my truths were dated and that any one of my findings might cease to be true tomorrow. But this, I pointed out, is not an argument against umbrellaology, but rather an argument for keeping it up to date, which exactly is what I propose. Let us have surveys monthly, weekly, or even daily, to keep our knowledge abreast of the changing facts. His next contention was that umbrellaology had entertained no hypotheses and had developed no theories or laws. This is a great error. In the course of my investigations, I employed innumerable hypotheses. Before entering each new block and each new section of the city, I entertained a hypothesis as regards the number and characteristics of the umbrellas that would be found there, which hypotheses were either verified of nullified by my subsequent observations, in accordance with proper scientific procedure, as explained in authoritative texts. (In fact, it is of interest to note that I can substantiate and document every one of my replies to these objections by numerous quotations from standard works, leading journals, public speeches of eminent scientists, and the like.) As for theories and laws, my work represents an abundance of them. I will here mention only a few, by way of illustration. There is the Law of Colour Variation Relative to Ownership by Sex. (Umbrellas owned by women tend to a great variety of colour, whereas those owned by men are almost all black.) To this law I have given exact statistical formulation (See vol. 6, Appendix 1, Table 3, p. 582.) There are curiously interrelated Laws of Individual Ownership of Plurality of Umbrellas and Plurality of Ownership of Individual Umbrellas. The interrelationship assumes the form, in the first law, of almost direct ratio to annual income, and in the second, in almost inverse relationship to annual income. (For an exact statement of the modifying circumstances, see vol. 8, p. 350.) There is also the Law of Tendency Toward Acquisition of Umbrellas in Rainy Weather. To this law I have given experimental veri-fication in Chap. 3 of Volume 3. In the same way I have performed numerous other experiments in connection with my generalizations.

Thus I feel that my creation is in all respects a general science, and I appeal to you for substantiation of my opinion ....'"

Dear reader, please compose a reply to this letter based on the chapter you have just read.

Based on considerable experience with my former students in science I predict that it will not be easy for you to develop a strong, rational argument for catego-rizing "umbrellaology" as science or as non-science.

2.4 The Aims and Claims of Science

Earlier I referred to the often stated distinction between science as *creating* knowledge (what is generally referred to as "pure science") and science as *applying* knowledge ("applied science"). Clearly, these have different ends.

The aims associated with pure science can be said to deal primarily with the aims of the individual scientist and with that of science as an institution, rather than with its utility in the form of contributions to technologies and products.

If we consider first the aims of the *individual scientist*, we can say that these most likely include at least the following: a) the pursuit of knowledge—to become a more knowledgeable, more competent scientist, b) the satisfaction that is derived from using one's intellectual powers—to grow as a person and to receive recognition from one's colleagues, and, hopefully, c) to contribute to the well-being of society and of the world.

When we consider the aims of *science as an institution*—and here I emphasize the natural sciences—we tend to focus on establishing true and generalizable statements about the world, i.e. on improving the existing descriptions, explanations and predictions as to physical reality. For example, according to (Kitcher 1982 in Klemke et al. 1998; 79): "Scientific investigation aims to disclose the general principles that govern the workings of the universe. These principles are not intended merely to summarize what some select groups of humans have witnessed … Science offers us laws that are supposed to hold universally, and it advances claims that are beyond our power to observe. The nuclear physicist who sets down the law governing a particular type of radioactive decay is attempting to state a truth that holds throughout the entire cosmos and also to describe the behaviour of things that we cannot even see."

But this apparently straight-forward and clear aim "to disclose the general principles that govern the workings of the universe" is perhaps not as clear, or at least not as generally applicable, as we tend to think it is. According to the internationally recognized founder of "the Copenhagen School" in atomic physics, Nobel Prize laureate Niels Bohr (1885–1962), it is "wrong to think that the task of physics to find out how nature *is*. Physics concerns what we can *say* about nature." (Pais 1991; 427) In other words, according to Bohr, at least at the micro-level of the atom and sub-atomic particles, science does not aim at presenting a true picture of nature, but only of improving our ability to *speak about* nature.[21]

This was in line with Bohr's philosophical reflections on our ability as language-users to describe events in a quantum reality that is so far removed from our experience as humans. We will return to this subtle distinction in Sect. 2.7 of this chapter when we discuss the concepts of realism and anti-realism.

Over and above aims as to disclosing or speaking about the principles that govern the workings of the universe, science can also be considered as having the aim of improving its *own ability* to describe, explain and predict. We generally think of applied science as having the aim of improving our ability to control, plan and utilize resources (physical and social) for practical purposes, in contrast to pure or basic science's more fundamental aims of generating knowledge for the sake of

---

[21]From this perspective, at the quantum level, the laws of nature no longer deal with the elementary particles but with our knowledge of them such that, for example, an electron does not have properties such as its position or momentum unless it is observed via measurement. According to (Morrison 1990; 41), "… the central question of wave-particle duality is not 'can a thing be both a wave and a particle?' Rather, the question is, 'can a thing be *observed* behaving like a wave and a particle in the same measurement?' Bohr's answer is no: in a given observation, quantum particles exhibit *either* wave-like behaviour (if we observe their propagation) *or* particle-like behaviour (if we observe their interaction with matter). … Note that by restricting ourselves to observed phenomena, we are dodging the question: 'what is the nature of reality behind the phenomena?' Many quantum physicists answer, 'there is no reality behind the phenomena.'"

knowledge. But since science also aims at improving its own capabilities to generate valid knowledge, it aims not only to improve our knowledge of the world but also *to improve our ability to improve our knowledge*. In this way, the borderline between pure and applied science becomes amorphous. The development of electron microscopes, laser technology, cloud chambers, nanotechnology and radio telescopes are examples of how scientific theories indirectly extend our perceptual capabilities, just as the development of computer models, for example as to development in climatic conditions, indirectly extend our cognitive abilities.

Furthermore, if a primary aim of science is to develop general laws from specific observations, a number of branches of science have difficulty in living up to this aim. Physics is the domain where generalizations based on empirical investigation are relatively easy to make (not that performing research in physics is an easy matter!). This is due to the fact that, with many exceptions of course, the data that physicists work with has relatively small variances compared to analyses in other fields. In addition, physics has a foundation in first principles dealing with the fundamental nature of matter and energy. This ability to build upon first principles is reduced as one moves further away from the well-structured domain of physics, with its apparent isomorphic relationship with mathematics (whereby the physical universe appears to be subject to precise description using the language and logic of mathematics), to that of chemistry and other natural sciences, to the life sciences such as biology and botany, to the behavioural sciences such as psychology and anthropology, and to the social sciences such as sociology and economics. In each case, as we move further away, so to speak, from physics, the data tend to have larger and increasing variances, such that when we reach the behavioural and social sciences, generalizations in the form of precise and widely accepted theories/laws do not exist—even though these branches of science with living organisms, organisations, and cultures as their objects of study, have their own norms and methodologies.

I note as well that there is also a very large variance in the data in some areas of scientific investigation where controlled experimentation is not possible, and where perhaps only one or a very limited number of observations is possible, and perhaps only very indirectly. Consider for example the investigation of such historic processes and phenomena as the origin of life on Earth, organic macroevolution, and the development of mountain ranges—and more modern processes and phenomena such as the effects on people and the environment of global warming and of massive catastrophes such as melt-downs at nuclear power stations (e.g. Chernobyl in Ukraine, 1986), the emission of dioxin into the air (Bhopal, India 1984), or the tsunami that devastated parts of southeast Asia in 2004. Since the relevant data generated by such phenomena are not obtained via experimentation they tend to have relatively large variances, whereby generalization becomes extremely difficult in these fields of study. Which is not the same as saying that scientists cannot and should not study such phenomena; in fact, just the opposite is true, and the catastrophes referred to have provided scientists with a wealth of data that enable research into their causes and effects.

Returning now briefly to applied science in general, the closer the scientific endeavour is focused on application the greater is the focus on empirical investigation rather than on first principles. According to (Siever 1968; 38): "... we have fields, such as engineering, where empiricism is the order of the day simply because there is no generally valid group of first principles from which to operate."

Thus, although the overall aims of science as an institution may be the same, no matter which branch of science we focus on, there are major differences in the ability of the various domains of science to provide accurate descriptions, explanations and predictions of physical reality.

Let us now delve a bit deeper into the claims of science and follow the arguments of (Gauch 2003; 28–41, 72), who argues that all natural scientists implicitly make four principle claims and that it is the simultaneous assertion of all four of these interconnected claims that fully expresses science's boldness.

Rationality: Rationality is good reasoning that regulates belief and guides action. Rational persons base their beliefs on evidence and reasoning. Scientists are rational and they use such beliefs to guide their actions so they are in accord with the goal of science of developing true statements (theories and laws) regarding the universe. I note however that stories told by creative scientists throughout the ages indicate that scientific discovery is often kick-started by *a*-rational processes (Koestler 1989; 208). We will consider the limits of rationality in science in greater detail in Sect. 2.5.3.

Truth: From a scientific perspective, truth is a property of a statement; true statements correspond with reality and truth is determined by the objective state of nature. In other words, according to this so-called "correspondence theory of truth" there is a correspondence between the external physical world of objects and events and the internal mental world of perceptions and beliefs—and where reality has priority over belief. Such truth claims can be made with various levels of confidence but not with absolute certainty. As we have reflected on earlier, as observers, analysers, and mediators we are fallible; scientific knowledge is tentative and subject to revision.

Objectivity: The dominating idea of objectivity in science, what we might refer to as classical objectivity, concerns observer-independent knowledge, i.e. knowledge about physical objects that exist independently of our observations. Such knowledge can be tested and verified so that consensus will emerge among knowledgeable persons. Thus there is a strong link between objectivity and inter-subjective agreement. It is this link that characterizes the consensual activity in what we earlier referred to as the "scientific community" where objective knowledge is considered to transcend political, religious and cultural divisions.

I note however that, as we recently touched on, at the micro-level of atoms and quarks, the indeterminacy of the quantum world implies that the classical idea/ideal of an objective reality must be replaced by the concept of an observer-dependent reality where what one measures depends on the apparatus used and on what has been chosen to observe. As strange as this may sound to observers who do not investigate quantum weirdness, one cannot speak of events in the micro-world without observing them.

I note too that the focus in the natural sciences on observer-independent truths about external physical objects is perhaps the major characteristic that differentiates the natural from the social sciences. In the social sciences understanding and decision making are often more in focus than truth, and the "external objects" are humans or collectivities of humans (groups, organisations) that may be affected by, and affect, the researcher's behaviour, rather than observer-independent physical entities.

Realism: Realism is the contention—the metaphysical presupposition—that physical objects exist independent of human thought and that our minds and senses enable access to the physical world so that it can be reliably known—that there is a correspondence between human thought and an external, independent reality. This is clearly closely related to the concept of objectivity considered above. We will return to the notions of realism and its counterpart, anti-realism, in Sect. 2.7 of this chapter.

2.4.1 Science and Democratic Development

Finally, over and above such aims of science, there are also aims that are far more general and involve society as a whole. In order for a democracy to function well, citizens have an obligation to be informed on issues that face the government. Since many such issues have a scientific component, for example, those dealing with environmental and educational policy, it follows that an aim of science is/should be to help citizens to develop a scientific literacy so as to be better able not only to make their own decisions, but also to contribute to the decisions made by their political representatives (Lee 2000; 7).

This aim with regard to democratic development is far from being achieved and the ability to achieve it is perhaps decreasing, rather than increasing, due to the accelerating increase in scientific knowledge. On the one hand, increased specialization within the sciences makes it increasingly difficult for ordinary citizens, even those who are "scientifically literate", as well as for professional scientists too for that matter, to understand the results and implications of highly specialized scientific investigations. On the other hand, when we face more complex, interdisciplinary investigation, even greater demands may be placed on us if we are to be able somehow, directly or indirectly, to contribute to the decision making processes of our political representatives.

In order to obtain the level of scientific insight that can empower a citizen to contribute to democratic development may require both a reasonably strong background in one or more fields of science as well as considerable investments of time and energy. Consider e.g. the demands on our understanding if we are to be able to meaningfully contribute to societal debate on such issues as global warming, the long term implications of using nuclear power for the generation of electricity, the effects of permitting patents to be taken out on genetic structures and computer software, the effects of building dams on local and regional economies as well as on the environment, the effect on our morals and belief systems of genetic engineering (e.g. cloning, stem cell research), etc. In each case, the ability of citizens to decide for themselves and to contribute to democratic/parliamentary decision making presumes an ability to deal with complex issues that far exceed the talents and knowledge of most ordinary human beings, including politicians and scientists, as such issues often involve not only scientific but also social, economic, ecological and ethical aspects!

An article in the major Indian daily newspaper, *The Hindu* (January 28, 2007), "Taking science out of the labs", presents a more optimistic perspective on democracy and science. Rather than focusing on an obligation of science to inform the citizenry, it deals with the participation of ordinary people in the conduct of experiments. The article commences as follows: "Radical biologist Rupert Sheldrake is working to change the way people think about science." It continues by citing Sheldrake, who is Perrick-Warwick Scholar at Trinity College, Cambridge University:

> I think we need to change not only the content of science and what it is about but also the way science is done. Western science was shaped by the needs of the industrial revolution, the needs to produce machines, so the image of scientists was of an elite priesthood. This spread across Europe and to India too. Science excludes ordinary people. A democratic society like ours, where computing power is available to everybody, creates a condition for a new way of doing science. ... Participatory research will become very big, very fast, as this transformation will happen through the internet." The article notes that Sheldrake's research "involves ordinary people conducting experiments at home and through his website: www.Sheldrake.org. His highly successful participatory online experiments have led Microsoft, Google and, recently, AOL, to show interest in his work.

These thoughts as to the difficulties—and possibilities—involved in living up to major aims of science pave the way to the next issue to be considered in these reflections on science: its limitations.

2.5   The Limitations of Science

When we use a tool or technology in our daily lives we are usually interested not only in its capabilities, but also in its limits. A lack of such knowledge can lead to an ineffective use of the tool (the battery on my laptop will only supply me with power for a limited period of time), to its destruction (cameras may be sensitive to heat; frozen food can only maintain its freshness for a limited period of time)—or even to the user's own destruction (too much of a pain-killer can kill you). Certainly therefore reflections on research methodology should also include consideration of the limitations of science.

According to the American Association for the Advancement of Science (AAAS), which is the world's largest scientific society and the umbrella organisation for roughly 300 societies and academies, "Being liberally educated requires an awareness not only of the power of scientific knowledge but also its limitations." Furthermore, according to the AAAS, learning the limits of science "should be a goal in all science courses". (AAAS 1990; 20–21)

We have already briefly considered limitations that are a result of our limited abilities to objectively observe physical reality. These limitations arise due to our physical characteristics (the limits of the senses in perceiving), our limited experiences and our varying expectations. We will now consider some other principle limitations of science—limitations in its "special way of looking at the world" and its "method of discovering, generating, testing, and sharing true and reliable knowledge about physical reality."

2.5.1   Presuppositions—and Science as Faith

Let us start by considering the proposition that just as geometry is based on a number of axioms (statements that are not proven or demonstrated to be valid but whose truth is taken for granted), so is science based on a set of presuppositions or beliefs that cannot be proved by logic or firmly established by evidence. *IF* this proposition is correct, then it would be reasonable to conclude that since science rests on a foundation of statements that are assumed to be true but that cannot be proved/firmly established, *science fundamentally is a matter of faith*! A statement that many scientists would find shocking due to their lack of serious reflection on the foundations of their fields of endeavour and their strong faith (!) in science as providing true and objectively testable knowledge of physical reality.

But is the proposition true? This appears to be the case. There are certainly a number of what we might be tempted to call common sense propositions underlying science. For example that the physical world exists, that it is orderly, that our sense perceptions are generally reliable, and that rational thought can synthesize the ordered reality of the physical world and the observations of our senses into true and reliable knowledge—in other words, that the physical world is comprehensible. Later on, we will reflect in some depth on these most fundamental presuppositions of science in connection with the treatment of "realism".

Here, let us instead start by reflecting on another of the major presuppositions, part of the metaphysics one might say, of science: that we are (or at some time will be) able, either directly or indirectly, with the aid of technology that extend our perceptual and cognitive abilities, to observe or measure every physical object/phenomenon that exists. In other words, that our intellect and senses enable us to access all of objective reality.[22] By this we do not require that we have access to these objects and phenomena here and now, since some may at present be outside of the direct reach of our measuring instruments, for example sub-atomic entities or some objects/phenomena in outer space or at the middle of the Earth or that occurred earlier in time. What we are talking about here is the implicit assumption

---

[22]As we have briefly considered earlier, quantum physics in particular presents us with the following dilemmas as to what we even mean when we speak of "objective reality": (1) the properties of microscopic entities are not, in general, well defined *until* they are measured, and (2) since interaction between an observer and the observed is unavoidable, and since a microscopic system cannot be observed without the process of observation altering some its properties, it is meaningless, at least at the level of the microscopic world, to speak of an objective world that exists external to and independently of our perceptions and therefore to speak of physics as studying objective reality (Morrison 1990; 7–8).

of the natural sciences that whatever exists is *potentially* subject to our observation, experimentation and measurement.[23]

To commence reflection on this implicit assumption, suppose that there are in fact phenomena that we do not and cannot have access to via our senses and intellect, such as e.g. the generation of thoughts or so-called psychic/paranormal phenomena.

Let us consider first the generation of thoughts. Most scientists would agree that at present we are limited in our ability to observe and measure such processes, even though they could point to the increased focus by neuroscience on analyses of the brain via scanning technologies. They might very well contend that since (and here comes another major presupposition) everything that exists has a physical cause, thought generation, self-awareness and all of conscious activity will at some time be observable/measurable.[24] They might also argue that all conscious activity will eventually be reducible (here comes yet another major presupposition!) to chemical, electrical and biological activity in the brain and central nervous system. Therefore, according to this materialist and reductionist line of thought, at some time, when our understanding of the brain and our measuring instruments are sufficiently improved, consciousness and the generation of thoughts will be observable, measurable, and explainable.

However, these assumptions can be seriously challenged. For example, extensive analyses in *Irreducible Mind* (Kelly et al. 2007) challenge what the authors consider to be the apparent consensus among scientists that "Mind and consciousness are entirely generated by—or perhaps in some mysterious way identical with—neurophysiological events and process in the brain." Their analyses provide theoretical arguments and empirical evidence that the mind cannot be understood as the product of simple physiological sensations or processes and that it is itself a "fundamental elementary and causal principle in nature" (Ibid.; 56). In *Beyond Physicalism* (Kelley et al. 2015), the theory-oriented sequel to the huge 2007 book, these analyses are further developed and synthesized within an overarching framework of the relationship between science and spirituality.

Let us now also consider psychic/paranormal phenomena, for example so-called "near-death experiences", typically occurring to individuals close to death (e.g. cardiac arrest, near-drowning, or a motor vehicle accident) or in situations of

---

[23]In some domains of natural science, such as e.g. astronomy and geology, this potential is extremely limited due to limitations on the ability to experiment. Here scientists must primarily rely on simple observation to generate their data. So this can limit their ability to focus on a limited number of independent variables and thus to generate data with low variances. I note however that the distinction between simple observation and experiments is not that clear. For example, in 2005 a rocket was sent into the middle of a comet in order to obtain better observations of its contents. This was neither passive observation, nor a controlled experiment.

[24]Buddhists, for example, would contest this, arguing that there is a continuum of consciousness but no physical basis in that continuum for an individual, solid, permanent and autonomous self. From this perspective, we are only aware of consciousness inasmuch as it is qualified by an object, and only contemplative practice—not scientific investigation—enables us to see the nature of the mind (Revel and Ricard 1998; 29–35).

intense physical or emotional danger. They are often characterized by over-whelming feelings of peacefulness and well-being, out-of-body experience (the impression of being located outside one's physical body), a flash-back of one's life, and meeting and communicating with deceased relatives (Moody 1975), (Charland-Verville et al. 2014). A scientist unfamiliar with the considerable liter-ature on such experiences[25] and who met someone who said that she had such an experience might very well reply that the person most likely suffered from a hal-lucination or that any number of such scenarios can be created by the imagination—although strongly felt, the experiences could not be "real" and the person's state-ment cannot be considered to be a true/valid/reliable observation statement (a "fact") from a scientific perspective. Suppose then that the scientist was presented with similar statements by a large number of otherwise reliable, trustworthy and respected people, including other scientists, all of whom told of their own near-death experiences? Once again the scientist would most likely reply that such experiences cannot be considered to be objective observations and that no con-trolled experiment can be devised that could confirm or falsify their statements. And since the scientist herself did not have such experiences, she would reject such reports since they would challenge a number of the implicit presuppositions underlying the profession of science.

This illustrates a methodological challenge: whether "scientific method" (see Chap. 7) which emphasizes objectivity and replicability of observations can be supplemented with subjective and non-replicable observations of individuals—and if so, how this could/should be done. A striking example of the challenge is pro-vided by (Alexander 2012). While clearly subjective and personal, this experienced neurosurgeon's story of his near-death experience appears to provide evidence as to the existence of consciousness independent of the brain and that the brain essen-tially performs a filtering function that enables us to limit the amount of information to be processed, which would otherwise be overwhelming, thereby facilitating a normal, conscious daily existence.

However, his non-replicable experience would not be accepted as valid evidence in a traditional scientific investigation, in spite of the details provided by a scientist who has considerable knowledge of the brain, having performed thousands of brain operations in the course of his career. We have earlier (in Sect. 2.3 in our discus-sions of non-science/pseudo-science) reflected on such limitations on our ability to expand our knowledge base by including information based on personal experi-ences due to the demands from scientific method as to objectivity and replicability as well as well-grounded theoretical explanations; personal experiences cannot provide objective, replicable data.

Therefore, the assertions of people who have had near-death experiences do not in general threaten scientists' convictions as to the "truth" of the implicit meta-physical assumption that whatever exists is potentially subject to observation and

---

[25]See e.g. the website of IANDS, The International Association for Near-Death Studies: http://iands.org/nde-stories.html.

measurement; most scientists would simply deny the relevancy/truth of the testimony. Since the personal experiences of individuals cannot be observed by others (their "observations" cannot be replicated by carrying out experiments whereby "neutral" and "objective" scientists can experience the same observations or carry out tests), a reverse logic is employed: The experiences do not exist as other than figments of the imagination. As mentioned in Sect. 2.2 ("What is science?") of this chapter, empirical science has a far more difficult task when it tries to demonstrate that something does not exist, than that it exists.

However, there are an increasing number of respected scientists who are trying to amass and analyse data that support what otherwise have been considered anecdotal evidence as to e.g. parapsychological phenomena, including telepathy, perception at a distance, mind-matter interaction and mental interactions with living organisms. Radin (1997) refers to research at highly recognized institutions as to the existence of such phenomena, and relies in particular on the use of meta-analyses.[26] Not only may this development potentially contribute to a deeper understanding as to such phenomena, but also contribute to an expansion of research methodology!

So if we assume, in contrast to the traditional presuppositions of the natural sciences, that reality is more inclusive than we can observe and measure with our senses, intellect and technologies, then we must conclude that the natural sciences as we know them are limited in their ability to describe, explain and predict. It would also mean that the sciences unknowingly have erected barriers that limit our ability to meaningfully deal with phenomena that are outside of the reach of our senses (even when these are "extended" via technology). One might even say that if the scientific community denies that humans can have reliable access to phenomena by other than the methods of science, science would be totalitarian.[27]

Thus, although traditional science has for centuries distanced itself from metaphysics, one can nevertheless argue that it is a metaphysical choice—let us call it an act of faith!—of the natural sciences that humans, with the help of their cognition and senses (and language) are or will be able to observe, describe and measure all of the phenomenal universe, which is said to exist independent of our cognition and senses.

Note that I do not imply that even though science cannot justify its presuppositions by referring to its own methodology, that its metaphysical choices—its acts

---

[26]These are analyses of analyses where the units of observation are, so to speak, the independent studies rather than the individual observations of an individual study. Such analyses are becoming increasingly important as there may be individual studies that provide conflicting results or that may be based on too small a sample size. Ideally, the meta-analysis is equivalent to a single study with the combined size of all the original studies.

[27]It is perhaps appropriate here to recall the condemnation of the great Galileo Galilei, who was twice put on trial by the so-called "Inquisition" of the Roman Catholic Church for his support of the Copernican heliocentric view of the universe. In 1632, under threat of death, he was compelled to retract his views and to spend the remainder of his life under house arrest. The point being made here is that whoever appears to have a "patent" on the truth and the power to determine what is true and false, must use this power with great discretion and humility. We will return to this theme in Chap. 10, Ethics and Responsibility in Scientific Research.

of faith—are arbitrary or unstable. Rather, the point is that reflective scientists should be aware of the presuppositions that underlie their research and, when appropriate, they should raise them from the level of implicit to explicit assumptions, for example when they perform their analyses and publish their results.

2.5.2 Fundamental Questions as to Physical Reality

Another limitation of the natural sciences is their inability to provide strong answers to a number of fundamental questions regarding physical reality. Although scientists recognize the challenges involved, most tend to shy away from confronting them, both as subjects to be professionally researched and as questions to be dealt with at a personal, existential level. This is due most likely to the holistic nature of the questions that can be far more inclusive and methodologically demanding than one traditionally faces in one's own field of specialization. Included here are such questions as:

> How did life evolve from non-life?
> Has everything (including life and consciousness) existed somehow in a "seed state" or been programmed since Planck time (said to be the smallest measurable unit of time, roughly 10^{-43} s after the presumed "Big Bang")?
> Is the universe a random event?
> And perhaps the biggest question of all: Why is there anything rather than nothing?

Commenting on such questions, the American physicist Heinz Pagels (1939–1988), former executive director of the New York Academy of Sciences, wrote: "What is the universe? Is it a great 3-D movie in which we are all unwilling actors? Is it a cosmic joke, a giant computer, a work of art by a Supreme Being, or simply an experiment? … I think the universe is a message written in code, a cosmic code, and that the scientist's job is to decipher that code." (Pagels 1982; 343)

Closely related to the above questions regarding physical reality are fundamental questions dealing with our experiences as human beings. We have already briefly referred to the question as to where thoughts come from. We can extend this to other areas such as: where do morals come from? preferences? love? aesthetics? conscience? loyalty? faith? awareness and self-awareness? Can these all aspects of consciousness be observed and explained by science—perhaps reduced to molecular/genetic/chemical/quantum-mechanical explanations? According to Erwin Schrödinger (1887–1961), Nobel Prize laureate in Physics in 1937: "The image of the world around us that science provides is highly deficient. It supplies a lot of factual information, and puts all our experience in magnificently coherent order, but keeps terribly silent about everything close to our hearts, everything that really counts for us." (Schrödinger 1954 as quoted by Revel and Ricard 1998; 185)

The noted Zen Buddhist rishi (sage) Steve Hagen (2003; 229) notes that the way that the natural sciences look upon the relationship between consciousness and matter may be 180° out of phase and that we may have put the cart (matter) before the horse (consciousness): "… unlike consciousness, which is directly experienced, matter is always secondary—that is experienced indirectly via mind. This is our actual, immediate, direct experience—it's purely mental, not physical. In short, physical reality cannot be fully accounted for apart from consciousness. Yet it's not at all clear that matter is necessary to account for consciousness."

It has not been my purpose here to pursue this issue of whether consciousness has a material basis, a question we will also meet several times in the sequel. Rather, it has been to underline that just as there exist fundamental questions as to the nature of physical reality, e.g. "why is there anything rather than nothing?" so too are there fundamental questions that deal with such a basic phenomenon as consciousness—something we are all aware of and mean that we possess, yet do not know why we have it or what its source is. It is the inability of the natural sciences (at least at present) to provide strong answers to such fundamental questions that constitutes a limitation of science.

Closely related to the above is the fundamental question of the relationship between a physical reality "out there" and our perceptions of that reality. The sciences dealing with cognition make it clear that we do not and cannot experience the world directly. Our knowledge of the world emerges in the mind. And as we briefly have touched on earlier and will shortly consider in more depth, scientific investigation often has to deal with phenomena that cannot even be perceived by our senses, even with the aid of the most advanced technologies, e.g. quarks and black holes. So it is perhaps not too strong an ontological statement that the world that is investigated by the natural sciences is *not* just an external physical world, rather it is also a product of our senses and consciousness; we cannot perceive the world in any other way.

2.5.3   Rationality

Another limitation characterizing science deals with the relationship between means and ends. Just as natural science, with its emphasis on rational investigation and analysis, cannot justify its own presuppositions, neither can it help us to decide whether a goal ought to be pursued or not; that is a matter of choice, where values, not truth, determine the outcome. Science can, at best, only help to determine which means are well suited to fulfilling pre-established goals. Thus, the natural sciences have contributed to the development of knowledge and technologies that can be used to serve ends that can be considered both beneficial and destructive. An example is knowledge regarding nuclear fission that can be used both to create weapons of mass destruction and to generate electricity. Another example is from the medical sciences. Medical science has established the veracity of the conditional statement that if we wish to deliver an incurably ill person from prolonged physical suffering, then a large dose of morphine affords an effective means of so doing; it reduces pain but, depending on the dose, can also lead to the patient's peaceful death. However medical science may also indicate ways of prolonging the patient's life—and thereby his or her suffering as well. This leaves open the question of whether it is right to give the one goal (relief of suffering) precedence over the other (preserving life). Such value judgments are not amenable to scientific analysis; they are simply outside the realm of science. I myself was once compelled to make such a value judgment towards the very end of my mother's terminal illness; the information I received from the doctors attending my mother had a scientific foundation—but the decision as to what was the "best" or "right" thing to do could not, morally or scientifically (!) be made by practitioners of science.

According to the Nobel laureate in Economics in 1978 Herbert A. Simon (1916–2001) who wrote extensively on the limits of reason: "Reason, then, goes to work only after it has been supplied with a suitable set of inputs, or premises. If reason is to be applied to discovering and choosing courses of action, then these inputs include, at the least, a set of *should's*, or values to be achieved, and a set of *is's*, or facts about the world in which the action is to be taken. Any attempt to justify these *should's* and *is's* by logic will simply lead to a regress to new *should's* and *is's* that are similarly postulated. ... We see that reason is wholly instrumental. It cannot tell us where to go; at best it can tell us how to get there." (Simon 1983; 7)

In other words, rationality cannot rationally justify itself. Thus, the aim of rational justification that not only underlies science, as was demonstrated earlier in this chapter in Sect. 2.4, but also much of human behaviour, is itself a choice that cannot be rationally justified. On the other hand, such a choice is clearly not arbitrary. Science, at least as it developed since the time of the so-called Scientific Revolution, has been built on a foundation of rationality. And this has enabled it to provide descriptions, predictions and explanations that are consistent with observations—and as a result, knowledge and technologies that could not possibly have been developed in the absence of such rationally-based descriptive, predictive and explanatory power. This, in turn, suggests that a choice of rationality is consonant with the structure of the world; further reflections on this fundamental assumption are provided in Sect. 2.8 of this chapter dealing with mathematics and science.

Continuing this line of thought, consider next a controversy that has been framed under the provocative name of "Science Wars" and that deals with the traditional perspective on science that it is a rational method of inquiry that provides objective truth about physical reality. Perhaps the origin of this "war" can be traced back to some of the past century's most influential reflections on the meaning and limits of science by the British philosopher of science Sir Karl Popper. As we saw earlier, Popper (1957 in Klemke et al. 1998) offered his demarcation criterion of "falsifiability" to separate science from non-science—but at the cost of separating science from truth. In a nut-shell, Popper argued that those meanings we can justifiably call knowledge are not verified or proved meanings—have not been *shown* to be true. Rather they are meanings that have been *challenged* by critical examination and have not yet been falsified—that is shown to be false—when confronted with facts, with reality. From this perspective, science does not provide us with rationally established truth. Rather, it offers an effective way for us to interpret events of nature and to cope with the world.

Later on, such noted scientists/philosophers of science as Thomas Kuhn (1922–1996), Imre Lakatos (1922–1974) and Paul K. Feyerabend (1924–1994) emphasized concerns such as:

- Empirical data can neither confirm nor falsify a theory. In other words, science cannot prove that a theory is either true or that it is false.
- Observations are theory-laden (the way we see the world is to some extent determined by the theories we accept) and theory choices are underdetermined (there are always rival theories that can also be supported by the empirical evidence).
- Different scientific worldviews or paradigms are incompatible. (Earlier we discussed how Newtonian mechanics gave way to Einstein's relativity when experiments went against the former and supported the latter theory for velocities approaching that of light.) So science cannot be said to progress towards truth, in the ordinary understanding of progress as a more or less "smooth" process.
- What makes a belief scientific is not the underlying methodology of investigation, but what scientists say about the belief. Science is simply what scientists say it is.

They concluded essentially that science is *a*-rational (not based on or governed by logical reasoning)—and that nature does not significantly constrain the choice of theory!

If they were presented with such radical perspectives and conclusions, the vast majority of scientists would most likely reply that our ability to develop technology based on scientific findings clearly demonstrates that (a) the methods of science find much that we consider to be true, (b) data help confirm or falsify theories, and (c) progress and truth are intertwined. In addition, they would argue that although science certainly rests on many beliefs and presuppositions, the strength of its statements lies in its methods and the vigilance of the scientific community in controlling its results. Therefore, most authors who write about science do not subscribe to a conclusion that science is a-rational. This is certainly the case with most of the literature I have referred to in this book. In particular, I can refer to the strong positions taken in some of the publications often referred to here, in particular the books (Chalmers 1999; Hacking 1983; Gauch 2003; Gooding et al. 1989).

Like those authors just referred to, Arthur Koestler (1903–1983), renowned as a keen observer of science and Fellow of the Royal Astronomical Society, does not observe that science as such is a-rational. However, his empirical investigation indicates the limits of rationality with respect to the development, one might even say the emancipation of creativity in science. In his opus work *The Act of Creation* (Koestler 1989) he presents results of his studies of the behaviour of a number of highly renowned scientists known for their creativity and observes: "... true creativity often starts where language ends. ... the evidence indicates that verbal thinking, and conscious thinking in general, plays only a subordinate part in the decisive phase of the creative act. ... On the testimony of those original thinkers who have taken the trouble to record their methods of work, this seems to be the rule in other (than mathematics; my comment) branches of science. Their virtually unanimous emphasis on spontaneous intuitions, unconscious guidance, and sudden leaps of imagination which they are at a loss to explain, suggests that the role of strictly rational thought processes has been vastly overestimated since the Age of Enlightenment." (Ibid.; 177, 208)

2.5.4   Innate Limitations

Without going into any great depth here, let us conclude this section on the limitations of science by briefly considering three fundamental discoveries that demonstrate certain inherent limitations on scientific reasoning and observations: Heisenberg's uncertainty principle (quantum physics), Gödel's theorem (mathematics) and more recently, chaos theory (non-linear dynamics). Each of these demonstrate that there are limits on our ability to describe reality and that these limits cannot be removed, or even reduced, by developing better technologies; in other words, they are intrinsic and due to the essential nature of reality.

Werner Heisenberg (1901–76) was a founder of quantum physics and is particularly known for his formulation of the so-called "uncertainty principle". In a paper written in German from 1927, (English translation: "On the Perceptual Content of Quantum Theoretical Kinematics and Mechanics"), he pointed out that the more precisely the position of a subatomic particle is determined, the less precisely the momentum is known at that instant, and vice versa.[28] So no matter how precise the measuring instruments used, there is an inherent limit to the precision of the measurements when these two properties are measured simultaneously. In other words, the outcomes of physical processes cannot be predicted with certainty simply because nature determines its future states through a process that is fundamentally uncertain! This has had profound implications, not only for the ability of physicists to describe/predict the future behaviour of subatomic particles, but also for fundamental concepts of causality.

(Bohm and Peat 1989; 79–87) provide a fascinating discussion of the effect of this principle on the informal language of physics whereby terms which had well defined meanings within Newtonian physics became ambiguous within quantum physics. This makes it difficult to discuss quantum theory using our ordinary language. In addition, it led to the separation of Einstein and Bohr whose use of the informal language of physics made it impossible for them to continue their dialogues. "This separation has had particularly serious consequences for the development of relativity and quantum theory, for there is now no common, informal

---

[28]It should be noted that the uncertainty relation does not apply to a single measurement on a single particle; it is a statement about a statistical average over many measurements of position and momenta (Pagels 1982; 90). A result of the relation is that the laws of classical physics are valid only at distances much larger than atomic scales—i.e., involving distances $\gg 10^{-8}$ m.

language that covers them both. As a result, although both theories are regarded as fundamental, they exist in an uneasy union with no real way of unifying them" (Ibid.; 85–86).

In 1931 Kurt Gödel (1906–78) at the age of 25 published his now famous paper in German: (English translation: "On Formally Undecidable Propositions of Principia Mathematica and Related Systems"). In it he developed two theorems of great significance for our understanding of the limitations of formal or logical systems of axioms and rules of procedure. Very briefly, Gödel demonstrated in his first theorem that *any* formal system of axioms and rules capable of containing/supporting certain axioms for the natural numbers, such as ordinary arithmetic, and that is consistent (the property of a logical system whereby there are not statements which the system regards as both true and false), is necessarily incomplete or "undecidable"—it contains statements that are neither provably true nor provably false by the approved procedures. A perhaps simpler way of stating this is that "he proved that in any sufficiently complex formal system, such as arithmetic, there must exist statements that make sense but can neither be proved nor disproved within the system". (Davies 2010; 88)

In his second theorem he showed that such an axiomatic system cannot prove itself to be consistent and complete from within without also proving itself inconsistent. He did this by showing that the statement of the consistency of the system, when coded into the form of a mathematical proposition, must be "undecidable". Although we may be able to prove any conceivable statement about numbers within the system by going outside the system (in order to come up with new rules and axioms), in so doing we create a larger system with its own undecidable statements. This implies that all such systems are incomplete; they contain more true statements then can possibly be proved according to the system's own defining set of rules.[29]

According to the renowned mathematical physicist Roger Penrose (Penrose 1989; 111–12): "...it seems to me that it is a clear consequence of the Gödel argument that the concept of mathematical truth cannot be encapsulated in any formalistic scheme. Mathematical truth is something that goes beyond mere formalism. This is perhaps clear even without Gödel's theorem. For how are we to decide what axioms or rules of procedure to adopt in any case when trying to set up a formal system? ... Any particular formal system has a provisional and 'man-made' quality about it. Such systems indeed have a very valuable role to play in mathematical discussions, but they can supply only a partial (or approximate) guide to truth. Real mathematical truth goes beyond mere man-made constructions."

In his lecture, "Gödel and the End of Physics", (Hawking 2002), Stephen Hawking builds upon this concept of mathematical truth by asking "how far we can

---

[29]Although the details of the proof are very difficult to follow for those without considerable mathematical training, (Nagel and Newman 1958) enables readers with limited, but basic, mathematical and logical training to understand the basic structure of Gödel's demonstrations and the core of his conclusions.

go in our search for understanding and knowledge" and considers the relationship between Gödel's theorem and whether we can formulate the theory of the universe in terms of a finite number of principles. Towards the end, he argues that "One connection is obvious. According to the positivist philosophy of science, a physical theory is a mathematical model. So if there are mathematical results that cannot be proved, there are physical problems that cannot be predicted. ... Some people will be very disappointed if there is not an ultimate theory that can be formulated as a finite number of principles. I used to belong to that camp, but I have changed my mind, I'm now glad that our search for understanding will never come to an end, and that we will always have the challenge of new discovery."

Together, Gödel's two theorems demonstrated the existence of limitations on what one can seek from a logical system—that any attempt to produce a paradox free mathematical system is bound to fail if the system is reasonably complex. His theorems indicate as well that what many consider to be an ideal of science—that we can devise a set of basic assumptions from which all phenomena in the external world can be deduced—cannot possibly be achieved.

Finally, in this brief discourse on innate limitations on scientific reasoning and observations, let us turn now to chaos theory. This theory had its origins not in logical deductions, systematic observations or experiments, but in a chance observation in 1961 by a meteorologist working with a system of equations that he hoped could contribute to weather forecasting. In brief, what is now referred to as chaos theory can, more correctly, be said to be a set of ideas that attempts to reveal structures in what are apparently disordered, aperiodic, unpredictable systems—i.e., to reveal an underlying order behind what appear to be random data. In a nut shell, such so-called "chaotic systems" are deterministic, non-linear models (systems of equations) that demonstrate unpredictable behaviour.

What is particularly significant for our reflections here is that chaos theory demonstrates the sensitivity of certain phenomena (more correctly: of the system of non-linear equations that have been chosen to describe the phenomena) to their initial conditions. Even minute changes in the initial conditions characterizing such a system (for example due to experimental errors, background noise, the lack of precision of measuring equipment or inaccuracy with which the measurement is made) can lead to significant changes in the long term behaviour of a chaotic system. Examples of such systems are the (models describing) the turbulent flow of liquids such as white water in a river, cloud formation, and plate tectonics. At a more down-to-earth level, our ability to give accurate descriptions of an apparently simple everyday occurrence as mixing milk in tea is amazingly limited. Although we may have excellent theories as to fluids, molecular motion and the like, the ability to describe what actually takes place when a particular portion of milk is poured into a particular cup of tea is extremely limited and no reliable prediction can be made as to the dynamics of the system! Reference is made to the "classic" exposition (Gleick 1987).

What is truly fascinating is that scientists have been able to discover or uncover such inherent limitations on scientific reasoning and observations as have been referred to above. It may be argued that the limitations are not inherent in science as such, but are a result of the way the "real world" is—that the limitations are of ontological rather than epistemological origin. In any case, we may regard it to be a triumph of science that its methods not only are effective in describing, predicting and explaining natural phenomena and relationships, but also in uncovering its own limitations.

Finally, on a more personal note, rather than considering such limitations as something negative, I strongly suggest that we consider them to reflect the magnificent complexity of what we refer to and experience as reality.

2.6   Description, Causality, Prediction, and Explanation

These four terms, description, causality, prediction, and explanation, are fundamental concepts in science. They are also, more or less, parts of our everyday vocabularies, so we tend not to really think much about their meaning. Having, so to speak, two sets of vocabularies, where many of the words are used by both the proverbial "man-on-the-street" and an academic working in a field of scientific investigation, is not unusual. For example, in everyday language people speak of their "philosophy", while the term has a far deeper meaning for students of philosophy (its actual etymological meaning is "love of wisdom"). Similarly, we speak of traffic systems, and computer systems, and the communist system, and simply "the system" ("you can't beat the system", here referring to a bureaucracy of some sort)—while a student of systems science attaches a more general and abstract meaning to the word "system" (e.g.: interacting elements that manifest as a whole). And as we saw earlier, even the strange word "paradigm" that Kuhn developed in his now-classic work on *The Structure of Scientific Revolutions* soon became used in common every-day language and in a far broader sense than originally intended.

Since the natural sciences aim at establishing statements (facts, hypotheses, theories and laws) with a higher degree of precision than our ordinary common sense and language enable us to do, as students and practitioners of science we have an obligation to dig a bit deeper and consider the meaning of fundamental terms we use in science. In so doing, we hopefully develop a greater appreciation of what science is really about, and what it is not about.

2.6.1   Description

Let us start with "description", which is in many ways the most fundamental of these terms. To begin with, I refer to our earlier reflections on observation, since description, at least in the natural sciences, presupposes observation. We saw that e.g. visual observation is not simply a mechanical process whereby e.g. light waves from an illuminated object causes retinal changes and these cause changes in the optic nerve which in turn lead to changes in brain cells. Observation is also a process involving memory, comparison, recognition and selection. There are many things we do not "see" when we look say at a room with its windows, doors, furniture, lamps, books, etc. What we see is a product of choice, of a filtering process whereby we are aware of certain images, and unaware of a great number of other potential images that have been recorded/observed. For example, although we may have seen the books in a bookcase, we cannot recall their titles, placements in the bookcase etc., nor can we recall whatever was on the desk, whether there was a bit of dust on the floor, the motif of the paintings on the wall, and so on, even though we in fact saw all these things. In general, even before we describe what we see/hear/feel/smell/taste, a filtering process takes place. And when we translate what we observe into a form that is available to others via a verbal statement or other form of description, additional filtering processes take place. Therefore, no matter how "objective" we may try to be, description is always selective, incomplete and can be fallible.

All branches of the natural sciences employ and are based on description, while they do not all focus on the other topics to be considered in this section: causality, prediction and explanation. For example, due to a lack of suitable theory, geologists may tend to place greater emphasis on describing the historical developments of the crust of the earth in a particular region, than on predicting its future development. Furthermore, particularly in the early stages of a field of study, there may be a tendency to emphasize description for its own sake, such that the collection, structuring and presentation of data are far more in focus than analysis and attempts at causal explanations.

Yet most scientists do not consider it to be "science" when investigations are simply aimed at observing everything that can be said to characterize an object, activity, or phenomenon, and then recording, categorizing the data, and mediating it —recall our exercise about umbrellaology! There appears to be a tacit agreement in science that the goal of data collection is analysis (leading hopefully to under-standing, explaining, and predicting), and not just straightforward description. For example astronomy, which has extremely limited opportunities for direct experi-mentation and is primarily characterized by more passive forms of observation rather than experimentation, nevertheless aims at going beyond description to make cosmological predictions and to explain and theorize as to the development of say a particular galaxy or nebula or the existence and properties of planets beyond our solar system.

I emphasize too, that even rather straightforward description requires some analysis and reflection as to the purpose of the description. Otherwise it would not be possible to rationally determine the desired accuracy of one's measurements, how to categorize the data, and how to present it in some well-organised manner.

In other words, in general, scientific investigations, even when performed in "pure science" without explicit motivation regarding some future application, must be based on ideas as to how the knowledge gathered will be used, why, and by whom. Astronomers studying sun storms, physicists studying quarks, chemists studying the properties of U_{238} and biologists studying the migration patterns of barn swallows, may say that they are studying these objects for no other reason than to add to our knowledge of the world. Nevertheless, in all such investigations, decisions are being made as to how to perform the observations and experiments, how to make measurements, how to allocate the available time and other resources to the various activities involved, how to determine which observations are of interest to one's peers and which are not, how best to communicate one's findings to others, and so on. Description in science automatically implies reflection, no matter how implicit, as to the aims and limitations of one's investigations.

In addition, what we actually observe under an investigation depends on our understandings of the phenomena we want to describe. And as considered earlier, our culture, our previous training, and the theories we have learned all influence what we perceive and what we deem to be worthy of our consideration.

Therefore description for the sole sake of description is a misleading concept, as is objectivity in observation. This is not to say that one description cannot be evaluated as better than another description in a given context, or that we should not emphasize objectivity in our observations, but rather that we must be aware of what it is that affects our judgment as to what a good description is—and what good science is. It is postulated that the more we are aware of the aims and limitations of our research and of our (implicit) biases and "cultural baggage", the better we will be able to carry out our investigations, make our observations, descriptions and explanations, and contribute to the development of science.

2.6.2 Causality

Causality is a tricky word. We are so used to saying things like: "my lack of sleep last night is the cause of my headache today," or "the drought caused the crops to wither," or simply, "every effect has a cause". But things are not that simple. My headache may be seen as related to or caused by a lack of sleep last night—but may also, perhaps, be related to or caused by how well I slept the last few days, as well, perhaps, by the food I ate before going to bed, the number of hours I sat looking at my computer screen before going to bed, the stress I felt due to a forthcoming examination, the humidity in the air, and so on. And maybe one can even argue that the relationship between a headache and poor sleep is 180° reversed—that a headache I had yesterday led to my having poor sleep—which may in turn have led to my again having a headache today. So it is not always clear as to what is cause and what is effect in a given situation.

So if we are to more meaningfully speak of A as a cause of the effect B, we see that the relationship may be characterized by:

- A being one of many possible causes of B (one possible cause of the headache is stress felt about an examination, another possible cause is overeating, a third possible cause is lack of sleep).

- *A* being a single step along a causal chain (stress due to being worried about an examination can lead to overeating which can lead to a lack of sleep which can lead to a headache).
- *A* being one of a number of factors which together lead to *B* (overeating in a very unpleasant and noisy restaurant and worrying about one's work all act together to lead to a headache).

Continuing with this train of thought, crop failure—however we may define this —may certainly be affected by draught. But the amount and quality of what is harvested may also be a function of the quality of the seeds that were sown, the characteristics of the soil, the weather conditions that prevailed when the seeds were sown, the timing and amount of fertilizers and pesticides used, the pattern of temperature and humidity during the growth season, the weather conditions on the day the crop was harvested, how the harvesting was performed, how the harvest was stored, and so on. So once again there are three possibilities: (a) draught may be one of many possible causes of the poor harvest; (b) the poor harvest may be due to several factors working in a sequence: poor seeds followed by poor weather at the time of planting, followed by ...; and (c) the poor harvest may be due to a combination of several factors.

The above reflections look upon *A* as being a/the cause of *B*. But as we have touched on earlier, sometimes it is difficult to determine what is cause and what is effect. Scientists working in the field of ecology tend to work with a concept of "systemic causality" rather than simple (or linear) cause and effect. By this they mean that all the phenomena they study are interrelated, and therefore there is in some sense a cyclical causality at work, rather than a linear—"*A* is the cause of *B*"—causality.

Before proceeding, let us summarize the implications of the above reflections in a more precise terminology:

- If *A* is said to be **a** cause of *B*, then *A* must always be followed by *B* and *B* cannot precede *A*; $A \Rightarrow B$. In other words, *A* is a sufficient condition for *B*.
- If *A* is said to be **the** cause of *B*, then *A* is both a necessary and sufficient condition for B; $A \Leftrightarrow B$; *B* occurs if and only if *A* occurs.

It follows, for example, that it is incorrect to say that smoking cigarettes causes lung cancer. Even though there may be a strong statistical correlation between smoking and the occurrence of lung cancer, many smokers do not get lung cancer, and many who have lung cancer have not smoked; smoking alone is neither *a* cause nor *the* cause of cancer (although we cannot ignore the possibility that smoking, together with some other phenomena, e.g. a particular DNA pattern, can be **a** cause of cancer—and that given the strong statistical correlation it would be foolish to smoke).

I also note that while most scientists subscribe to the so-called scientific method (the subject of Chap. 7) and consider experiments as a means of determining causality in the physical world, some scientists tend to downplay causality. They argue that there is no causality in nature; what we tend to regard as cause and effect is simply a high degree of consistency with which events of one kind are followed by events of another kind.[30] From this perspective, all that we can observe are correlations, not causations, and based on these we tend to make inductive inferences.

Such scientists assert, for example, that when people speak of the moon's gravitational pull being the cause of tidal movement, this is not a correct statement, not a fact. By this they mean that gravity is simply an expression of a constant observable relationship among masses, and the movement of the tides is an example of that relationship. From this perspective, the moon does not exert a pull that precedes and causes the tide to rise; the relationship between the masses involved is completely symmetrical. According to (Stace 1967 in Klemke et al. 1998; 354): "Gravitation is not a 'thing', but a mathematical formula … as a mathematical formula cannot cause a body to fall, so gravitation cannot cause a body to fall. Ordinary language misleads us here. We speak of the law 'of' gravitation, and suppose that this law 'applies to' the heavenly bodies. We are thereby misled into supposing that there are *two* things, namely the gravitation and the heavenly bodies, and that one of these things, the gravitation, causes changes in the other. In reality nothing exists except the moving bodies. And neither Newton's law nor Einstein's law is, strictly speaking, a law of gravitation. They are both laws of moving bodies, that is to say, formulae which tell us how these bodies will move."

An example of how, depending on our aims and focus, one can draw misguiding inductive inferences as to cause and effect, deals with lightning and thunder. Ordinary people (without a scientific background) tend to speak of lightning as causing thunder since we observe a constant relationship between them, whereby lightning is always followed, with a time delay dependent on the distance of the lightning from us, by thunder. It may be more correct to see both lightning and thunder as being caused by—or simply two results/perceptions of—the same physical event, an electric discharge that we perceive first with the aid of our eyes as light, then with the aid of our ears as sound, since the speed of light is far greater than that of sound.

Since the notion of causality is closely related to that of "determinism" , it can be fruitful now to briefly consider the relationship between these two concepts in order to demonstrate the potentially wide-reaching consequences of causality. According to determinism, since there is a cause underlying every effect, everything is necessarily predetermined—therefore everything has a primal cause. From a religious or spiritual perspective, such a cause is typically considered to be the "Creator";

---

[30]This is not a newer reflection. The reflections of David Hume (1711–76), referred to earlier, when we considered the concept of positivism, formed the basis for future empirical analyses of cause and effect. In his "regularity theory" he argued that no amount of experience in observing that A precedes B permits us to say that A is a cause of B (in our terminology: induction cannot provide us with proof). When we think in this way it is simply a reflection of the habits of our mind; we appear to have a need to make sense of our observations that A always appears (at least up to now) to occur before B; (Hume 1748/1955; 26–41, 64–67).

from a scientific perspective, such a primal cause is typically considered to be the "laws of nature".[31]

Continuing with these reflections, let us very briefly consider the arguments of Popper (1983; 1–2), in which he sets forth his reasons for being an indeterminist and where he discusses the relationship between the concepts of determinism and causality. "My central problem is to examine the validity of the arguments in favour of what I call '*scientific*' determinism; that is to say, the doctrine that the structure of the world is such that *any event can be rationally predicted with any desired degree of precision, if we are given sufficiently precise descriptions of past events, together with all the laws of nature*".

Popper argues that examining the validity of arguments in favour of scientific determinism is relevant mainly because exponents of quantum physics tend to argue that classical physics entails determinism, while quantum physics compels us to reject both classical physics and determinism. Popper argues that this position is not correct and that even the validity of classical physics would not impose a deterministic doctrine about the world. It can also be argued that quantum physics does not undermine determinism as such, only that it leads to a new form of determinism where the laws of nature determine the probabilities of future and past states rather than determining the future and past with certainty (Hawking and Mlodinow 2010; 72).

In the arguments he develops against determinism Popper presents an intuitive idea of determinism by using a metaphor of a motion picture film. He likens the World to such a film, where the picture being shown at any given time is the present, where those parts of the film already shown constitute the past, and where those parts not yet shown constitute the future. Building on this metaphor, he argues that the future coexists with the past, and that it is known to the producer of the film/the Creator of the World. This is a religious, not a scientific, perspective of determinism, connected as it is with the idea of omnipotence (the Creator created the film/physical reality) and omniscience (the Creator as film producer knows what has not yet been shown on the screen). What Popper refers to as *scientific* determinism is the above, but with the idea of the Creator (God) replaced by the idea of nature, and with divine law replaced by the so-called laws of nature. He essentially argues that while God may be known by revelation, determinism assumes that the laws of nature may be discovered by human reason and experience—and that one can use these laws to predict the future state of a physical system based on the present data (or past data) that characterize it and by using rational methods. In other words, if only we know the laws of nature and the state of the system at any time, we can rationally calculate any future event in advance of its occurrence. It follows as well according to Popper that if only one future event could not in principle be predicted using the natural laws and data as to the present or past state of the world, then scientific determinism must be rejected.

---

[31]However, an argument can also be made that if these laws first 'emerged' after the Big Bang, then they cannot be a primal cause.

I will not continue with further reflections on Popper's analyses, which are based on a physical interpretation of probability theory. The purpose of the presentation was simply to indicate that working with a concept of causality can compel us to consider far more inclusive questions, e.g. prediction and determinism.

2.6.3   Prediction

Prediction in everyday language is a statement that a particular event will occur in the future. In a scientific context, prediction is a precise statement, given specific starting or initial conditions, about the future state of a system. Prediction is a key concept in scientific method, where it is a precondition for testing hypotheses. It is by applying deductive reasoning to our hypotheses that we generate predictions, and it is by analysing the results of our observations/experiments that we evaluate whether our predictions are confirmed or contradicted, and thus whether our hypotheses are supported or not by our observations/experiments.

Note that an extended concept of prediction would permit inferences not just regarding future events, but as to past events as well; the physical system is characterized by reversibility. We might, for example, be interested in 'predicting' earlier astronomical events (eclipses, the trajectory of comets) in just the same way as we make inferences as to future eclipses, trajectories of comets, etc. In this sense, prediction is the application of logical/mathematical reasoning to the dynamics of a physical system, and the mathematical system that describes the dynamics of the physical system is totally indifferent as to future and past. Clearly this understanding of prediction is closely related to the concepts of causality and determinism that we have just treated.

Note too that predictions do not need to be evaluated by the use of experiment in order to be "scientific". For example, consider the prediction in 1705 by Edmund Halley (1656–1742) that a bright comet would return some time in December, 1758. His prediction was the result of his applications of Kepler's theory of elliptical orbit and Newton's laws of planetary motion, together with his own studies of existing reports of comets that appeared in 1456, 1531, 1607 and 1682. His analyses led him to hypothesize that the earlier comets were in fact the same comet since they appeared to follow similar paths. His prediction was shown to be true on Christmas day, 1758, 16 years after his death. Since then the comet has been named after him.

Another example of predictions not being evaluated via experiment is the many experiences I had while being on the board of an international consulting company in the late 1960s and early 1970s. The primary resource of this company was the talent of its employees in building mathematical models that could be used to predict the outcomes of an organization's decisions. In particular, simulation techniques were used with considerable success to enable the leaders of private, public and governmental organisations to consider in advance the potential results of alternative actions. There was no ability to run controlled experiments, so the models became a kind of "management's laboratory".[32] Although there was no capability of comparing the actual outcome to what might have occurred had other decisions been made, the leaders were able to evaluate the "goodness" of the models and thereby, indirectly, of the predictions, by simulating the outcomes of earlier decisions and comparing the results to what actually occurred, as well as by evaluating the logic (as expressed via the mathematical relationships) of the models employed.

Tying together the concepts of causality and prediction, as well as placing them in a historical context, consider the ideas of the French astronomer and mathematician Pierre-Simon Laplace (1749–1827). Based on his work with Newton's mechanics, Laplace was convinced that everything in the universe is predetermined by the laws of motion and that the universe could simply be considered to be a big machine—a deterministic mechanical system. The following quote from (Truscott and Emory 1951; 4), the translation of Laplace's publication *Essai philosphique sur les probabilities* (*Philosophical Essay on Probabilities*) sums up his argument: "We may regard the present state of the universe as the effect of its past and as the cause of its future. An intellect which at a certain moment would know all forces that set nature in motion, and all positions of all items of which nature is composed, if this intellect were also vast enough to submit these data to analysis, it would embrace in a single formula the movements of the greatest bodies of the universe and those of the tiniest atom; for such an intellect, nothing would be uncertain and the future just like the past would be present before its eyes."

In other words, according to the Danish physicist Niels Bohr (1957; 118, my translation): "The idea was that all interactions between the machine's parts were subject to the laws of mechanics and that therefore an intelligence with knowledge of all these parts' relative positions and velocities at any given time would be able to predict any future event in the world, including the behaviour of animals and humans."

To round off these reflections that link prediction, causality and determinism, the following are six concise arguments, primarily based on our reflections so far, as to why deterministic prediction is not, neither in principle nor in practice, achievable:

1. First of all, at a more philosophical level, there is the logical problem of recursion: If the "intelligence" (say something like a huge super-computer) is assumed to be inside the universe, it has a physical structure (particles) that affects and is affected by everything else, and therefore must include itself in its description which can lead to an unlimited recursion. It also means that the "super-computer" is subject to the laws of nature and these may limit its ability to find the laws; for example, its ability to transfer information would be limited by the finite speed of light. On the other hand, if the "intelligence" is not part of the universe, the universe is not complete and therefore the calculations cannot take into account the influence of the intelligence on the universe.

---

[32]The title of my very first publication, written together with several classmates at Harvard Business School in 1959 was: *Simulation: Management's Laboratory*.

2. Determinism is based on the underlying assumption that the universe is a mechanical system characterized by its elementary particles and the forces that act upon them/their movements. This assumption is not supported by quantum physics where, at the microscopic level of atoms, quarks, bosons etc., it is not even meaningful to speak of the universe as composed of particles.

3. It is also presumed that classical mechanics applies; yet as we know from Heisenberg's uncertainty principle, in quantum systems one cannot specify precisely the required data as to position and momentum. According to (Hawking and Mlodinow 2010; 72) "... the outcomes of physical processes cannot be predicted with certainty because they are not *determined* with certainty. Instead, given the initial state of a system, nature determines its future state through a process that is fundamentally uncertain."

4. The reductionist assumption as to a universe being built up of individual units of matter that obey the laws of nature is also challenged when we consider the nature of consciousness. It can be argued that human consciousness (and that of other sentient beings) cannot be reduced to a physical system, and therefore consciousness itself can lead to behaviour that is not predictable—this will also be considered in this book's concluding chapter.[33]

5. As we referred to earlier, if in fact physical reality can be represented, at least to some extent, by systems of nonlinear equations, then it may demonstrate chaotic behaviour which would not permit precise prediction.

6. Finally, on a very practical level: The number of particles in the universe is amazingly large (there are on the order of 10^{23} molecules in just one cubic centimetre of gas), and even to be able to calculate their future states would require precise data on their present positions and momenta, a task which no computer can achieve. So even if we assume that Laplace's assumptions and reasoning are correct, since we as humans are not the "intelligence" referred to and have a limited ability to know everything about the material structure of the universe as well as a limited ability to make all the computations that would be required, we are only able to predict events with a certain probability.

---

[33]A differing opinion is held by the distinguished scientist Stephen Hawking who, based on the implicit assumption that consciousness is a biological system, argues that since "... biological processes are governed by the laws of physics and chemistry and therefore are as determined as the orbits of planets. ... we are no more than biological machines and that free will is just an illusion." But "... since we cannot solve the equations that determine our behaviour, we use the effective theory that people have free will." (Hawking and Mlodinow 2010; 32, 33)

2.6.4   *Explanation*

We conclude this section by reflecting briefly on the concept of explanation, which will be seen to be closely related to that of causality as well as to that of reductionism.

In science, to explain the events or phenomena we observe is to answer the question "why" rather than simply the question "what", that we met in connection with our reflections on the concept of description. Answering such "why-questions" is generally considered to be one of the primary aims of scientific inquiry. For example, although a person without scientific training may be able to observe and to provide reasonably accurate descriptions of the phenomenon of natural materials reflecting light in one direction—when you place a straw in a glass of water, it appears to bend toward the surface—a scientific explanation will be required in order to provide an answer to the question *why* the straw appears to bend.

While a typical dictionary definition of the verb "explain" is: "to make understandable ... to give the meaning or interpretation of ... to account and state reasons for", the philosophy of science provides us with a far more nuanced approach to explanation. I will frame the major part of our reflections on "explanation" by referring to the seminal work of Carl. G. Hempel (1902–1997), in particular to (Hempel 1948). Since Hempel's model of explanation assumes a process whereby the event or phenomenon to be explained is subsumed under a general law, it is often referred to as a "covering law" model of explanation.

To illustrate the model, consider an example from (Hempel 1948 in Klemke et al.; 206–7, 210):

> A mercury thermometer is rapidly immersed in hot water; there occurs a temporary drop of the mercury column, which is then followed by a swift rise. How is this phenomenon to be explained? The increase in temperature affects at first only the glass tube of the thermometer; it expands and thus provides a larger space for the mercury inside, whose surface therefore drops. As soon as due to heat conduction the rise in temperature reaches the mercury, however, the latter expands, and as its coefficient of expansion is considerably larger than that of glass, a rise of the mercury level results. ... This account consists of statements of two kinds. Those of the first kind indicate certain conditions which are realized prior to, or at the same time as, the phenomena to be explained; we shall refer to them briefly as antecedent conditions. In our illustration, the antecedent conditions include, among others, the fact that the thermometer consists of a glass tube which is partly filled with mercury, and that it is immersed into hot water. The statements of the second kind express certain general laws; in our case, these include the laws of the thermic expansion of mercury and of glass, and a statement about the small thermic conductivity of glass. The two sets of statements, if adequately and completely formulated, explain the phenomenon under consideration; they entail the consequence that the mercury will first drop, then rise. Thus, the event under discussion is explained by subsuming it under general laws ...

Hempel then adds that so far "... we have only considered the explanation of particular events occurring at a certain time and place. But the question 'Why' may be raised also in regard to general laws. ... the explanation of a general regularity consists in subsuming it under another, more comprehensive regularity, under a more general law ... the validity of Galileo's law for the fall of free bodies near the earth's surface can be explained by deducing it from a more comprehensive set of laws, namely Newton's laws of motion and his law of gravitation, together with some statements about particular facts, namely, about the mass and radius of the earth."

Thus, according to Hempel's model, explanations are arguments offered to establish that the event or phenomenon to be explained had to occur given the prevailing initial conditions and one or more laws of nature (which he also refers to as regularities). In brief, his model, which is in essence deterministic but which can be extended to treat probabilistic events, can be represented as follows:

There are statements of initial conditions:	$C_1, C_2, \ldots C_k$
And scientific laws:	$L_1, L_2, \ldots L_r$
Logical deduction based on these leads to:	E

which is an explanation—a statement describing the event/phenomenon to be explained.

Hempel notes that the same formal analysis applies as well to scientific prediction, where the difference between the two is that if E is given (the event/phenomenon has occurred) and a suitable set of statements $C_1, C_2, \ldots C_k$ and $L_1, L_2, \ldots L_r$ is provided afterwards, we speak of an explanation—while if the statements are given and E is derived prior to the occurrence of the actual event/phenomenon, we speak of a prediction.

The type of explanation considered here is often referred to as a "causal explanation" (reference is made to our previous discussion of causality), since the regularities expressed by the laws $L_1, L_2, \ldots L_r$ imply that whenever conditions of the kind indicated by $C_1, C_2, \ldots C_k$ occur, an event of the kind described in E will occur.

Since Hempel's seminal work, a number of relevant criticisms have been offered of his equating the logical characteristics of explanation and prediction. For example, there may be good predictors of events that we would not regard as explanatory when provided after the event occurred; for example, barometers serve as reliable predictors of storms yet we would not explain a storm's occurrence by referring to the readings of a barometer. And a good explanation may not be accepted as a good predictor of an event should it be known prior to the occurrence of the event; theoretical knowledge as to why a roulette wheel stops on a particular number doesn't serve as a basis for predicting the number the wheel will stop at prior to spinning the wheel (Ibid.; 202).

It should be noted as well that there are other characterizations of explanation in the natural sciences than those provided by causal models. For example the concept of motivation has been used in the biosciences, as well as in the behavioural and social sciences. In models that employ this concept, a person's actions (or perhaps the actions of less complex forms of life, even those of a cell) are explained based on the living creature's motives, purposes or intentions.

Before concluding this section on explanation it should also be noted that not all scientists agree that explanation is a reasonable aim of science. In spite of the observation that it appears to be a characteristic of human nature to seek explanations, that is, to come up with answers to "why" questions, arguments can be made that challenge the ability of science to provide explanations. It can be argued that science simply provides us with generalized statements as to *what* happens, but not as to *why* it happens. For example, as we briefly touched on earlier, we tend to explain the movement of the planets by referring to "gravitation" or "forces". But "gravitation" and "forces", although apparently providing answers to such questions, can be said to be human constructions—not things out there in the so-called real world that can be studied by using the methods of science. Similarly, one can argue that although the mathematical formulae which comprise atomic theory are simply formulae for calculating what effects/sensations will occur under certain conditions, we tend to demand that something corresponding to these formulae exists and we refer to that something as atoms. According to (Stace 1967 in Klemke et al. 1998; 356), who takes a strong anti-realist stance: "The 'existence' of atoms is but the expiring ghost of the pellet and billiard-ball atoms of our forefathers. They of course had size, shape, weight, hardness. These have gone. But thinkers still cling to their existence, just as their fathers clung to the existence of forces, and for the same reason. The reason is not in the slightest that science has any use for the existent atom. But the imagination does. It seems somehow to explain things, to make them homely and familiar. ... strictly speaking, *nothing exists except sensations* (and the minds which perceive them). The rest is a mental construction or fiction. ...Their truth and value consist in their capacity for helping us to organise our experience and predict our sensations."

This so-called "instrumentalist" or "anti-realist" perspective is in direct opposition to a "realist" position as to the existence of phenomena that we regularly speak of in science but that are not perceivable by our senses. It is to the differences between such fundamental perspectives regarding the nature of reality that we now turn.

2.7 Realism and Anti-realism

We have previously introduced the term "realism" when we considered the aims and limitations of science. Realism was presented as the contention—the metaphysical presupposition—that physical objects exist independent of human thought and that our minds and senses enable access to the physical world so that it can be reliably known—that there is a correspondence between human thought and an external, independent reality. In other words, realism is the position that the physical world has objectivity that transcends subjective experience such that science can present an objective, true picture of nature, including those parts of nature that are not available to our senses, and that the universe really is as described and explained by our scientific statements.[34] (Husted and Lübcke 2001; 108)

Before continuing, a brief clarification is called for about the statement "those parts of nature that are not available to our senses". Consider for example the concept in astronomy of a black hole. Such an object cannot be directly observed since its gravitational pull is so powerful that light cannot escape it. An "objective, true picture" of such a black hole can however be deduced from the fact that we can observe an acceleration of stars in their orbit and this indicates the presence of a powerful gravitational force. In fact, such observations permit astronomers to both infer its position and mass.

The perspective of realism is not that straightforward as it might appear. As touched on earlier, according to Bohr, at least at the level of the atom, science does not aim at presenting a true picture of nature, but only of improving our ability to *speak* about nature. Similarly, one of the most celebrated scientists of our more recent times, Stephen Hawking clearly takes the position that one need not worry too much about whether or not it is meaningful to assume an isomorphic relationship between theory and reality, since the term "reality" has little meaning for him: "I don't demand that a theory corresponds to reality because I don't know what it is. Reality is not a quality you can test with litmus paper. All I'm concerned with is that the theory should predict the results of measurements." (Hawking and Penrose 1996; 121) He elaborates on this in a more recent book: "...it is pointless to ask whether a model is real, only whether it agrees with observations. If there are two models that both agree with the observation ... then one cannot say that one is more real than the other. One can use whichever model is more convenient in the situation under consideration" (Hawking and Mlodinow 2010; 46). From this perspective, the Copernican heliocentric model that replaced the geocentric model of the nature of the universe cannot be said to be a more "real" model, but certainly a more useful model; the earlier model can in principle be developed to provide accurate descriptions of the heavens, but the equations of motion would be far more complex and the resultant model extremely difficult to apply. We return to such ideas in Sect. 3.2 when we consider the concept of parsimony in connection with the choice of criteria for determining "good" theories.

The distinction between a true picture of nature and our ability to practically describe nature is the distinction between realism and anti-realism. This latter, more pragmatic, perspective does not aim at "truth"—at developing a truer description/understanding of reality, including its unobservable parts—but only at

---

[34]Other observers of science have a narrower definition of realism whereby only the ontological aspects are included—that the world exists independently of our representations of it. For example, (Searle 1995; Chaps. 7–9) considers in depth: "Does the real world exist?" and "Truth and Correspondence". Searle's conclusions are very much in line with this narrower definition of realism; "... if we had never existed, if there had never been any representations—and statements, beliefs, perceptions, thoughts, etc.—most of the world would have remained unaffected. Except for the little corner of the world that is constituted or affected by our representations, the world would still have existed and would have been exactly the same as it is now." (Ibid.; 153).

developing theories and models that provide us with knowledge of reality that is sufficient for the purposes of explaining, predicting, and controlling events and phenomena. Thus, for a realist, electrons and gravitational forces exist, while for an instrumentalist they are practical concepts that may or may not exist. According to (Hawking and Mlodinow 2010; 44) "Anti-realists suppose a distinction between empirical knowledge and theoretical knowledge ... theories are no more than useful instruments that do not embody any deeper truths underlying the observed phenomena."

The natural sciences, at least as they are taught and understood by most scientists, are tied to realism. In general, the natural sciences claim to have rational methods that provide humans with objective truth about an independent physical reality. Another way of saying this is that the realist view of the natural sciences is that their *epistemological* methods (epistemology deals with what is, or should be, regarded as acceptable knowledge) aim at *ontological* knowledge (knowledge of the nature of reality).

Note that this aim is not characteristic for the social sciences, where many adhere to the idea that the (social) world is a projection of the mind and that we create rather than discover reality. This social constructivist or social constructionist perspective (I will use the two terms interchangeably, although some distinctions can be made between them) is anti-realist. Although its perspective on reality is therefore far removed from the traditional realist perspective of the natural sciences, with their emphasis on an objective, independent external reality that it is the task of science to develop true knowledge about, the constructivist approach to reality offers all scientists thought-provoking challenges. Let us therefore briefly consider this perspective from the social sciences, which also has supporters in the natural sciences.

Social constructionism was first introduced into the field of sociology in the book: *The Social Construction of Reality* (Berger and Luckmann 1966). Its underlying idea is that social reality—social phenomena and objects of consciousness—does not exist independently of our attempts to know and describe it. Our theories therefore are a product of the social aspects of scientific processes. Some of social constructivism's major perspectives are:

- We (both scientists and the participants in the social processes they observe) make sense of our realities based on our historical and social perspectives.
- We are born into a world of meaning bestowed on us by our culture.
- Knowledge is constructed by humans (both the observers and those being observed) as they engage with the world they interpret.
- Knowledge is not only personally constructed—it is also socially mediated and therefore is shaped by our political and social contexts.
- Social reality is a constantly shifting emergent property of individual and collective creation.

A radical expression of this perspective, a so-called "strong social constructionist" view on the nature of science, is provided in (Gross and Levitt 1994; Chap. 3): "Science is a highly elaborated set of conventions brought forth by one particular culture (our own) in the circumstances of one particular historical period; thus it is not, as the standard view would have it, a body of knowledge and testable conjecture concerning the real world. It is a discourse, devised by and for one specialized interpretive community, under terms created by the complex net of social circumstance, political opinion, economic incentive and ideological climate that constitutes the ineluctable human environment of the scientist. Thus, orthodox science is but one discursive community among the many that now exist and that have existed historically. Consequently its truth claims are irreducibly self-referential, in that they can be upheld only by appeal to the standards that define the scientific community and distinguish it from other social formations."

It should be noted that this quote does not represent the opinion of the book's authors, who are opposed to this strong constructionist concept of science.

The social constructivist perspective on truth, whereby knowledge is tentative and a result of not only objective and rational processes but also of social processes, is rather challenging for most natural scientists. The traditional realism perspective whereby science can present an objective, true picture of nature, including those parts of nature that are not available to our senses, is far more in line with common sense. But, as we have already noted earlier, common sense has severe limitations as regards scientific investigation and rationality. I note in this connection that the debates about realism are primarily based on those objects and phenomena that are *not* observable—and therefore not amenable to our common sense.

Perhaps then the main question to face when considering the perspective provided by scientific realism is whether the entities and relationships that are postulated to exist by the natural sciences, and in particular by physics, are real, or whether they are simply our perceptions of reality, constructs of the human mind that enable us to order, classify, experiment and develop meaningful knowledge. Realism contends that the entities, phenomena, and processes described by correct theories really exist, no matter whether they are the smallest of the small or in the most distant reaches of the universe. According to (Hacking 1983; 21): "Protons, photons, fields of force, and black holes are as real as toe-nails, turbines, eddies in a stream and volcanoes. The weak interactions of small particle physics are as real as falling in love. Theories about the structure of molecules that carry genetic codes are either true or false, and a genuinely correct theory would be a true one."

The anti-realist/instrumentalist takes the opposite position. According to Hacking, the position here is that there are no such things as photons, DNA, gravitation, dinosaurs, and viral infections—light bulbs emit light, but not photons; inheritance takes place, but there are no genes; heavenly bodies move together, but there are no gravitational forces; fossilised bones exist; but not dinosaurs; colds exist, but not viruses.

But anti-realism is not only based on objects and phenomena, observable or not. It also deals with our theories and explanations. The anti-realist accepts that the theories we develop may certainly be useful tools for our thinking and may help us to develop technologies. But the theories cannot be said to be true or false; we cannot obtain true knowledge of the physical world; theories are logical constructions, instruments, for reasoning about reality. They can be useful and adequate for their purposes, but we have no compelling reasons to believe that they are *true*. And the entities/phenomena they deal with may or may not exist, but there is no need to assume their existence in order to develop a meaningful understanding of the physical world.

I conclude this section on realism and instrumentalism/anti-realism by suggesting that there cannot be any final, conclusive argument for or against realism in the natural sciences. However, there is ample evidence that no matter whether one takes a realist or an anti-realist/instrumental position (or for that matter a constructivist or a positivist position), as a basis for investigating and understanding the world, science works! From this pragmatic epistemological vantage point, scientists develop theories and technologies that describe the world reasonably precisely— including its unobservable parts. Just how these theories and descriptions are developed in the natural sciences is considered in the next section on the relationship between mathematics and the natural sciences.

2.8 Mathematics and Science

Mathematics, which is deductive in nature, is not a natural science, at least given the perspectives on the natural sciences provided here with their emphasis on empirical investigation. Nevertheless, mathematics plays a vital role in the natural sciences. Its significance for the natural sciences is that it is essentially their language. The so-called laws of nature are in general expressed in, that is reduced to, a mathematical form.

The following brief series of reflections on mathematics and science is structured around four basic questions that I consider to be of great significance for the natural sciences. The first of these questions serves as the leitmotif of this section:

2.8.1 *Is Mathematics* Created *by Humans, or Is It Immanent in Nature and* Discovered?

Neither practitioners of the natural sciences, nor mathematicians, can provide a definitive answer to the above question. In spite of the fundamental nature of this question, it is surprising to note that educational programs in science do not attempt to answer it—or even to raise it. The main reason for the question's significance is that if mathematics, like ordinary language, is an artefact, a creation of human endeavour, and not something inherent in physical reality, then we are compelled to reflect on the extent to which it is sufficient for describing relationships between objects and phenomena in the physical world. After all, the natural sciences rely greatly on mathematics for describing physical reality and essentially all the "laws" of nature are expressed in, that is reduced to, a mathematical form. So if there are structures, processes and relationships in the physical world that cannot be precisely translated/described via the language of mathematics, scientists face the question as to how they can gain access to and communicate the "secrets" of physical reality that are *not* reducible to mathematics.[35]

Let us continue with this line of thought as to what we are able to "say about nature". It is not possible to have a one-to-one direct translation between languages that have different syntaxes, say from Sanskrit to English. In the same manner, if mathematics is not "embodied in nature"—and by this I mean isomorphic with structures in nature—it may not be possible to "translate" directly, even from observable physical reality, to mathematics—and translation may be even more out of the question if we consider those aspects of reality that are not directly observable. If this is the case, we must rethink fundamental concepts in science such as truth, objectivity and scientific laws!

Thus, it may be very optimistic and unrealistic to assume, as most scientists do without further reflection, that mathematics is sufficient to provide truthful/isomorphic depictions of nature's elements, processes and relationships.

If, on the other hand, mathematics is embodied in nature, i.e. if all of physical reality has a mathematical structure, then we can perhaps feel secure in assuming that all of nature can be reduced to, expressed in, mathematical statements—that mathematics is the language, so to speak, *of* physical reality, observable or not.[36] Another way of expressing this is to say that if physical reality has a mathematical structure, then humans in their scientific investigations will discover (not invent, as is the case with artefacts) mathematical realities.

2.8.2   What Is the Ontological Status of Mathematics?

Closely related to the question of the relationship between mathematics and reality therefore is the question of the ontological status of mathematics: whether the elements of mathematics have an existence of their own, whether they are not only

---

[35]When I refer to 'mathematics' here, I refer to the study and the methods of study of deductive relationships dealing with quantities (arithmetic), structures (algebra), space (geometry), and change (analysis). A humanistic perspective is provided by emeritus professor of mathematics E. Brian Davies (2010; 101) who refers to mathematics as an aspect of human culture, just like ordinary language, music and architecture. According to Davies, it is because its vocabulary is so highly specialized that its domain of applicability excludes much of importance in our daily lives.

[36]Professor Claus Emmeche, Niels Bohr Institute, Denmark, sent me the following example that illustrates the complex nature of the question as to whether there an isomorphic relationship between mathematical and physical structures. If one looks at a head of cauliflower it appears to have a fractal structure—but is fractal geometry imbedded in the cauliflower? It appears that the answer can be both yes and no. Yes, since we may be able to describe its form using fractals. No, since fractals are limitless (are self-similar at arbitrarily small scales) while the number of times that a form on a cauliflower head can recursively repeat itself is limited since its basic building blocks are cells, which themselves are not fractal in nature.

useful in describing reality, but in fact are part of the reality they describe? This is a question that was central to the reflections of the schools of thought that developed around Pythagoras (roughly 570–495 BCE) and Plato (429–347 BCE) in ancient Greece.

After presenting the famous so-called Mandelbrot set,[37] (Penrose 1991; 94–5) poses the question: "How 'real' are the objects of the mathematician's world? From one point of view it seems that there can be nothing real about them at all. Mathematical objects are just concepts; they are the mental idealizations that mathematicians make, often stimulated by the appearance and seeming order of aspects of the world about us, but mental idealizations nevertheless. Can they be other than mere arbitrary constructions of the human mind? At the same time there often does appear to be some profound reality about these mathematical concepts, going quite beyond the mental deliberations of any particular mathematician. It is as though human thought is, instead, being guided towards some external truth—a truth which has a reality of its own, and which is revealed only partially to any one of us. ... The Mandelbrot set is not an invention of the human mind; it was a discovery, like Mount Everest, the Mandelbrot set is just *there*!"

Even though they lack spatiotemporal properties, Penrose attributes ontological existence to mathematical objects themselves: "... my sympathies lie strongly with the Platonistic view that mathematical truth is absolute, external and eternal, and not based on man-made criteria; and that mathematical objects have a timeless existence of their own, not dependent on human society or on particular physical objects."[38] (Penrose 1991; 116) Penrose does *not* here say that all physical objects have a mathematical structure, only that there are mathematical objects, like the Mandelbrot set, that exist eternally, unchangingly and independently of anything else.

But if we take the position that mathematics exists, i.e. is part of reality (even though its objects may not have physical properties such as mass or location) and that, like physical reality, it is discovered and not created by us, then we face a series of significant metaphysical questions: Is there/was there an "Intelligence" that embodied mathematics in nature? If so, did this "Intelligence" have a plan—is it a predetermined part of evolution that the physical world should be amenable to human investigation such that humans should discover mathematics so as to be able to use it to crack the "cosmic code"? To pretend that one can answer such questions conclusively would be the height of audacity.

According to the Hungarian American physicist and mathematician, Nobel laureate in physics in 1963, Eugene Wigner (1902–1992): "...the enormous usefulness of mathematics in the natural sciences is something bordering on the mysterious and there is no rational explanation for it." (1960; 2) ... "The miracle of

---

[37]The Mandelbrot set is defined as the set of complex values of c for which the mapping of c under iteration of the complex quadratic polynomial $z_{n+1} = z_n^2 + c$ remains bounded.

[38]In contrast to Penrose, the renowned mathematician E. Brian Davies (2010; 97–113) provides a critical assessment of Platonism.

the appropriateness of the language of mathematics for the formulation of the laws of physics is a wonderful gift which we neither understand nor deserve. We should be grateful for it and hope that it will remain valid in future research and that it will extend, for better or for worse, to our pleasure, even though perhaps also to our bafflement, to wide branches of learning." (Ibid.; 14)

Finally, before leaving these reflections as to whether there is an isomorphic relationship between mathematics and physical reality I would briefly like to reflect on the following question: If mathematics is embodied in nature is it only in the nature of this Earth or is it embodied in the entire universe?

It is a basic presupposition of the natural sciences that the regularities we describe via the so-called "laws of nature" are invariant and universal. In other words, it is presumed that since the advent of the Big Bang the laws and the phenomena they describe are independent of time and space; gravitational or electromagnetic forces are not just regarded as phenomena that exist in this world of this solar system, but everywhere and at all times—that is, after "time" began. This basic presupposition as to the regularity of the universe appears to be well-founded and documented; e.g. investigations of stellar activity support the generalizability of the relevant laws of nature as they have been uncovered here on Earth.

While this last statement appears to be obvious, Wigner (1960; 6) raises the question as to whether what appear to be invariant, universal laws can be in conflict with each other:

> We have seen that there are regularities in the events in the world around us which can be formulated in terms of mathematical concepts with an uncanny accuracy. ... The question which presents itself is whether the different regularities, that is, the various laws of nature which will be discovered, will fuse into a single consistent unit, or at least asymptotically approach such a fusion. Alternatively, it is possible that there always will be some laws of nature which have nothing in common with each other. At present, this is true, for instance, of the laws of heredity and of physics. It is even possible that some of the laws of nature will be in conflict with each other in their implications, but each convincing enough in its own domain so that we may not be willing to abandon any of them. ... We may lose interest in the "ultimate truth," that is, in a picture which is a consistent fusion into a single unit of the little pictures, formed on the various aspects of nature.

I conclude this sub-section with the observation that the entire universe appears to be "designed and constructed" with an amazing precision—based on a few constants. If for example the gravitational constant (or the speed of light or absolute zero on the Kelvin scale or Planck's constant or...) had been just slightly different, the universe would not have the conditions that could support the development of life on earth—and perhaps the universe might never have come about (Ward and Brownlee; Chaps. 2, 3, 12). So has *everything*, including life and consciousness, existed embodied as a seed—that has been "programmed"—since "Planck-time" (roughly 5.4×10^{-44} s after the Big Bang)? Or is the universe simply a random event? This leads to a third basic question regarding the relationship between mathematics and science.

2.8.3 If Mathematics Is Embodied in Nature, Is the Universe Deterministic?

In connection with reflection on whether the world is mathematical, I raised the question: "Is there/was there an 'Intelligence' that embodied mathematics in nature?" In connection with our earlier reflections on the concept of prediction in science, we considered the thought experiment of the French astronomer and mathematician Pierre-Simon Laplace that involved a hypothetical master-intelligence or intellect. Laplace reasoned that if the intellect somehow was able to know the precise location and momentum of every element in the universe then, since Newton's laws of motion were assumed to hold universally, it would be possible, in theory at least, to calculate the position and momentum of every object in the universe at any time. In other words, he argued that the universe is a deterministic mechanical system.

When we considered this argument earlier, I strongly challenged it:

(a) The findings of quantum physics challenge the underlying assumption that everything that exists consists of particles whose position and motion can be described by classical mechanics and therefore as well that all physical processes are reversible.

(b) Human behaviour (as well as that of other sentient beings) is conscious behaviour, and since consciousness may not be a deterministic mechanical system—at least there is no compelling evidence that it is—and since humans consciously intervene in the system, the universe cannot be a deterministic system in the sense that its past, present and future can be completely and precisely described by mathematics.

(c) Even if we assume that Laplace's assumptions and reasoning as to the universe being a deterministic mechanical system are correct, this does not imply that scientists are able to completely describe this system where all there is are particles interacting in a completely deterministic manner. Scientists, even employing the most powerful computers, are not the "intelligence" referred to by Laplace. We have a limited ability to know at any given moment of time everything about the material structure of the universe, and have limited capacity to compute the evolution of the universe. Thus, at least in practice, scientists are only able to predict events with a certain probability.

I conclude these brief reflections on mathematics and determinism by referring to (Hofstadter 1989) who touches on this question in a similar manner in his section on Formal Systems and Reality: "... it is natural to wonder about what portion of reality can be imitated in its behaviour by a set of meaningless symbols governed by formal rules. Can all of reality be turned into a formal system? In a very broad sense, the answer might appear to be yes. One could suggest, for instance, that reality itself is nothing but one complicated formal system. Its symbols do not move around on paper, but rather in a three-dimensional vacuum (space); they are elementary particles of which everything is composed. (Tacit assumption: that there is an end to the descending chain of matter so the expression 'elementary particles' makes sense) ... The 'typographical rules' are the laws of physics ... the theorems of this grand formal system are the possible configurations of particles at different times in the history of the universe. The sole axiom is (or perhaps *was*) the original configuration of all the particles at 'the beginning of time'. This is so grandiose a concept, however, that it has only the most theoretical interest; and besides, quantum mechanics (and perhaps other parts of physics) casts at least some doubt on even the theoretical worth of this idea. Basically, we are asking if the universe operates deterministically, which is an open question." (Ibid.; 53–54)

2.8.4 Are We Able to Describe All Relationships in the Universe with Mathematics?

It should be clear that if we do *not* assume that mathematics is embodied in nature, then we cannot a priori answer this question affirmatively.[39] But what if we assume that mathematics is embodied in nature—are we then able to describe all phenomena and relationships using mathematics?

It is argued that if we adhere to a presupposition that the *physical* universe *is* mathematical—that there is an isomorphic relationship between physical reality and mathematics—we risk being blind with respect to phenomena that cannot (reasonably) be reduced to physical elements and thus cannot be described using mathematics. For example, we risk a mathematization of areas of investigation such as economics, psychology, anthropology, consciousness, aesthetics and ethics whereby the resulting theories and models may be (most probably are) unable to accurately and operationally describe the phenomena being investigated. Even more threatening: we risk that such areas of investigation become so dominated by their mathematization that the phenomena and relationships they investigate are transformed to "fit" the models and theories that have been formulated mathematically—instead of the opposite.

The risk is real. Educational programs in science, including the cognitive-, bio-, behavioural- and social sciences, are becoming ever more enamoured with such mathematization and the supposed "order" it enables. However, and as was argued earlier in Sect. 2.5.4 of this chapter where we considered chaos theory, reality is not necessarily characterized by order, at least the kind of "order" we are accustomed to refer to. So-called chaotic systems are disordered, aperiodic and unpredictable even though their mathematical representation is deterministic, i.e. without any random elements.

---

[39]We will many times, particularly in Chap. 8 dealing with uncertainty in science, refer to the concepts a priori and a posteriori: The truth or falsity of an a priori statement can be established without reference to observational evidence (e.g. as in determining the truth of a theorem in geometry or in a tautological statement such as "All bachelors are unmarried") while such evidence is required to establish the truth or falsity of an a posteriori statement.

Another way to consider the question as to whether we are able to describe all relationships in the physical world with the aid of mathematics is to reflect on the role of probability theory in science. For over two thousand years following the development of mathematics as an axiomatic-deductive system, science was considered as dealing with causes, not with chance. Then, with the development of probability theory from the mid-17th century (and later on, the development of statistics as a related area of study), science and chance gradually became reconciled. In the 18th century probability theory expanded its focus from gambling problems to law, to the analysis of data, and insurance—and from there to sociology, to physics to biology, and to psychology in the 19th century, and to agronomy, polling, medical testing, sports and so on in the 20th century.[40] The theory developed as its applications developed. Today, the study of probability is not solely confined to quantitative matters; concepts of chance also are predominant in considerations of philosophical and scientific matters such as free-will, causality, explanation and inference.

---

[40]For a fascinating exposition on what the authors refer to as "the probability revolution" that is said to have taken place between 1830 and 1920 "and began with a statistical revolution, that is with a mania for collecting and analysing quantifiable data about people and their doings and that got underway at the turn of the seventeenth century", leading to a "shift in the conceptual and methodological underpinnings of many sciences", see Chap. 8 in (Depew, D.J. and Weber, B.H. 1996), particularly pp. 202–208.

Chapter 3
Hypotheses, Theories and Laws

Although we tend to speak of hypotheses theories, and laws as though they were clearly differentiated, this is not the case. In fact, as we shall see, some scientists argue that they are all synonyms, and that there is no essential difference between say a hypothesis on the one hand and a law of nature on the other hand. And many scientists use these words interchangeably in their speaking and writing.

I will argue that although there are in fact meaningful distinctions between these three concepts, the distinctions are not clear and are primarily a matter of belief and faith regarding how well relationships that have been conjectured are supported by evidence.

I emphasize that when I use the words "belief" and "faith", I *am not* criticizing science and its methods by introducing concepts ordinarily considered to be alien to science. Rather, I am simply attempting to provide a more complete picture of scientific methodology than is generally provided to students of science; belief, faith and metaphysical choice, contrary to popular scientific opinion, play important roles in science and are inherent elements of science.

And the faith of reflective practitioners of science in the activity, profession and institution we call science appears to be well-founded: Science "works"! Scientific theories and laws are far more than just matters of faith; they are also the product of a methodology that has developed via continual confrontation with reality. It is not an exaggeration to state that scientific theories and laws are the most reliable forms of generalization produced by rational human activity.

Faith and rationality co-exist in science in a most vital symbiosis!

3.1 Hypotheses

There are a number of definitions of "hypothesis", all of which essentially deal with how a proposition or conjecture can be formulated in such a manner that it can be tested and provide the "raw material" for the development of theory. I note that while I earlier referred to "hypotheses" and "research questions" as more or less synonymous, at this point it is appropriate to make a distinction. Research questions can be considered to be statements of what you want to learn, while hypotheses can

© Springer International Publishing Switzerland 2016
P. Pruzan, *Research Methodology*,
DOI 10.1007/978-3-319-27167-5_3

be considered as tentative answers to these questions—and as such can be tested. The following are several common definitions of "hypothesis":

- A proposed explanation for a phenomenon (e.g. global warming is a function of the amount of carbon dioxide, methane and other such gases that are released to the atmosphere)
- A conjecture that can be tested via experimentation and observation (e.g. when mercury is heated it expands)
- A preliminary generalization subject to empirical investigation (e.g. all swans are white)
- A basis for predictions that can be tested (e.g. light is subject to Einstein's theory of gravity)

The general idea that underlies the generation and use of a hypothesis is that by applying deductive logic, the hypothesis can lead to predictions as to observations, in particular of the outcome of experiments. If the results of these observations/experiments support the hypothesis, and if further testing—by others and under varying circumstances—continues to support it, we speak of the hypothesis as leading to knowledge generation in the form of a theory.

In order to illuminate the relationship between hypotheses and theories, let us now consider a completely constructed (i.e. make-believe) "case", inspired by my many teaching experiences in India.

A researcher in the biosciences who makes regular visits to an ashram in India has observed that the wild dogs and monkeys that visit the ashram appear to live relatively peacefully with each other, while his experience elsewhere tells him that they can be very aggressive with respect to each other. After reading literature that contains discussions of the potential beneficial effect on animals of being in peaceful areas, for example, where people meditate, he formulates a very preliminary hypothesis to guide his thinking: "Monkeys and dogs that frequent the ashram do not exhibit violent behaviour with respect to each other (although dogs may attack other dogs there, and monkeys may attack other monkeys there)." Should this hypothesis be supported by evidence, he would then continue to develop a new hypothesis that the lack of the violent behaviour is a result of the animals frequenting a place where many people meditate; this is in fact the underlying motivation for his investigation.

Suppose that after forming the hypothesis that dogs and monkeys that frequent the ashram do not exhibit violent behaviour with respect to each other, the researcher begins to carry out preliminary well-structured and documented observations and that these seem to support the tentative hypothesis.[1]

Certainly we would not say that this then justifies referring to the hypothesis as a scientific theory. First of all, before the hypothesis can be tested his terminology

---

[1]We will see later on when the focus is on the test of hypotheses that this is the so-called "alternative hypothesis", the hypothesis we are interested in and that motivates the study. The "null hypothesis" here would be that the behaviour of the dogs and monkeys in the ashram is not different from dogs and monkeys who do not frequent the ashram.

would have to be operationalized, i.e. defined in ways that would permit precise measurement and such that others could replicate his investigation. For example, the concepts of "peaceful nature" and "violent behaviour" would have to be defined in such a way that the definitions would correspond to the use of such terminology by others working in similar fields and permit measurement. In addition, tests would have to be designed such that the results could be said to support or not to support the hypothesis (including the use of statistical analyses that demonstrate that the predictions based upon it are supported by or not-supported by the result of the tests).

Assume that "violent behaviour" has been operationalized according to some accepted scheme from the behavioural or biosciences (this might include measures of physical contact, threatening behaviour, biting, barking etc.). Then, for example, one would have to see whether this apparent lack of violent conflict is enduring or is simply a more or less short-lived random occurrence that can be explained by one or more of a number of variables such as: the time of year (does the season affect the behaviour of the animals?); climatic conditions (the effect of such factors as temperature, humidity etc.); the types of dogs and monkeys (are some more prone to antagonism?); the surrounding conditions (does the number and density of dogs, monkeys and inhabitants of the ashram have an effect? And so on.

Let us take yet another giant step and assume that additional investigations reveal that conflict between monkeys and dogs generally occurs outside the ashram, independent of the time of year, climatic conditions etc., but that it does not occur (or occurs far less often) inside the ashram. At this point, the researcher might want to operationalize the concept of "peaceful surroundings" and make his hypothesis more specific. He might conjecture that it is the immediate presence of a large number of spiritual seekers in the ashram that is the cause of the lack of violent conflict—or perhaps that it is related to the daily meditation sessions, and so on. These ideas could be translated into measurable concepts (the number of people per square kilometre who reside in the ashram; the frequency with which there are meditation sessions; the number of people meditating; and so on).

In other words, at each step, depending on the results so far, new and more precise reformulations of the hypothesis could be generated. But, and this is crucial, the end goal of generating a hypothesis and confronting it with data is theory development—where the resultant theory is able to provide reliable explanations and predictions that are of significance and accepted by the scientific community. In addition to clear conceptual definitions, this requires considerable evidence in the form of observations, measurements, tests, and analyses—as well as, if possible, causal explanations that are based on existing well-established theory.

Note too that it is quite possible that the researcher's hypothesis is not sufficiently supported by the evidence. For example, detailed preliminary observations followed by statistical analyses may have indicated that the dogs and monkeys in the ashram behave very similarly to—or at least do not behave very differently from —dogs and monkeys outside the ashram. In this case, depending on the resources available and the motivations of the investigator, the hypothesis would most likely be rejected, the investigations terminated, and no new theory development would take place. On the other hand, new information and data would be available that indicate that e.g. the presence of spiritual seekers in an ashram does *not* affect the mutually aggressive behaviour of monkeys and dogs. Thus, although it would probably be far more remarkable if his investigations had led to theory development, or at least to results that would inspire others to test the hypothesis, science also progresses when conjectures are refuted.

However I note that such "negative" contributions generally are not made available to the scientific community; scientific journals tend primarily to publish "positive" results that can lead to theory development or at least provide a basis for further scientific investigation. Presumably there are far more hypotheses that are rejected than hypotheses that are supported/confirmed by rigorous investigation.

Of course the artificial case presented here, dealing as it does with phenomena involving the behaviour of humans and animals, is probably far more complex than the hypotheses that most younger researchers develop and test. It was constructed here so as to be understandable by researchers in any of the natural sciences (which might not have been the case had I chosen a case involving specific concepts and data from say quantum physics or organic chemistry or neuroscience or …). Hopefully however, the case has demonstrated that when formulating a hypothesis the researcher must seriously consider the following:

- the potential significance of the research
- how to operationalize the concepts used
- how measurements can/should be made
- how experiments can be designed to test the hypothesis
- the choice of statistical or other methods that can be used to analyse the data
- the resources (time, supervision, knowledge, etc.) that may be required

The following is another example of how hypotheses can lead to the development of scientific knowledge. This time the example will be based on a quick sketch of a development known to most students of science. It is of interest here due to the fact that theory development took place over a period of roughly two hundred years. It is also significant that the reason for the rejection of what was previously considered to be well-established theory was that an underlying presupposition was shown to be false (or at least not correct under all circumstances).

After a careful study of the motion of bodies, Newton formulated his theories that have since been designated as the laws of motion, laws that provide the basis for what is referred to as classical or Newtonian mechanics. As in the above case of monkeys and dogs in an ashram, when Newton formulated and tested his hypotheses, operationalization was required; the terms used had to be defined in such a manner that they could be measured. In order to record motion, one has to use time intervals, and measure distances covered in those intervals. An implicit presupposition underlying Newton's analyses was that different observers who observed the motion of moving bodies measured time in the same way, using what can be referred to as *absolute time*. To be more precise, the intervals of time elapsed during motion of an observed body were assumed to be the same for all observers regardless of whether they were at rest with respect to one another or whether they were themselves in uniform motion with respect to one another. Given this pre-supposition, the results he derived from his hypotheses were found to be in excellent agreement with observations not only of terrestrial bodies but also of astronomical objects. What were originally hypotheses became recognized as theories—and with time, when a huge number of tests and observations corroborated their accuracy, they became elevated to the stature of laws with universal applicability.

However, more than two hundred years later, Einstein argued that the presupposition of absolute time regardless of the state of motion of the observer is not tenable. His analyses also led him to conjecture that there is a maximum (finite) speed, the velocity of light (roughly 3×10^8 m/s), at which information can be conveyed. Further, he hypothesized that this velocity of light is the same for all observers regardless of their state of motion. It is because of this finite (although large) velocity of light, that the assumption of exact simultaneity in time for two observers regardless of their state of motion has to be abandoned. According to Einstein's hypotheses, each observer carries his own time, just as they are in different spatial locations from which they observe the motion of the body. Time is not absolute, he argued; it is relative, just as space is relative.

This new, and startling, hypothesis provided by Einstein had to be subjected to intense scrutiny to see if it accounted for motion at all speeds including those close to the speed of light.[2] Many experiments have since demonstrated the validity of what is referred to as Einstein's special theory of relativity and the theory is now accepted by the scientific community as one of the fundamental laws of physics. Even though Newton's "laws" of motion are of practical importance when velocities are well below the speed of light, they were shown to be incorrect in the case of objects/observers whose velocities approach the velocity of light. Therefore Newton's "laws" were replaced by what later became known as "Einstein's laws", which encompassed all states of motion, including motions close to the speed of light. Newtonian mechanics can be derived as the limiting case of Einsteinian mechanics when the velocity of light is set to infinity.

This discovery, which took place some two hundred years after the time of Newton, illustrates something very important about the development of knowledge in science. It shows how, when facts were shown to be in disagreement with what were considered to be accepted presuppositions and scientific laws (in the case of Newtonian mechanics: the absolute nature of time and Newton's laws of motion), the presuppositions and the laws based on them were modified.

It may now be instructive to consider various characteristics of hypotheses. We will distinguish between two types of hypotheses: descriptive and relational (Cooper and Schindler 2003; 50–51). *Descriptive hypotheses* typically state the existence, size, number, form or distribution of a variable (e.g. the diameter of the

---

[2]It should be noted here that Einstein did not just formulate his theory (which, in fact was a hypothesis that later became accepted as a theory) on the basis of intuition; his formulation of relativity theory was preceded by the Michelson-Morley experiment on the velocity of light.

double helix of DNA is between 19 and 21 Å). Such a descriptive hypothesis does not (directly) aim at theory generation, only at the generation of knowledge. Note that such a hypothesis can also be rephrased as a research question: What is the diameter of the double helix of DNA?

Relational hypotheses on the other hand describe the relation between two or more variables. Here we can make another distinction, this time between correlational and causal hypotheses. A *correlational hypothesis* states merely that certain variables occur together in some specified manner, without implying a causal relationship. An example could be: "The rate of radioactive discharge of radioactive graphite is at least 1.00462 times as great in New York as it is in London" (a nonsensical hypothesis that most likely would not meet the criterion of relevance presented later in this section, nor would it be supported by systematic observations/experiments). An *explanatory* or *causal hypothesis* not only describes a relation between two or more variables, it also implies that the existence of a change in one variable causes or leads to changes in another variable. For example, "The rate of discharge of radioactive graphite increases 0.002 % for each kilometre one travels west from London" (another nonsensical hypothesis). Such hypotheses are closer to the concept of theory development than are correlational hypotheses, which in turn are closer to theory development than purely descriptive hypotheses. Therefore, in general, causal hypotheses provide a firmer basis for operationalization and experimentation than do correlational hypotheses or descriptive hypotheses. Once again—the motivation here is not to develop definitions but to stimulate reflection.

Note that some research projects are not based on hypotheses at all and just aim at carrying out measurements to provide new or more detailed information in connection with an-going investigation. For example, this is often the case when a Ph.D. student is asked by an advisor to test a hypothesis that has already been formulated. Although the training that the researcher obtains when performing such measurements may be of considerable value for his or her career development, such measurement activities are rather far removed from the creativity and ingenuity involved in generating and testing one's own hypotheses and development of theory.

However, there is an important exception here. Once scientists have published what appear to be new theoretical developments based on their hypotheses and experiments, these must be confirmed by others before they are broadly accepted by the scientific community. So replication of investigations carried out by others, even though this is not based on one's own hypotheses, can be important both for the development of scientific knowledge and of the scientist performing the investigation.

What is generally not discussed in the literature is the question of *how* a researcher can or should develop hypotheses—i.e. how the underlying idea evolves. By this I do not refer here to the content of the specific formulation but to the methodological question of how to formulate hypotheses that are of scientific relevance/interest.

I imagine that most researchers have heard the myth of the apple and Newton's hypothesizing what eventually became known as the universal law of gravitation. But this is not the way that the vast majority of hypotheses evolve. Rather, typical underlying mechanisms seem to include one or more of the following:

(a) Your previous studies, in particular your reading of relevant literature, inspire you to conjecture as to a relationship between certain variables;

(b) you have had empirical observations that stimulate your thinking as to possible causal relationships—for example, you have noticed that when you put an oar into the water in a lake it appears to bend, and you speculate whether the degree of this apparent bending is independent of the temperature of the water or the elevation of the lake above sea level;

(c) your knowledge of the literature leads you to consider questions that do not appear to have been satisfactorily answered and/or to consider apparent inconsistencies or disagreements between results published by different authors. This can be a significant outcome of a Ph.D. student's review of the literature.

Since the development of specific hypotheses is so dependent on the particular field of study, more general guidelines cannot be provided here. However, the following reflections on criteria for the relevance of hypotheses should provide food for thought no matter what one's field of research.

3.1.1 Criteria for the Relevance of Hypotheses

The following are a small set of reasonable criteria for ascertaining the relevance of a hypothesis for a research project:

(a) *Fruitfulness*: The hypothesis appears to be a statement that may, if accepted, or at least if not rejected, lead to new hypotheses and eventually to useful knowledge and/or theoretical development. In the above make-believe case of the monkeys and dogs in an ashram it could be of considerable interest to biologists and behavioural scientists if empirical observation appeared to confirm a hypothesis that the presence of meditators affects the behaviour of dogs and monkeys—and thereby, perhaps, of animals in general. For example, this could potentially have significant value for farmers raising livestock.

(b) *Clarity, precision, and testability*: The hypothesis makes it clear as to which facts are relevant. It also enables clarity and precision as to which outcomes of testing will support/confirm the hypothesis and which will not support/refute the hypothesis.

(c) *Provides a framework for organising the analyses*: The hypothesis provides a framework for the performance of the empirical investigations, including the appropriate experimental design, as well as for organising the presentation of the analyses and conclusions.

(d) *Relation to existing knowledge*: The hypothesis can be related, directly or indirectly, to existing scientific knowledge. In our make-believe case, there may have been literature that pointed towards beneficial influences of peaceful surroundings on the potentially aggressive behaviour of sentient creatures.

(e) *Resources*: The availability of resources (time, money, supervision, technology) so that the research project (which might include far more than the testing of the original hypothesis) can be carried out within the existing constraints as to these resources.

(f) *Interest*: The hypothesis is one that interests you and your advisor/superior!

3.2 Theories

A theory in the natural sciences is more "powerful" and abstract than a hypothesis. It provides generalizations about relationships between variables in the form of an interrelated, coherent set of ideas that have been supported by considerable empirical investigation.[3]

Another way of expressing this is that a theory is a hypothesis that has been tested (supported by considerable empirical evidence/not falsified) and generalized so that it can be useful in explaining, predicting and controlling phenomena.

I note once again it is not correct to say that a theory has been proved, even though it has been verified (supported by evidence) or not been rejected (shown to be false). While a theorem in mathematics can be proved, a theory regarding physical or social reality cannot be proved. A theorem is a statement of mathematical fact which logically follows, is deduced, from a set of axioms and from theorems that have already been established based on these axioms. In contrast, a theory in the natural sciences is a generalization, based on existing knowledge and empirical evidence—and observations and experiments can never prove anything; we never can assume that we know all there is to know or have observed everything that can be observed. According to Hawking (1988; 10), "Any physical theory is always provisional, in the sense that it is only a hypothesis; you can never prove it. No matter how many times the results of experiments agree with some theory, you can never be sure that the next time the results will not contradict the theory."

---

[3]The term "model", which we have met several times until now, is often used interchangeably with "theory", although it most often refers to a representation of a theory. In the case of many of the natural sciences, this representation is most often in the form of a mathematical model, i.e. a system of equations representing the interrelationships between variables. The term "model" often refers as well to physical or pictorial representations. Examples could be a physical structure representing a geocentric model of the universe or a visual model created on a computer (and generated by a mathematical model) of the double helix of DNA with its two polynucleotide strands woven around each other and running in opposite directions.

I underline here that Hawking considers a theory to be essentially a hypothesis ("... it is only a hypothesis; you can never prove it"). The equally renowned mathematician and physicist Sir Roger Penrose agrees: "Even quantum mechanics is a provisional theory, which has to be replaced by a greater theory. This is not just my own contention. It has also been stated by Einstein, Schrödinger and Dirac." (From the convocation address, National Institute of Technology, Durgapur, India; cited in the major Indian daily newspaper *The Hindu*, Nov. 5, 2006).

Although theories are generalizations based on empirical evidence, sometimes they are based on objects that cannot be directly observed. This may be because the objects are not available to perception, for example due to their size, their distance from the observer or due to their having existed earlier in time. An example of the latter is finding fossils of marine animals on mountains. Logic tells us that since these animals lived in the sea, the soil on the mountain must have been lower earlier and/or the sea must have been higher—in any case, the soil where they were found must have been covered by the sea.

The observant reader might ask: Couldn't there have been other explanations? Might not other animals, say some unknown ancient bird that loved to carry fossils in its beak, have carried the fossils from lower altitudes to the mountains? Or couldn't the sea animals somehow have crawled up the mountain before they died and eventually became fossils?[4] And so on. In any scientific investigation one can generate any number of such (apparently irrelevant) hypotheses. This is just another example of the earlier reflections as to relevance and irrelevance. I note though, that at times, however seldom, what appears to be irrelevant hypothesizing may turn out to be brilliant creative thinking that leads to major contributions to science!

Continuing with these characteristics of theories note that not all experimental results can be explained or foreseen by existing theory. When such unexpected results are obtained, they tend to lead to new hypotheses, new investigations, and potentially new theory development. On the other hand, theories, once firmly established, tend to remain as part of science's building blocks. They remain accepted until shown rather conclusively to be incorrect. This will typically be the case when the results of observations/experiments performed independently by a number of scientists and reported in respected scientific journals are in conflict with the theory. At this point such an apparently negated/falsified theory can either be

---

[4]Having just considered an example involving fossils, I am tempted to mention that what is considered by many to be one of, if not the most successful and fundamental theories in the biosciences, that of evolution, is regarded by a number of leading scientists as a research programme rather than a theory. Aside from some of the programme's underlying partial theories (dealing with such matters as macro- and microevolution and the molecular and genetic basis for the form and function of organisms), it has not been formalized, and there exist significant controversies as to its content and form (e.g. as to whether the development of new organisms has been more or less continuous or whether revolutionary changes such as the sudden development of completely different organisms have occurred)—as well as to whether the programme's representations of organisms' abilities to adapt, compete, cooperate and survive in fact represent a teleological perspective on the development of organisms. See e.g. (Depew and Weber 1996) and (Ward and Brownlee 2004).

modified and serve as a "new" hypothesis that can be tested, or it can simply be rejected. However, as we touched upon earlier in Chap. 1 when we considered "Scientific revolutions and paradigms", this process may not be as straightforward as it sounds. Sometimes it took many years for what eventually became well-established theories to be accepted. And sometimes discussions about a hypothesis can go on for years without any firm conclusion being arrived at as to whether or not it is accepted as a theory. This is e.g. the case when there is experimental evidence that can be interpreted as both supporting and refuting the proposed theory.

An example of such a complex process is the on-going discussions as to so-called "cold fusion". In 1989 an amazing press release was issued by the University of Utah in the USA. Its headline was: "Breakthrough process has potential to provide inexhaustible source of energy". During the press conference on March 23, two scientists, Stanley Pons and Martin Fleischmann, presented the results of what was essentially a very simple experiment. This was amazing for several reasons, primarily of course due to the results themselves. But also due to the fact that what was pronounced had not been subject to peer approval, as is almost always the case in the natural sciences; no articles had been sent into a recognized journal and evaluated by the journal's so-called "referees", and only the two scientists, Pons and Fleischmann, were able to provide any data as to the experiments and the results. In a nutshell, the theory they proposed was that low-level fusion reactions leading to the generation of heat could be obtained via a process, corresponding to the electrolysis of water; all that was required were two electrodes connected to a battery and immersed in a jar of heavy water (deuterium oxide). Although this apparently amazing new discovery appeared to be in conflict with fundamental knowledge in chemistry and physics, it was originally met with great expectations. These expectations soon appeared to vanish as a number of other scientists reported that they were not successful in their attempts to replicate the results reported by Pons and Fleischmann. But not all such attempts failed, and a number of other scientists, some from prestigious institutions, reported that they were able to replicate the results.

What is noteworthy here is that these debates are still continuing. According to Wikipedia: "A small community of researchers continues to investigate cold fusion, now often preferring the designation low-energy nuclear reactions (LENR). Since cold fusion articles are rarely published in peer reviewed scientific journals, the results do not receive as much scrutiny as more mainstream topics." (http://en.wikipedia.org/wiki/Cold_fusion; accessed March, 2015) There is a quarterly journal devoted exclusively to the subject of cold fusion: *Cold Fusion Times—The journal of the scientific aspects of loading isotopic fuels into materials and the science and engineering of lattice-assisted nuclear reactions* (see the website: http://world.std.com/ ∼ mica/cft.html). A bi-monthly electronically published magazine: *New Energy Times: Low-Energy Nuclear Reaction (LENR) Research and Next-Generation Nuclear Power* provides "Original reporting on research in leading-edge energy technologies, with a focus on low-energy nuclear reactions (part of the field of condensed matter nuclear science historically known as "cold fusion")";

http://www.newenergytimes.com/. There is also a bi-monthly journal with a major focus on cold fusion (*Infinite Energy—The Magazine of New Energy Science and Technology* (http://www.infinite-energy.com/). For a historical overview until 1998 including interviews with a number of scientists actively engaged in relevant research, see (Platt 1998).

There are still scientists from renowned scientific institutions who support the thesis that the theory as to cold fusion is in fact correct and can provide the basis for unheard of inexpensive and controllable fusion reactions. For example, in 2009 MIT held a Colloquium on Lattice Assisted Nuclear Reactions (LANR) whose theme was "Successful Mathematical Formulae and Engineering Concepts in LANR (Cold Fusion)"—and this colloquium followed up on earlier Cold Fusion colloquia at MIT in 2005 and 2007. In February 2011, the 16th International Conference on Condensed Matter Nuclear Science was held in Chennai, India (the conference is held in rotation between Russia, US, Europe and Asia). According to an article based on the conference in the major daily newspaper, *The Times of India*, February 4, 2011, p. 14, "Our dream is a small fusion power generator in each house" with a capacity to produce 20–100 KW of energy. "Mankind needs a new source of energy and this could be a major source to meet the ever-increasing demand for power. The findings of this research could change the face of science."

So the debates have gone on now for more than 25 years and there is no sign of their stopping! This again underscores the comment made earlier that there is no fine line of demarcation between a hypothesis and a theory—and for that matter, when a hypothesis or theory can be said to be falsified.

With basis in these definitions and reflections, let us now consider what it is that characterizes "good" theories in the natural sciences. In other words, which criteria can be used by scientists when choosing from among alternative hypotheses/theories, all of which appear to fit the data more or less equally well? Not surprisingly, this is closely related to the earlier discussion on criteria as to the relevance of hypotheses. All such criteria are, of course, subjective, a product of human thought; there are no absolute criteria for what are good criteria or good theories.

3.2.1 Criteria for "Good" Theories

1. *Power of explanation*

   A theory's explanatory power is related to: (a) the sources of the data employed, (b) the specificity of the theory, and (c) its generalizability—its ability to connect a wide range of phenomena.

   Suppose that, based on some earlier observations in Mexico, a researcher hypothesized that supplementing the diet of poor pregnant woman in and around Mexico City with 200 g of lima beans a week would lead to a reduction in unprovoked abortions compared to other pregnant women in the area who do not receive the dietary supplement. Lima beans are native to Central America and are an inexpensive source of healthy food. Suppose too that her observations over a period of several years strongly supported this hypothesis and led to the development of a (very limited) theory that dealt only with women in regions with characteristics similar to that of Mexico City. In contrast, had the investigation been expanded to cover far more general circumstances (for example, geographically, time wise and regarding the characteristics of the women who participated such as age, height, health, nutritional state, etc.), it could have led to a far more general, i.e. a theory having more explanatory power—such as: 'Supplementing the diet of underweight pregnant women who suffer from Type 2 Diabetes with 200 g of lima beans a week from the 16th week of their pregnancy leads to a significant reduction in unprovoked abortions.' Note too in this make believe case that the explanatory power of this more general and more precise theory could be significantly increased if investigations also indicated *why* lima beans have this beneficial effect on such women and this could lead to the theory being far more general, indicating perhaps how certain nutrients affect muscular spasms in underweight diabetic woman. That "final" theory would be the result of a chain of hypotheses and investigations (controlled experiments), most likely extending over a period of years and covering diverse populations (with respect to geography, culture, physical characteristics, genetic makeup etc.).

2. *Irrelevant details are left out and relevant details are included*

   To give a simple example of what is meant by "relevant" and "irrelevant" here, in ancient times people found out that they could create fire by rubbing pieces of dry wood together. Gradually individual experience as to the generation of fire became collective knowledge. Assume here that a primitive theory was in fact formulated at some time. Irrelevant details in this primitive form of theory generation might have included the day and the time of day that the pieces of wood were rubbed together, the phase of the moon, the characteristics of the person who did the rubbing, and where the rubbing took place. The relevant details might have included e.g. the dryness of the pieces of wood, how they were rubbed together and for how long, whether certain kinds of wood were better suited than others, and so on. A better theory would leave out the irrelevant details and operationalize the relevant details.

3. *Parsimony*

   The criterion above of leaving out irrelevant details and including relevant details is closely related to the criterion of parsimony, which is in fact an epistemological principle.

   This principle, often referred to as Ockham's Razor (called after William of Ockham/Occam, circa 1285–1347), states that the simplest account which explains a phenomenon is to be preferred. That is, an account with a smaller number of arbitrary assumptions that explains the phenomenon is to be preferred to an account with a larger number of arbitrary assumptions—a simple theory that fits the data is preferable to a more complicated one.

The underlying pragmatic assumption is that simplicity can lead to increased insight and require less data collection and analysis than more complicated theories or models. But the criterion also has an aesthetic aspect; just as we prefer beautiful buildings to less attractive architecture, we prefer "beautiful theories" to less beautiful, more "messy" theories; we prefer simplicity to complexity, everything else being equal. In general, scientists prefer theories that combine simplicity and elegance with inclusiveness.

A simple example that most scientists are familiar with regards curve-fitting. Any number of mathematical models can generate curves that go through or are close to points on a graph, for example a graph relating the inputs to a process to its outputs. The criterion of parsimony would choose a model that is simple, fits the data well, and that is supported by existing theory/logic, thereby providing a more readily comprehensible description.

But this invites reflection as to what to do when one model or hypothesis fits the data better than another, but where the other model is simpler and has greater appeal to say common sense or our ability to understand the model. According to Frank (1954 in Lemke et al. 1998; 465–6) "Everybody would agree that a linear function is simpler than a function of the second or higher degree … For this reason physics is filled with laws that express proportionality, such as Hooke's law in elasticity or Ohm's law in electrodynamics. In all these cases there is no doubt that a nonlinear relationship would describe the facts in a more accurate way, but one tries to get along with a linear law as much as possible." Frank indicates that a choice is required between precision and simplicity that cannot be determined by science. But this applies of course not only to balancing two criteria, such as to accuracy and simplicity, but to the principle question of how to make the best compromise between all the criteria that characterize a good theory—or, for that matter, any good decision. This is the subject of the fascinating field of "multiple criteria decision making"; see e.g. (Bogetoft and Pruzan 1997; Chaps. 1 and 4).

Earlier, in Sect. 2.2.5 of Chap. 2 (Scientific revolutions), we considered how, in the 16th century, Copernicus showed that the existing geocentric model was incompatible with the radically different heliocentric model he had developed whereby the planets revolve around the sun. Contrary to what one might expect, his arguments were not based on improved accuracy. Furthermore, both his model and the geocentric model employed epicycles. Instead, he demonstrated that his model was simpler, involving fewer epicycles. According to Gauch (2003; 273–274): "From the perspective of science, Copernicus was revolutionary for placing the sun in the centre of the cosmos; but from the perspective of scientific method, Copernicus was revolutionary for elevating parsimony to a prominent position."

According to the noted English philosopher, mathematician and Nobel laureate (1950) Bertrand Russell (1872–1970): "To seventeenth-century astronomers it *seemed* that more than simplicity was involved, that the earth *really* rotates and the planets *really* go round the sun, and this view was reinforced by Newton's work. But in fact, because all motion is relative, we cannot distinguish between the hypothesis that the earth goes round the sun and the hypothesis that the sun goes round the earth. The two are merely different ways of describing the same occurrence, like saying that A marries B or that B marries A. But when we come to work out the details, the greater simplicity of the Copernican description is so important that no sane person would burden himself with the complications involved in taking the earth as fixed." (Russell, B. 1961; 30)

Finally, the following more modern illustrative example, though not from the natural sciences, is provided by experiences I had in the early 1970s. I had developed a computer model to aid in the decision making regarding the transportation of large quantities of cement from the site of its production to a few major cities that had harbours. This transportation was performed at the time using a small fleet of ships, and the cement company was planning to make a major investment in both new ships and improved harbour facilities. The analysis could have been made by developing a detailed simulation model of the new operations in such a way that the total costs associated with the new system could be compared to the costs with the existing system. Instead, a far simpler model was developed which made a number of simplifying assumptions, each of which would be in favour of the new system, i.e. lead to reduced costs, compared to more realistic assumptions as to the new system's efficiency. Having made these simplifications, it was possible to develop a far simpler optimization model than the simulation model that would otherwise have been required. The analyses then showed that even under these favourable assumptions the new system would be far more costly than the existing transportation system. Thus, the simple model enabled the company to avoid making costly and highly inefficient investments.

So here parsimony in the form of a simple model that, though less accurate than a more complex and realistic model, resulted in the same decision that the more complex model would have led to—and did so with far greater efficiency in terms of the resources used to collect, process and analyse data. Of course things might have turned out otherwise. The simplified model might have been inconclusive and shown that under the favourable assumptions made, the new system would be less expensive. In this case the more complex model would have been required, which would have led to greater total costs of model building etc., more time than would otherwise have been required compared to the situation where the complex model had been developed from the very start, and so on. Here, just as in many situations in science, it was a good hunch/intuition that led to the decision to start out with a simple model.

4. *Predictive accuracy*

By this I mean that a theory: (a) avoids making predictions that do not hold true, and (b) survives critical tests that could have shown it to be false.

A famous example, illustrating both aspects of this criterion, deals once again with planetary motion, since readers can easily follow this no matter what their field of specialization. By the early 1800s, Newtonian mechanics had enabled scientists to make calculations that were in excellent correspondence with their observations of heavenly bodies. There was though one exception; the anomalous behaviour of Uranus. Uranus, the outermost planet, followed an orbit that differed from the calculations. Scientists were unable to explain this apparent error. So it appeared that either there was an error in treating the solar system as a Newtonian gravitational system, or else there was something special that did not permit the method to apply to Uranus. If the scientists of the time had heard of falsification they might have argued that one could not consider the solar system to be a Newtonian gravitational system, which was the central claim of Newtonian mechanics, since the underlying theory appeared to be falsified by the observations of Uranus. But fortunately they did not make such claims. And with good reason! The scientists of that period had experienced that when they applied the methodology of Newtonian celestial mechanics (by specifying the relative positions of the heavenly bodies studied; using the "law of gravity" to calculate the forces acting on the bodies; using the laws of dynamics to compute the equations of motion; then solving these equations to obtain the motions of the bodies) they had always arrived at conclusions that appeared to be corroborated by the observations.

Two scientists, independently of each other, sought an answer to the apparent anomaly. They proposed that there was an unobserved planet beyond Uranus whose gravitational force affected its orbit. Using Newtonian mechanics, they determined the orbit that such a planet would have to have if it were to account for the apparent perturbations in the orbit of Uranus. And this enabled astronomers to search the skies for such a new planet. Within a few years, in 1846, Neptune was found!

The above is by no means unique. There is a long history that documents that scientists do not simply give up an established theory simply because it appears to meet difficulties in explaining or predicting—which is a reason why so-called "scientific revolutions" can take many years to be completed. Scientific theories are in general characterized by a considerable stability since they only achieved "theory status" after having been subject to considerable testing.

5. *Testability/falsification*

This is the criterion that the better a theory enables predictions to be made that can lead to a demonstration that the theory is *not* correct, the better is the theory. As was very briefly referred to earlier in our discussions of science versus pseudo-science, in 1919 the famous scientist Eddington carried out a strong test of Einstein's gravitational theory and was able to provide the first important confirmation of that theory. This was based on a fascinating "experiment". Einstein's theory led to the conclusion that light must be attracted by heavy bodies in the same way that material objects are attracted to each other. Therefore, stars with light rays that pass near the sun should appear to be slightly shifted since their light would be curved by the Sun's gravitational field. Eddington tested this theory by photographing the apparent relative position of a distant star to that of a group of stars behind the sun and whose light passes very close to the edge of the sun. Ordinarily the gravitational effect of the sun on the light from these stars cannot be determined since the stars are invisible in daytime. However, he was able to take such photographs during a total eclipse of the sun and compare them to photographs of the same constellation taken at night six months later when the stars were not in the region of the sun. He was thus able to measure the apparent distances between the stars on the different photographs and in this way test Einstein's theory.

According to (Popper 1957 in Klemcke et al. 1988; 41), "...the impressive thing about this case is the *risk* involved in a prediction of this kind. If observation shows that the predicted effect is definitely absent, then the theory is simply refuted. The theory is *incompatible with certain possible results of observation* —in fact with results which everybody before Einstein would have predicted." In other words, the theory permitted a strong test to be made—"strong" in the sense that it enabled predictions and measurements to be made that could have clearly refuted it (or at least challenged it strongly).

It follows as well that in order for theories to be testable by others, the experiments that underlie a theory must be replicable. That is, others must be able to perform the same experiments and assuming similar conditions, obtain the same results. It follows—and I underline this once again—that good science requires clear and detailed descriptions of the observational and experimental processes employed.

6. *Consistency with pre-existing knowledge—even if the theory may show that an earlier theory was wrong*

   Certainly the ability of a theory to be accepted by one's peers depends to a great extent on its being in harmony with what is already accepted as legitimate (true) scientific knowledge. As we have seen in our earlier discussion of the concept of scientific paradigms, any natural science, like all other "professions" has its own traditions, presuppositions, standards, vocabulary, behavioural norms, "classical" texts and the like. So even though a theory may lead to startling new results, in order for it to be accepted it must be consistent with the bulk of the knowledge that is accepted in one's field. In contrast to the time of Galileo Galilei who in 1633 was forced by the Catholic Church's court of the Inquisition, at the threat of death, to retract his statements about his astronomical observations, the depository of this knowledge today is the amorphous body we refer to as the scientific community—one's peers. And it is they who provide the judgment as to the legitimacy of one's theory, not an inquisitor and not the ordinary man-on-the-street. In particular such judgment is performed by those who lead and edit the journals where new scientific results are published.

   To briefly exemplify the criterion of a new theory being consistent with an earlier theory, even though it may refute that theory, we refer once again to Einstein. We have previously discussed how his theory of relativity is consistent with Newtonian mechanics for situations where $v/c \ll 1$, while it showed that Newtonian mechanics is not correct in situations where v/c approaches 1. In other words, the new theory was not only consistent with existent theory but also extended it and demonstrated under which circumstances the existing theory would not lead to accurate/correct predictions. In so doing, it demonstrated that the existing theory was not universally "true" (although usable from a pragmatic perspective).

3.3 Laws

Laws can be said to be theories that:

- have been subject to extensive testing,
- have been found to apply over a wide range of time and space and to be correct in every possible situation—or at least are judged to have an extremely low likelihood of being found to be incorrect,
- appear to be in coherence with existing knowledge, and
- are widely accepted by the scientific community.

In other words, laws are well established generalizations about regularities between phenomena (objects, events, relationships) in the physical world in the form of an interrelated, coherent set of ideas. Thus laws have a status similar to that of a proved theorem in mathematics; since they are considered to be exceptionless generalizations, they take on the status of a presupposition. According to Bertrand Russell (1961; 147), "The discovery of causal laws is the essence of science ... the maxim that men of science should seek causal laws is as obvious as the maxim that mushroom gatherers should seek mushrooms."

But, continuing the line of thought introduced when considering "theories", the fact that the theories underlying laws are: (a) well established, (b) appear to be highly generalizable and tested, (c) are supported by existing knowledge, and (d) are widely accepted, provides no proof of their "truth". No matter how strong the apparent congruence between phenomena in nature and a law in science, laws are always fallible since they are the product of human intellectual endeavour. Consider the reflection in (Revel and Ricard 1998; 88): "...there's nothing that proves they (laws) exist as permanent principles underlying phenomena. Knowledge of them can only come through the mind, and it's a metaphysical choice that science makes when it states that with the help of our concepts we can discover the ultimate nature of a phenomenal world that exists independently of our senses." Recall too the statement by Stephen Hawking as to the temporality of theorieslaws (Hawking 1988; 10) "Any physical theory is always provisional, in the sense that it is only a hypothesis; you can never prove it. No matter how many times the results of experiments agree with some theory, you can never be sure that the next time the results will not contradict the theory."

In other words, since scientific laws transcend finite observable evidence they can never be proved in the sense of being logically deduced from the evidence/facts. There is no provable link between internal mental beliefs (theories/laws) and an external physical world. No matter how many observations we have, we cannot extrapolate from statements about "some" to statements about "all". From this perspective, scientific laws simply express *observed* regularities between events, where for example events of type A have invariably been accompanied or followed by events of type B. I note that these reflections on "proof" apply not only to scientific laws, but to all scientific statements based on empirical observation and induction: When scientists use the word "proof" they really are making a probabilistic judgement with a high level of confidence, based on evidence. In other words, research in the natural sciences should not be considered as seeking to provide proof, but, more modestly, to provide good answers to carefully asked questions—answers that can *improve* (not prove) our knowledge as well as to improve our ability to generate such knowledge.

However, even though laws can be considered to be well established generalizations about regularities, a regularity exhibited by a phenomenon is not sufficient to constitute a law, the reason being that there is more to law-like behaviour than regularity. For example, while it may be common for a person not trained in scientific methodology to argue that the sun will rise tomorrow morning since all her personal experience as well as what she has read and learned about the experience of others indicates it has always done so in the past, science provides a far more powerful explanation; given particular initial conditions (e.g. our location, the date) and based on Newton's laws of planetary motion, we can predict such an astronomical occurrence as the sun rising tomorrow at a certain time with compelling arguments and with considerable descriptive precision. "Scientific laws and theories have the logical form of general statements, but they are seldom, if ever, simple generalizations from experience." (Salmon 1967; 17–18)

In section IV of *An Enquiry Concerning Human Understanding* that has served as the classical starting point for inquiry as to our ability to draw scientific inferences, the Scottish philosopher David Hume (1748/1955; 24–25) argues that no matter how great our evidence is of a matter of fact, such as the rising of the sun, its contradiction is still possible: *"That the sun will not rise tomorrow* is no less intelligible a proposition, and implies no more contradiction than the affirmation, *that it will rise*. We should in vain, therefore, attempt to demonstrate its falsehood." (Hume's italics) In other words, in spite of our belief in the uniformity of nature it might change and be radically unlike the past whereby regularities that have been observed could prove to be inapplicable to unobserved cases. Commenting on Hume's analysis, (Salmon 1967; 10) argues: "We have found by experience, of course, that nature has exhibited a high degree of uniformity and regularity so far, and we infer inductively that this will continue, but to use an inductively inferred generalization as a justification for induction, as Hume emphasized, would be flagrantly circular." Of course, such logic does not in any way establish that empirical evidence cannot and should not provide a most fundamental guide to inference and learning, only that it cannot constitute a basis for proof!

While regularity is not a sufficient condition to constitute a law, it may not be a necessary condition either. For example the well-known principle of Archimedes that objects denser than water sink is refuted by needles that float (surface tension of the water inhibits the sinking), and leaves that fall to the ground rarely fall with a constant acceleration (due to wind, air resistance, the shape of the leaves etc.)—but this does not cause us to challenge the (Newton's or Einstein's) law of gravity.

In any case, just as with all products of human endeavour, what may at any given time or in any given culture be considered to be a law of nature, may turn out at some later time or some other context to be incorrect. To emphasize this point, consider the following example of a theory that once was so widely accepted (at least in one part of the world) that it was considered to be a law, but now is primarily of interest for students of the history of science. In Sect. 3.3 of this chapter where we considered "Science and non-science/pseudo-science", we referred to the theories of Lysenko. These theories were accepted as laws in the former Soviet Union from the 1930s to the 1960s. They rejected much of modern biology, especially Mendelian genetics and strongly affected the development of agricultural and biological sciences in the Soviet Union—with disastrous consequences. It was only after the acceptance by political leaders in the Soviet Union of the potential dangers of maintaining political dogma as a criterion for judging scientific progress that Lysenko's "laws" were removed from teaching materials.

Note in this connection that laws are mainly found in domains where mathematics can be employed to express the regularities. For example, one almost never hears of laws in the "softer sciences" since the regularities one meets in such fields cannot be characterized by mathematical precision and the variances that appear in data are significantly greater than one typically meets in the "harder sciences". Nevertheless, in scientific domains such as biology, economics, psychology, political science, sociology, and anthropology, generalizations play an explanatory role just as they do in the "harder" and more mathematically oriented branches of science. This is so in spite of the fact that they fail to live up to a characteristic such as: "laws have a status similar to that of a proved theorem in mathematics … they take on the status of a presupposition". So the concept of "law" is also relative to the individual branch of science.

Let us close this section on "scientific laws" by reflecting on their nature. We introduced the topic with a statement by Bertrand Russell: "The discovery of causal laws is the essence of science… the maxim that men of science should seek causal laws is as obvious as the maxim that mushroom gatherers should seek mushrooms." This apparently self-evident statement assumes implicitly what I earlier referred to as a fundamental presupposition of science—that the physical universe is characterized by regularities and that it is the aim of science to describe and explain these regularities. While Russell's statement deals with what the universe is, it does *not* touch upon the even more fundamental issue: *why* is the physical universe as it is? Stephen Hawking (1988; 174) reflects on this question by considering the role of theories in science: "What is it that breaths fire into the equations and make a universe for them to describe? The usual approach of science of constructing a mathematical model cannot answer the questions of why there should be a universe for the model to describe. Why does the universe go to all the bother of existing? … Up to now, most scientists have been too occupied with the development of new theories that describe *what* the universe is to ask the question *why*. On the other hand, the people whose business it is to ask *why*, the philosophers, have not been able to keep up with the advance of scientific theories." He concludes his reflections on the possibility of a unified theory with the provocative statement: "...if we do discover a complete theory, it should in time be understandable in broad principle by everyone, not just a few scientists. Then we shall all, philosophers, scientists and just ordinary people, be able to take part in the discussion of the question of why it is that we and the universe exist. If we find the answer to that, it would be the ultimate triumph of human reason—for then we would know the mind of God."

Chapter 4
Scientific Statements: Their Justification and Acceptance

In Chap. 2 we considered natural science to be "…a special way of looking at the universe—a rational approach to discovering, generating, testing, and sharing true and reliable knowledge about physical reality." In order to provide such knowledge, science must have means and methods for justifying its statements—its facts, hypotheses, theories and laws.

Scientific knowledge is distinguished from belief, no matter how true or strongly felt that belief may be, by its processes of justification and by its being accepted as such by the scientific community. And this applies both to limited factual statements (e.g. there were 37 varieties of butterflies on the island of Moen in Denmark during the summer of 2014) and to law-like statements that are universal in the sense that they are said to hold true for all phenomena, everywhere, at all times (e.g. the law of universal gravitation or the statement that the velocity of an object cannot exceed that of light).[1]

Before we begin the discussion of justification and acceptance by considering how scientific statements can be verified or validated, a few words are to be offered on "truth", since science is said to deal with "true and reliable knowledge". It should be emphasized here that we have mainly used the term "truth" to refer to a special property of statements: that they correspond with "reality".[2] In other words, that there is a correspondence between the external physical world of phenomena and the scientist's internal world of perceptions and beliefs as they come to expression in statements. The truth of a scientific statement is thus determined by the state of nature—and not by democracy or laws or feelings or beliefs. In other

---

[1]I note that some laws, such as those said to govern heredity, may not be universal in the sense used here since these laws may not hold for life elsewhere in the universe; they were developed based on investigations of life in the form it has on the planet Earth.

[2]Like "truth", the concept of "reality" itself is also subject to interpretation and definition; reference is made to the earlier discussions on "realism" and "anti-realism/instrumentalism". So once again we experience that language, which is a closed system where words are defined by other words, establishes inherent limitations on our ability to truly understand. Nobel Laureate in economics Amartya Sen reflects on similar concerns, though in an expanded context of public reasoning and objectivity: "This is a kind of dual task, using language and imagery that communicate efficiently and well through the use of conformist rules, while trying to make this language express non-conformist proposals (Sen 2009; 122).

© Springer International Publishing Switzerland 2016
P. Pruzan, *Research Methodology*,
DOI 10.1007/978-3-319-27167-5_4

words, the justification of a scientific statement deals primarily with the arguments that can confirm (or refute) its truth.

But as we shall see in more detail later on, there is a difference between justification and acceptance. You may have carried out great research and justified your results—but this will not make a difference unless the results have been communicated in such a manner that they are considered as reliable and are accepted by "the scientific community"! The following quote is from a letter from the editor of a respected physics journal to one of my former Ph.D. students in 2008: "You should also realize that the editors and referees are judging a manuscript, not only its results. Even if the results are great, but not clearly presented, the chances of getting the paper accepted are slim."

4.1 Verification (Deductive and Inductive Reasoning in Science)

Verification deals with the confirmation (or disconfirmation) of a statement via its being tested. But many scientific statements cannot be tested via direct observation. For example the statements that the earth is (roughly) spherical, or that all electrons have a charge of -1, or that genes transmit hereditary characteristics. Here science has to resort to indirect methods to verify its statements—methods that have the same basic rationale as those used when attempting to verify or validate observable phenomena. In those cases where phenomena are not directly amenable to observation, other statements that can be directly inferred from the original statement and that describe observable phenomena must be tested. The statement that the earth is (roughly) spherical in shape permits the inference that circumvention of the earth is possible by following a straight course, that the distance of such circumvention is roughly the same independent of the direction one follows, that photographs from satellites should show the curvature of the earth's surface, that if one take into account the altitude of one's location with respect to sea level, gravitational force should be the same no matter where on earth one measures it, and so on. And these inferences permit us to formulate new scientific statements that can be verified by observation and testing. So as discussed earlier in Chap. 2, where we discussed the concept of science in relation to the aim of generalization, a fundamental challenge to a reflective scientist is to justify conclusions concerning the unobserved from what has been observed (Salmon 1966, Hume 1955; 25).

As we also considered in the very first chapter, facts are the most fundamental of the scientific statements. We cannot speak of verification of hypotheses, theories, and laws without assuming the existence of "facts". I note that a "fact" can be considered to be both a *thing*—an event or phenomenon referred to and that has actual existence, or the *statement* about what is referred to as having actual existence. In the sequel, we will use the latter interpretation; facts are verified, accepted statements about events, relationships or phenomena.

Let us therefore begin by reflecting on the question of how scientific knowledge can be derived from facts assuming that the relevant facts can be established. It is common for both the layman and the scientist to speak of "proof", which of course would be the ultimate form of justification. Can we prove scientific statements based on facts?

I have earlier in Chap. 2 (the section *Science as facts*) and Chap. 3 (the sections *Theories* and *Laws*) argued that the strongest possible claim—that statements in the form of hypotheses, theories and laws can be proved—cannot in general be substantiated. The arguments rested in part on the conclusion that we cannot prove that facts, which are the basis for scientific statements, are true, but also on limitations to evidence; no matter how many observations we have, we cannot extrapolate from statements about "some" to statements about "all".[3] Let me build on these earlier reflections regarding proof and generalization by considering some fundamental concepts from the field of logic.

In logic, one can distinguish between two primary forms of reasoning: deduction and induction. We are all familiar with both of these in practice. Deduction is the method used in establishing proofs, e.g. in geometry, while inductive reasoning is the primary method used to draw inferences in the natural sciences; it begins with observations/data, and then determines what general conclusion(s) can logically be drawn from those observations/data.

Deduction

The underlying idea of deduction is very simple: one deduces a statement from other, given statements. If the given statements, the premises, are true and the reasoning is valid, so are the conclusions of a valid deductive argument; deduction is truth preserving.

Another way of phrasing this would be that a statement is a valid deductive consequence of a group of other statements if and only if it would be self-contradictory to assert all the statements in the group and to deny the statement (Kitcher 1982 in Klemke et al. 1998; 81, Salmon 1968; 24).

Note that when we here speak of validity, we refer only to the logic involved; if the reasoning is valid, the deduction is valid—but not necessarily true, since the truth also depends on the premises. In other words: A statement S is both a valid *and* a true deductive consequence of a group of statements if all its premises are true and if the reasoning that leads from the premises to conclusion is correct.

---

[3]Note that we here implicitly assume that rationality is a precondition for establishing the truth of a statement. But other approaches to truth are not based on such a presupposition. For example, mysticism is an approach to truth, in the sense of achieving oneness with reality or the Divine through a-rational means such as faith and spiritual practices. Similarly, the truth of a statement by a mother as to her love for her child cannot be rationally established, neither by herself nor by others who observe her; love here is an experience that is personal. It may change with time and circumstances, and cannot be proved or disproved, even though the mother's behaviour may provide some indication of the veracity of her oral expressions of love.

Before proceeding to examples that illustrate deduction, it is important here to distinguish between "truth" within the context of deductive logic and "truth" with respect to the natural sciences. Recall that in the natural sciences, judgments as to the truth of statements are determined by their correspondence to the state of nature, i.e. by what we are able to observe, directly or indirectly. According to Reichenbach (1949; 38), "Whereas the formulas of (deductive) logic are tautologies, science is constructed from synthetic statements. The aim of the scientist is to make not empty statements, but statements that inform us about the physical world; and the assertion that certain synthetic statements are true is the very task of science."

In other words, while observation and experience provide the basis for the assertion of the truth of a scientific statement, they play no role whatsoever in deductive logic. In addition, while valid deductive inference is necessarily truth-preserving, it is impossible to deduce from accepted premises any conclusion whose content exceeds that of the premises themselves; in contrast, science aims at generalization (Salmon 1968; 23–24).

The following examples illustrate these concepts of deductive reasoning and truth:

Example 1

1. All books on Research Methodology are boring.
2. This book is on Research Methodology.
3. This book is boring.

Based on the premises (1) and (2), the deduction/conclusion (3) is valid. But perhaps it is not true; perhaps even though this book is about research methodology you do not find it boring. The deduction is both valid and true only if both of the premises are true, i.e. that this book is on research methodology (one could reasonably argue that it is, at least that is its title), and that *all* books ever written on research methodology and ever to be written are boring. This last premise is a rather nonsensical and certainly is not a demonstrable statement; how could one ever establish its truth? Thus, although you must agree with the validity of the conclusion given the premises, you may (hopefully) disagree with the truth of the conclusion.

Note too that the concept "boring" is quite subjective and therefore also for this reason it will be difficult to argue as to the truth of the premises and conclusion. This would not be the case if e.g. "boring" were replaced by a more objective description such as "more than 100 pages", but then one might question what a "page" is, since pages can be "large" or "small", contain more or less information and so on. The point being made here, once again, is the need to operationalize the concepts we use—to define them so that the researcher can make measurements that can be replicated by others; operationalization of concepts will be treated in greater detail in the next chapter.

An even clearer example of how a deductive argument can be valid while its conclusion may be false is the following:

Example 2

1. All teachers who have taught research methodology have two heads.
2. Peter Pruzan has taught research methodology.
3. Peter Pruzan has two heads.

Clearly (3) is a perfectly valid deduction since it follows directly from the two premises. But both the premise (1) and therefore the conclusion (3) are clearly false from a scientific perspective, as simple observation can confirm. So the conclusion of a deductive argument may be valid, yet the conclusion may be false.

The following example illustrates a situation where the conclusion *can* be both valid and true:

Example 3

1. All students Peter Pruzan has taught are human beings.
2. Deepak is one of Peter Pruzan's former students.
3. Deepak is a human being.

If the premises (1) and (2) are true, then the conclusion (3) must be valid and true; if (2) is not true, the conclusion is still valid but may or may not be true (Deepak may be a horse, in which case (3) is false; and Deepak may be a baby boy, in which case even though (2) is not true, (3) is true).

If we slightly modify Example 3, we can obtain an invalid conclusion, which may however be true:

Example 4

1. All of Peter Pruzan's students work hard.
2. Deepak works hard.
3. Deepak is a student of Peter Pruzan.

Even if the premises (1) and (2) are true, the deductive conclusion (3) is not valid, although it may or may not be true! The argument is invalid since the premises (1) and (2) fail to establish that Deepak is a student of Peter Pruzan. If in fact he is a student of Peter Pruzan, even though he may not be hard working, the invalid conclusion is nevertheless true. But this has nothing to do with the reasoning; it is simply a matter of definition.

Deductive reasoning is thus a process of logic that leads to valid and true conclusions *if* all the evidence/facts/premises provided are true (which was not the case in Example 2) and *if* the reasoning used to arrive at the conclusion is correct (which was not the case in Example 4).

Thus, in spite of its truth-preserving character, we cannot in general use logical deduction to prove the truth of a scientific statement. So we can deduce (!!) that we must give up the search for truth in empirical science by using deduction.[4]

Therefore the question we now face is how can we verify or validate scientific statements if we are not able to prove them using deduction? This leads us to consider the other primary form of logical argument, induction.

Induction

Induction is radically different from deduction. It deals with drawing conclusions from data/observations and is the primary logical method upon which science is based. A fundamental presupposition that underlies induction is that the universe is ordered and governed by general laws.

The classic example of an inductive argument is that since every swan we may have seen (or heard about) is white, all swans are white (an argument of 17th century biologists). Yet this movement from a statement about "some" to a statement about "all" (from a finite number of observations to a universally valid conclusion) cannot be logically deduced since it goes beyond the information contained in the premises. Thus it provides a much weaker bond between reasons and conclusions than deduction. In fact, if someone somewhere at some time observes just one swan that is not white (which in fact took place in Australia where black swans, members of the species *Cygnus atratus*, live), this observation refutes the conclusion. It still might be true that the vast majority of the swans in the world are white, and even that all swans alive today are white, but it is not necessarily true that all swans at all times are white.

The same reservation as to absolute truth can be said to hold for any scientific statement, even those we call laws of nature. And experience over the course of history illustrates this with the development and acceptance of new, more inclusive theories/laws having an explanatory power that exceeded those of their predecessors. A classic example of this, referred to earlier, is the law of relativity, whose explanatory power exceeds that of the classical equations of Newtonian mechanics —even though in most situations where predictions or decisions are to be made and explanations are to be provided, Newtonian mechanics is a reliable guide.

Chalmers (1999; 46–7) attempts to provide a pragmatic answer to the question: When is it legitimate to assert that a scientific theory or law has been "derived" from some finite body of observational and experimental evidence? He suggests that if an inductive inference from observable facts to a scientific generalization is

---

[4]In this context one may want to consider the concept of "fuzzy logic" which is derived from so-called "fuzzy set theory" and which deals with reasoning that is approximate rather than precisely deduced from classical predicate logic. It is not to be confused with probabilities. Fuzzy truth represents membership in vaguely defined sets (not the likelihood of an event or condition as with probabilities) and permits membership values between 0 and 1(where 0 could be considered to correspond to "false" and 1 to "truth"); in its linguistic form, it permits the meaningful use of imprecise concepts such as "slightly", "very likely" etc. Thus it permits partial membership in a set rather than crisp membership ("truth") or non-membership ("false").

to be justified, the following apparently reasonable and straightforward conditions must be satisfied:

1. "The number of observations forming the basis of a generalization must be large.
2. The observations must be repeated under a wide variety of conditions.
3. No accepted observation statement should conflict with the derived law."

Based on these conditions, he arrives at a statement of what he refers to as the principle of induction:

> If a large number of A's have been observed under a wide variety of conditions, and if all these A's without exception possess the property B, then all A's have the property B.

But there are problems associated with such a seemingly straight forward principle.

Regarding condition 1: The demand for a "large number" of observations is vague and context dependent as well; are 10, 100, 1000, or perhaps 1,000,000 observations required? However, in many cases, statistical methods are able to provide guidance as we will see in Chap. 8 which deals with the role of uncertainty, probability and statistics in research methodology. Note too that in some situations even the demand of a "large number" is misleading: it is not necessary to put one's hands in the fire many times in order in order to conclude that fire can burn human flesh and cause pain; once should be enough, as many of us experienced in our childhood! And as we shall see when we consider the next condition, no matter how large a series of observations we use, we can only logically reach a conclusion based on observations if induction itself works!

Regarding condition 2: What counts as a wide variation in conditions is not clear. If the observations deal e.g. with the fertility of fruit flies in a well-defined region of India, is it necessary to perform observations where there are considerable variations in the following: (1) the type of fly? (2) the temperature in the region? (3) the time of year? (4) the time of day? (5) the colour of the shirt the scientist was wearing? (6) what she or he had for breakfast? (7) the value of the index on the Mumbai stock exchange? The answer to the first four questions is probably "yes", and most probably "no" to the remaining questions—unless of course one believes that the fertility of the flies may be affected by the colours in the surroundings, in which case condition 5 may not be as ridiculous as it may seem at first glance.

This second condition as to a wide variety of conditions also raises the fundamental question as to the choice of which variables to include in one's investigations. Unless we eliminate the infinite number of possible superfluous/extraneous variables from our investigations, we will never be able to completely satisfy the conditions under which an inductive inference can be accepted. When considering how to determine which conditions/variables are superfluous and which are to be included when investigating the generality of an effect, scientists logically draw on their knowledge. But this then leads to the question of how the choice of this knowledge is to be justified. If we demand that the knowledge itself is to be justified by induction, we arrive at what is called an infinite regress: each inductive argument will call upon additional prior knowledge, which then involves an appeal to even more prior knowledge, and so on. So in principle the choice of which variables to include and exclude introduces some degree of arbitrariness; we can never be certain that the variables we have excluded are in fact superfluous—or that we have included all the variables that are relevant.

Therefore, although in practice we always refer to our own and others' previous experiences and knowledge, conclusions based on inductive reasoning are based on an inferential jump beyond the evidence observed; *we cannot use inductive reasoning to justify induction*. No matter how many observations we have that a particular metal expands when heated, it is only logically valid to conclude that the next time we heat such a metal it will expand *if* induction works. And whether induction works can only be concluded inductively, and we have an infinite regress. Note however that scientific explanations tend to combine elements of both induction and deduction. In the case of metal expanding when it is heated, not only may we have observations that support such a statement, but also supporting theory in terms of heat being a function of molecular motion and knowledge as to coefficients of expansion in different types of materials. But note too that the theory that supports the conclusion as to the thermal expansion was arrived at via induction—via observations, experiments and reasoning.

Regarding condition 3: Also this condition is problematic since there is much scientific knowledge today that would not meet the demand that there cannot be exceptions; we will consider this in greater detail in the section below dealing with "falsification".

It follows that when we use induction, although the conclusion we draw may appear to follow from our observations, there is no guarantee for this. Induction does not provide us with truth in an absolute sense, but it provides us with the opportunity to learn. And this is what science is really about. As we mentioned earlier, scientific statements can never really be proved in the sense of being logically deduced from observable evidence. They are all in some sense "hypothetical"—what distinguishes hypotheses from theories, and theories from laws, is how strong, general, well-documented, and accepted the statements are.

Let us briefly summarize the discussion so far about deduction and induction:

1. The conclusion of valid *deductive* reasoning is contained in its premises; if its premises are true, the conclusion of a valid deductive argument is true with certainty. And it is impossible to deduce from its premises any conclusion whose content exceeds that of the premises themselves.
2. Even given the truth of all of its premises, the conclusion of *inductive* reasoning cannot be proved. The conclusion of inductive reasoning is in the form of a generalization that extends beyond the information available in its premises.
3. *Deduction* reasons from the general to the specific, whereas *induction* reasons from the specific to the general (from specific observations to general conclusions).

4. *Deduction* reasons from a mental model to expected observations, whereas *induction* reasons from actual observations to a mental model.

Thus, a major challenge to researchers when they attempt to justify their statements is how to design the interplay between deduction and induction. Since the time of Aristotle, scientific inquiry has been considered as alternating between inductive and deductive steps and reasoning. Given observations (data), often supported by theoretical reasoning based on existing knowledge, induction leads to a tentative generalization in the form of a hypothesis (conjecture, model). Based on this hypothesis as a premise, deduction leads to prediction as to observable phenomena/facts that should follow from the hypothesis and that can be tested— leading to further generalization, and so on. This iterative process, often referred to as "hypothetico-deductive", whereby there is interplay between deduction and induction is at the heart of the so-called "scientific method". We will return to this in greater detail in Chap. 7.

4.2   Falsification

Up until now we have primarily considered justification of scientific statements from the perspective of confirmation or verification. From this perspective, we test our hypotheses and justify our theories and laws by demonstrating that they are supported/confirmed/corroborated by the data we obtain from our observations and experiments. But as we saw above, although seeking confirmation from data is a most logical way of thinking, logic itself also demonstrates the limits to such thinking; neither deduction nor induction are able to prove the truth of statements regarding physical reality. In fact it was just such reasoning that led Sir Karl Popper to develop the approach to science and to the justification of scientific statements that is referred to as "falsification".

Falsification is the paradoxical idea that a statement cannot be scientific if it does not admit consideration of the possibility of its being false. The reasoning is that for a statement to be falsifiable (and therefore scientific), it must be possible to make an observation which would refute the statement—that would show it to be false.

From the perspective of falsification, this is what distinguishes statements of empirical sciences from metaphysical statements, tautologies and faith-based theories such as creationism from science. For example, those who advocate creationist theories base their explanations of the existence of life on Earth on biblical statements that all life on Earth was created as it is today. Therefore they deny the theory of evolution (which may be better referred to as a "research program" than a "theory"; see Footnote 4 in Chap. 3). Since falsification requires that theories must be open to refutation, and since creationists refer to the scriptures as an absolute source of truth that cannot be challenged, scientists who employ falsification as a criterion for distinguishing between science and non-science/pseudo-science reject the claims made by creationists that their explanations are scientific.

As referred to in Chap. 2, when Popper attempted to distinguish science from pseudo-science, he raised the question: When should a theory be ranked as scientific? (Popper 1957 in Klemke et al. 1998; 38). In his deliberations, he came to the conclusion that the apparent strength of a number of "theories" that were very popular in the period following World War I and that claimed to have scientific status, was in fact their weakness! They appeared to be compatible with even the most divergent human behaviour—the theories were always interpreted by their supporters so they fitted new observations and in this manner were always confirmed. But when he analysed Einstein's gravitational theory, things were quite different. The theory could be (and eventually was) subjected to a strong test since it is incompatible with certain possible results of observation. If observation had shown that the predicted bending of light did not occur, the theory would have been refuted. Such considerations led Popper to conclude that confirmations should count only if they are the result of risky predictions, and that good scientific theories are prohibitions—they forbid certain things to happen. In other words, the more a theory forbids (the more falsifiable it is), the better the theory. He concluded that a theory that is not refutable by any conceivable event is not a scientific theory, and every genuine test of a theory must be an attempt to falsify it.

So, according to Popper, falsifiability is a stronger criterion of the goodness of a scientific theory than verifiability. Therefore, the aim of scientific investigation should not be to generate hypotheses that are highly supported by evidence, rather it should be to develop and test hypotheses that have significant implications for our knowledge of the world.

In his book on the foundations of scientific inference, Salmon (1967; 22) supports this. He argues that the aim of empirical science is not to confirm hypotheses; rather it is to subject them to every possible serious attempt at falsification. "Scientific theories are hypotheses or conjectures ... they are never to be regarded as final truths. Their status is always that of tentative conjecture, and they must continually face the severest possible criticism. The function of the theoretician is to propose scientific conjectures; the function of the experimentalist is to devise every possible way of falsifying these theoretical hypotheses. The attempt to confirm hypotheses is not part of the aim of science."

It follows that falsification directs us to seek bold hypotheses/theories for they will run a great risk of being controverted by the evidence provided by empirical investigation. It is the duty of the scientist to seek evidence that falsifies such a bold conjecture—and the more serious are the attempts at falsification that it survives, the more highly corroborated or supported it is. This is clearly in contrast to the more traditional understanding of scientific method whereby a researcher starts with a hypothesis that is considered worthy of investigation and then seeks support for it via its positive confirmation.

Referring to the discussion in the preceding section on deduction and induction, this focus of falsification, on the ability to prove a statement to be wrong, was an attempt to re-introduce deductive reasoning into the debate about justification;

although a theory can never be proven by using inductive reasoning and empirical investigation, it was argued that it is possible from only one scientific observation to deduce/prove that a theory/law is false. It follows that those statements that have survived severe tests contribute to the development of science. But falsification, in spite of its avoidance of relying on inductive reasoning, has its difficulties.

First of all, although it provided a criterion for distinguishing between science and non-science, falsification did so at the cost of separating science from truth. In a nut-shell, Popper argued that those statements we can justifiably call knowledge are not verified or proved statements—they have not been shown to be true. Rather they are statements that have been challenged by critical examination and have not yet been falsified—that is shown to be false—by confrontations with reality. From this perspective, science does not provide us with truth. Although it offers a way for us to interpret events of nature and to learn, it removes the search for truth from the domain of science!

Secondly, as Popper himself pointed out, although the logic of falsification is valid it is also rather limited. To start with, it is not always clear whether the results of an investigation falsify a theory since the investigation itself may be faulty. For example, the measuring instruments employed may be improperly calibrated or improperly used. Even more important here is that nearly any statement can be made to fit the available data if compensatory adjustments are made to the statement (e.g. if the statement "all swans are white" appears to be falsified by the observation of a black swan in Australia, the statement could be modified to: "all swans are white except for those in Australia"). Furthermore, one can simply reject the observation of the black swan by arguing (making the hypothesis) that Australian ornithologists are colour blind. Or that they are poor scientists and that the large black bird found in Australia was not in fact a swan (a member of the genus *Cygnus*), but a member of some other genus. Thus, adding on ad hoc hypotheses can save any theory from falsification.

This is a serious weakness of falsification since scientists appear to have difficulty in rejecting theories that have been around for some time and that have appeared to be in harmony with our knowledge. However, in some cases, this is a most fortunate weakness and leads to new insights! An example is the serious challenge we referred to in Chap. 3 regarding the apparent anomalous behaviour of the planet Uranus. Instead of rejecting that the solar system was a Newtonian gravitational system, scientists sought and then successfully found an explanation that would maintain the strong belief in Newtonian celestial mechanics; the perturbations in the orbit of Uranus were due to the existence of the previously unknown planet Neptune.

Another example of the reluctance to reject a well-established accepted theory, that at some time appears to be falsified, is from radiation physics. In radioactive decays, the nucleus changes its current energy state to another, lower, energy state and in the process emits an electron. According to the law of energy conservation, the emitted electron, if it is the only particle emitted, must have an energy which is exactly the energy difference between the nuclear states, that is, it must be "mono-energetic". What was observed from a radioactive collection of nuclei was puzzling—the electrons instead of being mono-energetic, had a distribution of energies with only the most energetic electrons carrying exactly the nuclear energy difference. Based on this observation, the validity of the law of conservation of energy was questioned! However, not wanting to give up the law of conservation of energy, Wolfgang Pauli looked for other explanations. He suggested that along with the electron another particle was emitted so that the two together carried off the nuclear decay energy. Then it would be possible for the electron not to be mono-energetic while the invisible emitted particle could carry off the balance of the energy, and energy conservation would not be violated. This invisible particle was given the name neutrino by Enrico Fermi. Such adherence to the law of conservation of energy spawned a new field of research—the study of neutrinos!

The moral from both of these stories is that even though there may be some experimental results/observations that appear to be inconsistent with an established scientific statement, it may not be wise to reject the statement, at least not before considerable reasoning and investigations support the rejection. A failed prediction or the apparent refutation of a theory has often led to new insights as to the domain of the statement's applicability rather than to its rejection.

In addition, strict adherence to falsification as a criterion for justification can discourage scientific creativity and the progress of science. New ideas that have broad implications or challenge existing understanding may require long periods of gestation in order to be developed and refined; the risk is that they may be "falsified" and rejected at too early a stage in their development. According to Bohm and Peat (1989; 59), "New theories are like growing plants that need to be nurtured and cultured for a time before they are exposed to the risks of the elements."

A striking example is provided by the reflections of the British physicist Paul Dirac (1902–84) who in 1933 shared the Nobel Prize in physics with Erwin Schrödinger. In an article on how Schrödinger discovered the wave equation of the electron, he states that this was a result of pure thought and that when Schrödinger obtained experimental results that did not agree with his new equation (due to his neglecting that the electron has a spin which was not known at that time), he supressed publication of the equation. Koestler (1989; 245) cites Dirac's conclusion: "I think there is a moral to this story, namely that it is more important to have beauty in one's equations than to have them fit experiment.... If there is not complete agreement between the results of one's work and experiment, one should not allow oneself to be too discouraged, because the discrepancy may well be due to minor features that are not properly taken into account and that will get cleared up with further developments of the theory ..." Koestler provides his own reflections: "In other words, a physicist should not allow his subjective conviction that he is on the right track to be shaken by contrary experimental data." (Ibid.; 245–46)

Finally, adherence to falsification can mistakenly lead one to believe that a failed attempt to falsify a hypothesis somehow adds to its corroboration (which, in this context, clearly is not the same as its confirmation). But there may also be other hypotheses with respect to the same phenomenon that investigation does not falsify.

So while falsification may provide a filtering process whereby hypotheses may be rejected, it does *not* provide a method for selecting among *un*falsified hypotheses.

Thus, although falsification is a powerful perspective on how to generate scientific statements and design tests as to their justification, once again, as with verification, falsification cannot be considered to be the only, or the dominating, criterion for determining whether or not a scientific statement is justified or not—or for demarcating science from non-science. Neither of these two dominating perspectives, verification and falsification, can provide us with precise means for arguing that a statement is justified. They are neither necessary nor sufficient conditions for justification. Although logical and most useful in practical discourse and decision making, they are too subject to interpretation. But this is only part of the explanation. Another, and often overseen, explanation is that science is *not* just a rational approach to understanding, explaining, and predicting physical reality. It also has a very strong social aspect, to which we now turn in the next two sections dealing with "Acceptance" and "Peer Review".

4.3 Acceptance

Justification, no matter whether via verification and/or falsification is only part of what is required in order for a scientific statement to be accepted by the scientific community. Before it is recognized as reliable scientific knowledge a statement in science is typically also judged according to a number of other criteria, such as usefulness, simplicity, meaningfulness, potential harm that can result if acceptance/rejection is mistaken, and compatibility with common sense, moral, political, cultural, and, depending on the context, even religious views.

Let us briefly consider several of these criteria, where the reflections below build upon earlier discussions in Chap. 3 of "Criteria for the relevance of hypotheses" and "Criteria for 'good' theories". Consider first the potential harm that can result if a hypothesis or theory is mistakenly accepted or rejected. To illustrate the meaning of this criterion one can imagine what must have been highly dramatic evaluations during what was an unprecedented experiment—the activation of the so-called Chicago pile (the term 'nuclear reactor' did not exist then) on December 2, 1942. The successful activation demonstrated that a nuclear chain reaction could be readily controlled, something of crucial significance in connection with research leading to the development of the atom bomb. The scientists had to consider the probability that a potentially devastating uncontrolled chain reaction or explosion would occur (in the midst of a major city) and weigh this with their aim of developing a controlled chain reaction so as to permit further work on developing the bomb itself. We know from historical records that similar discussions took place among scientists working on the Manhattan Project as to whether or not such a bomb should be developed at all, but in this case with respect to military and moral issues; we will return to these issues in greater detail when discussing Ethics and Responsibility in Research in Chap. 10.

But we do not have to look at such dramatic decisions. Decisions involving say the development of new drugs also illustrate the significance of this criterion. There are always side-effects associated with the consumption of medicine, some of which can be the cause of discomfort to a patient or even lead to serious health problems. Therefore the scientists who develop a drug, the manufacturer who will produce the drug, and the scientists who will evaluate the drug (at institutions such as the FDA —Food and Drug Administration in the US) have to pay serious attention to the potential harm that can result from permitting the drug to be marketed. They have to take into account the risks associated with the hypothesis of "no dangerous side-effects" being false (risks which include both damage to the health of users as well as possibly huge economic costs to the company producing the drug) and balance this with the potential benefits associated with developing a new, marketable drug.[5]

Another of the criteria referred to above for evaluating the acceptability of a statement was how meaningful the statement is. I emphasize that neither the criteria for distinguishing science from non-science that we considered earlier nor those considered above for justifying statements involved the concept "meaningful". Statements can be meaningful without being scientific—the presuppositions upon which so much of natural science rests are certainly meaningful, but such statements are not scientific in the sense that they cannot be verified or falsified. And similarly, statements can be scientific without being meaningful—remember the exercise regarding umbellaology.

Yet another criterion we referred to in connection with the acceptance of a scientific statement was "usefulness". Many scientists would reject such a criterion as being irrelevant, since they have learned that (pure) science deals with objective truths about physical reality. However, in our discussions of "realism and non-realism" in Chap. 2 we considered arguments as to why scientists cannot/should not simply ignore the usefulness of their statements—objective truth is an ideal and science is fallible. It is rare that there is only one theory as to a phenomenon or relationship where the theory is in total agreement with observable facts. Often several hypotheses are developed that are in partial agreement with the observations, and therefore the final choice as to accepting a scientific statement is a matter of compromise and choice. Not only does it have to fit the data well, it must also e.g. be useful in making accurate predictions, in deriving many facts from simple principles, or in developing technology. According to Klemke et al. (1998;

---

[5]In summer 2005, at the time of the writing the first version of the notes that this book is based on, a court in the USA had just ordered the major international drug company Merck and Co. to pay US$253 million (later significantly reduced) to the widow of a man who had taken the painkilling medicine Vioxx and died of heart failure. This was the first of a huge number of such law suits facing the company in connection with illness and deaths attributed to the medicine, which was removed from the market by the company in September 2004 when internal investigations in the company showed that its consumption increased the risk of heart failure and strokes. A class-action lawsuit dragged its way through the courts for years, eventually being settled for $4.85 billion in 2007! According to http://www.drugwatch.com/vioxx/recall/ (updated July 11 2013), by the time of its withdrawal, more than 38,000 deaths were related to the use of Vioxx.

473), "We chose the theory according to our purpose. ... The actual acceptance of theories by man has always been a compromise between the technological and the sociological usage of science."

In our considerations of criteria for evaluating the acceptability of scientific statements, we should also consider the criterion that statements should be compatible with common sense, or at least be communicable to non-scientists. This has some characteristics similar to those of the criterion of simplicity or parsimony (treated in some detail earlier in Chap. 3). Certainly the more that a hypothesis or theory can be expressed in common sense language and related to our daily life experiences, the more it can be understood and appreciated by laymen.

But why should this be of any significance from a scientific perspective? An answer is that when scientific statements are incompatible with common sense they tend to lose their ability to support positive attitudes in such areas as ethics, politics, and religion (Klemke et al. 1998; 469–71). Consider for example, the theory (which was shown to be a myth) of spontaneous generation, which lasted from ancient times until it was demonstrated to be false in the mid-1800s. According to this theory (myth), organisms originate from inanimate matter without descent from similar organisms. It was strongly rejected by many scientists, not on the basis of scientific investigation but primarily since such a belief was considered to weaken the belief in the innate dignity of mankind and in the existence of a soul, and was therefore considered to be harmful to moral conduct.[6] Another example from more modern times is quantum theory that was enthusiastically acclaimed by some scientists, but opposed by others on the (non-scientific) basis that it would introduce indeterminism into physics.

Finally the criterion of compatibility with common sense/communicable to non-scientists is relevant since, as we touched on in Chap. 2, an aim of science is/should be to help citizens to understand science. An increased understanding and appreciation of science and its methods will not only improve their ability to make their own decisions in a world increasingly dominated by products and services with a high level of advanced technology, but also to contribute to the decisions made by their political representatives.

4.4 Peer Review

Over and above the perspectives on justification and acceptance provided so far, it is also necessary to consider the influence of one's peers, the community of scientists. Since scientific knowledge is, at least in principle, public property, scientific

---

[6]The theory of spontaneous generation was finally discarded by most scientists following an experiment in 1859 by the French chemist, Louis Pasteur (1822–1895), who later became famous for his important theoretical and practical contributions to biology, including discoveries dealing with rabies, silkworm disease, vaccines and fermentation.

statements are evaluated by others, not just by one's self—meaning that the statements must be found to survive critical study and testing by others. Of course there are exceptions to this broad statement due to the fact that many findings dealing with practical applications, such as e.g. new technology or drugs, are kept secret and patented. But in general, science involves not just investigation but also communication with the scientific community—and, at times, with society. Science is not only a rational activity; it is a social, communicative activity as well.

Science as an (informal) institution with its networks of individual scientists, universities, professional associations, journals, regulatory bodies etc. has evolved a primary system for controlling the quality of scientific production and determining whether researchers live up to the norms of science: peer review. This applies all over the world, at least when the political climate does not prohibit this, as it does from time to time in dictatorships.

Consider the following four situations in science where such quality control and evaluation takes place: (a) evaluation of theses submitted for a Ph.D., (b) "refereeing" of articles submitted for publication in scientific journals, (c) determination of tenure for university researchers, and (d) evaluation of proposals for research grants. Each of these, which virtually all scientists will be involved in during the course of their careers, is based on peer review. In each case, a major aspect of the evaluations performed consists, explicitly or implicitly, of determining whether the researcher and the research live up to the norms of "good" science.

For this form of quality control to be effective, researchers must develop an ability to evaluate not only their own research, but that of others as well. Such a skill is seldom formally taught; it is almost always a matter of learning by doing and trial and error. For example, young scientists working on a Ph.D. project will, both during the research and in connection with the thesis defence, not only receive feedback from mentors that will affect their understanding of what good research is, but also witness how more experienced scientists perform and communicate such an evaluation. Similarly, when they submit articles for publication, they will receive written feedback from the journal editors; this feedback will most often include evaluations not only of the specific topic being written about, but also of one's methodology and the clarity of one's writing. And of course since more and more research is performed in collaboration with others, they will learn such evaluation skills by "osmosis" in their teamwork with other researchers.

Very often those performing a peer review (at least in the final phases) do this in collaboration with others. This is in general the case for all the four activities mentioned above with the exception of refereeing articles submitted for publication, which is essentially a private matter. But since peer review is both a personal and a social activity, it is by no means a perfect system. Evidence of this can be found in two totally opposite observations: articles are accepted for publication that should not have been accepted, and articles that should have been accepted were rejected.

Let us consider first the case where articles that should have been rejected are in fact accepted. Shocking evidence of this can be found in an article with title "Who's afraid of peer review?" and with subtitle: "A spoof paper concocted by *Science* reveals little or no scrutiny at many open-access journals". The article is based on an investigation carried out amongst several hundred open-access journals by the editor of *Science*, one of the world's most highly prestigious scientific journals, published by AAAS, the American Association for the Advancement of Science (Bohannon 2013). In a nutshell, the data from the investigation "… reveal the contours of an emerging Wild West in academic publishing. From humble and idealistic beginnings a decade ago, open-access scientific journals have mushroomed into a global industry, driven by author publication fees rather than traditional subscriptions. Most of the players are murky."

We briefly consider now as well the case where articles that, perhaps, should have been accepted were rejected. Many scientists who have submitted articles for publication to a recognized journal or who have sought funds for research projects may, at one time or another, have felt that the evaluation they received was not fair or perhaps included serious errors. Aside from ordinary problems arising from the fact that one's peers are humans and therefore fallible, a real, but often unrecognized aspect of peer review is that factors other than merit can exert influence on evaluations.

For example, it is quite possible that those (anonymous) referees who review an article you have submitted for publication or a research proposal you have sent into obtain funding are your "competitors"—competitors in the sense that they are writing in the same field or seek funds for their own research in the same field. There is keen competition in science today—for positions, for recognition, for rewards. And this competition is not just at the level of the individual scientists. Also scientific institutions such as universities, government-run laboratories, and journals seek recognition and funding; a Nobel Prize can have a great and lasting impact, not only on the individuals who receive it, but also on the institutions where they are employed. Many universities receive a major share of their research funding from external sources, and this is to a great extent dependent on the prestige of their employees and of the institution as a whole. The huge increase in evaluative research (see Chap. 8, Sect. 8.1.3) whereby universities are ranked according to the quality of their research plays a key role in their ability to attract top quality scientists as well as good students. Even journals are being ranked today as A, B, C, journals. And the individual researchers are increasingly being evaluated, for example, according to the journals they have published in, the number of times that articles they have written are referred to in publications by others, by the prizes they may have won, by the networks they participate in …and so on.

Since competition and the search for rewards and prestige are not just characteristics of the market, but also of science, this has also led to increasing demands that, just as in the market, scientists and their institutions live up to ethical standards. Scientific research is not simply a neutral and objective activity (the way many scientists tend to characterize their field), but also a profession characterized by the values, ethics and responsibilities of its institutions and its individual members; we will consider these important aspects of research methodology in Chap. 10 (Ethics and Responsibility).

Chapter 5
Measurement

5.1 Processes, Instruments and Operationalization

Why measure? From a scientific perspective, measurement can provide a consistent yardstick for gauging differences. It thereby enables precise estimates of the degree of relationship between variables.

Measurement is thus a fundamental element in science and it is an implicit assumption in the natural sciences that whatever exists in the physical world is potentially subject to direct or indirect observation, and therefore to measurement.

Before considering more technical aspects of measurement, consider the following examples that illustrate that even though something to be measured exists and is subject to observation, it is not at all clear as to how the measurement can/should be performed.

The first example is "down to earth" and one that I, and many others from my part of the world (Northern Europe), am familiar with. When it is cold, many people who have central heating in their homes supplement it with wood stoves. The problem is how to define the volume of firewood that is purchased as fuel. The firewood is produced by cutting trunks and branches of trees into pieces of roughly the same length and then cleaving these so that they are not too thick and have increased surface area so that they are easier to burn. The resulting pieces of firewood are then piled into containers for delivery to users. There are a number of different ways of measuring the volume of wood in such a container, none of which can be said to be very accurate since it is impossible to completely describe how much of a container's volume is wood and how much is space between the pieces of wood. In addition, the usefulness of the firewood depends on the type of wood (some types burn very quickly and do not give off much heat, while others burn slowly), how dry the wood is, how dense it is, etc. So any reasonably good measurement of the quantity and quality of firewood should in fact provide a vector of measures and not just a single number.

© Springer International Publishing Switzerland 2016
P. Pruzan, *Research Methodology*,
DOI 10.1007/978-3-319-27167-5_5

A second example is also inspired by my own situation as I live close to a beautiful lake. The challenge here is how one can measure a lake's circumference? There are a number of possible approaches. For example, if the required accuracy is limited, for example in connection with a tourist brochure, one could walk along the side of the lake and use a pedometer. If far greater accuracy is required, one could obtain detailed satellite photos and use these as the starting point for measurement using laser technology. But if, for whatever reason, a highly accurate measurement is required, how is it possible to take into account all infinitesimally small uneven characteristics of the lake's perimeter—even down to the size say of small pebbles or grains of sand? And what IS the lake? Its circumference changes with the weather (when it rains, it contains more water; when the sun shines, water evaporates) and the circumference also changes, however minutely, if one or more boats are placed in the lake. Clearly, the answer to the question of how to measure the circumference will depend on the reason for measuring; there is no standard answer.

A third and final example deals with an agricultural experiment. Suppose a producer of fertilizer or an agricultural advisory body wants to determine the effect of a new fertilizer on a crop. To do so, a controlled experiment is designed to be performed on two fields located adjacent to each other; one of the fields is treated with the fertilizer while the control field does not receive the fertilizer (or perhaps receives another fertilizer if the goal is to establish the relative effectiveness of the new fertilizer). The fields are located adjacent to each other so as to remove or at least significantly reduce the influence of weather and soil quality on the results. Assume in addition that detailed information is available as to the size of the fields and that the same seeds are sown at the same time, identical conditions are provided as regards the use of pesticides, the treatment of the fields, the harvesting, and so on. So there is considerable detailed information as regards the inputs. But the question arises as to how to measure the differences in the outcomes, the yields. For example, if the crop is a particular berry for human consumption, then not only the weight of the crop will be important, but also such matters as the taste/attractiveness/smell/ robustness of the fruit, each of which cannot be objectively measured, but can nevertheless be evaluated using more subjective evaluations.

Clearly, in all three of these illustrative examples, in order to be able to answer the question of how best to measure, additional information would be required— including at least information on the required accuracy of the measurements, how the measurements will be used and for what purposes. I note in this connection that the more *precise* the measuring instrument and the more *accurately* the measurements are performed, the narrower is the difference between human perceptions of the physical world and the physical world itself. It is to these and other similar issues we now turn.

Let us commence with the question what we really mean by "measurement". In science, measurement is a process of quantification that determines the *magnitude* (e.g. the size, quantity, weight, time, length, density, opaqueness...) of something *in terms of units*. To say that something weighs 37 is meaningless because 37 has no unit. To say that the weight is 37 kg does make sense. I note that we here will be dealing with the measurement of macro- rather than micro-phenomena; due to unavoidable interaction between the observer and the observed "... it is impossible to perform a measurement on a microscopic system that does not disturb the system in a significant, unpredictable, and uncontrollable way." (Morrison 1990; 592).

But what do we really know about the units we use? What is a "metre" and why is it as it is? In Shakespeare's famous play *Romeo and Juliet* from 1594, Act II, Scene II, while declaring her love for Romeo, Juliet says: "What's in a name? That which we call a rose by any other name would smell as sweet"—meaning, please excuse this less-poetic interpretation, what really matters is what something is, not what name we give it. If we use the definition provided in Appendix A and rephrase Juliet's famous statement in our context as "a metre by any other name would still be 'the length of path travelled by light in vacuum during a time interval of 1/299,792,458 of a second'", we can see that some additional thinking is required. "Meter" is a word, just like "rose" is a word—and just as the word "rose" doesn't smell, the word "metre" does not in and of itself provide a basis for measuring length. Therefore, and because the units of measurement are fundamental to communication in science, rules have been established for their use. This permits communication in such a way that scientists can understand each other's findings no matter where the scientists come from and what language they speak.

The units of measurement that science works with are a particular version of the metric system called the International System of Units, abbreviated as SI (for the French term "Le Système Internationale d'Unités"); for more details, see Appendix A, which also provides some historical background for the basic units of measure—including why and how a "metre" came to be what it is. The standards for these units of measurement are overseen by The International Committee for Weights and Measures, located near Paris, France. The system was originally developed following the French Revolution in 1789 when the government had the goal of establishing a coherent system of units, based on the properties of natural phenomena, which would employ the decimal system.[1]

Since the 1960s, SI has been internationally recognized and all countries have adopted the metric system. It provides both a useful standard by which measurements can be made and general rules for reporting on measurements, not only in science, but in all situations where measurements are made, although it is not used with respect to time. In particular, it provides the background for laws designed to prevent fraud as to measurement in commercial transactions and many nations have their own institutions for regulating commercial measurements; for example, in the USA the regulatory body is the National Institute of Standards and Technology under the Department of Commerce.

However, not all countries use SI in practice; even such major countries as the USA and the UK continue to use units of measure (such as inches, feet, yards and miles; pounds and tons; quarts and gallons, and so on), although the standards for

---

[1]The decimal system is often said to have originated and developed in the Indus River Valley about 5000 years ago. The form of the numbers, e.g. 0, 1, 2 ... are of Hindu-Arabic origin and are commonly referred to as 'Arabic numbers' (as distinguished e.g. from so-called Roman numbers').

these units are based on SI units. These countries are going through a very slow process of converting directly to the SI system.

As was demonstrated in the introductory remarks to this chapter, even though measurement can appear to be a straight forward matter, this is not always the case. To avoid problems that can arise due to the process of measurement, measuring instruments are usually specifically designed to compare observational data to standards. Such instruments can be simple devices such as a ruler or spring weight or a container calibrated as to volume (say as to the volume of firewood)—or highly complicated and expensive devices based on advanced technologies, for example such as a CT (computer tomography) scanner that uses highly specialized X-ray machines to make multiple, finely layered pictures of e.g. the heart and its surrounding blood vessels. The precision required of an instrument depends of course on the kind of measurements it will perform, the significance of the precision, and the accuracy and reliability of the measurements. The more precise (and unbiased) the measuring instrument and the more accurate the measuring process, the more uniform will the distribution of measurements be dispersed around their true values.

Before considering a number of criteria that can be helpful when judging the goodness of measurements, let us quickly recapitulate various criteria that are often used in the natural sciences and that we have considered earlier:

1. *In connection with the formulation of hypotheses*:

   Fruitfulness
   Clarity, precision, and testability
   Provides a framework for organising the analyses
   Relation to existing knowledge
   Resources available
   Interest

2. *In connection with evaluating the "goodness of a theory"*:

   Power of explanation
   Relevance of details
   Predictive accuracy
   Parsimony
   Testability
   Consistency with pre-existing knowledge

3. *In connection with determining the justification and acceptability of a hypothesis, theory, law*:

   Verifiability
   Falsifiability
   Meaningfulness
   Usefulness
   Simplicity
   Fruitfulness
   Potential harm that can result if acceptance/rejection are mistaken

Compatibility with common sense and with moral, political, cultural, and religious views.

I note that some of these criteria are overlapping; e.g. some of the criteria in connection with the formulation of hypotheses are clearly also relevant for the evaluation of theories—and vice versa. In Sect. 4.2 we will focus on criteria that are relevant as a guide to measurement. In particular we will focus our attention on three widely referred to criteria when the results of observations and experiments are discussed in scientific publications: *validity*, *reliability*, and *reproducibility/ replicability*. These are also terms that are used in our ordinary daily language where they have a less precise meaning than presented here.

However, in order to be able to present and reflect on these three criteria, it is necessary to briefly make the terms "concepts" and "operationalization" more precise.

Concepts

Simply stated, a *concept* is a label we give to elements having common features. We may make this more precise in the present context by defining a concept to be a category for the organisation of ideas and observations that can serve as a building block of theory. For example, when you think of a noun such as "house" or "book", or "hypothesis", what comes to mind is not just a single instance but collected memories of houses and books and hypotheses, abstracted to a set of specific characteristics. This is even clearer if you think of a non-observable entity such as an electron or a black hole. What comes to mind, once again, is not a specific object (impossible in this case) but a concept with specific characteristics (e.g. a charge of minus one and a mass of …).

Concepts may require a considerable degree of specification before they can be useful in science. For example, a biologist may speak of how the resistance of plants to a pesticide depends on the size of the plant. But what does "size" mean? It could include the weight of the plant, its height, number of branches, average diameter of branches, number/size of leaves, and so on and so on. Which leads us to consider now how concepts are to be operationalized.

5.1.1   Operationalization (Variables and Indicators)

A term I have used a number of times so far is *operationalization*. This refers to defining a concept in terms of criteria that specify how to observe, describe and measure the concept. In other words, operationalization is the process whereby concepts can be applied on an empirical level so that they can be transformed into *variables*, that is, into symbols to which we can assign numerals or values based on measurements of the concept's properties. But, as referred to earlier, we also frequently meet concepts that cannot directly be measured. This is clearly the case for the concepts we use to characterize things that are not directly observable (e.g. electrons or black holes) or theoretical/abstract concepts (e.g. intelligence, happiness[2] or quality of life.[3]), or events/phenomena that took place earlier in time. In these cases, in order to be able to carry out empirical research and draw conclusions based on quantitative analyses, scientists create *indicators*—indirect measures of concepts.

For example IQ is commonly used as a measure of intelligence—but IQ is not the concept "intelligence", although it is often used an indicator of intelligence. An IQ measurement simply provides the total score obtained by adding up the points associated with an individual's answers to a series of standardized questions. Based upon empirical investigations, the score is often used as an indicator of the rather abstract concept, intelligence, for example in connection with evaluating an individual's potentials as regards educational achievement or job performance.

Finally, before reflecting on the three measurement criteria (validity, reliability, and reproducibility/replicability), we must distinguish between *independent* and *dependent variables*. The first class of variables refers to variables that are considered to be possible causes or predictors of variables in the second class and whose measurements are to be explained or predicted. We can express their relationship in scientific jargon in different ways: presumed cause—presumed effect; stimulus—response; manipulated input—measured outcome. Clearly in any scientific investigation a major choice is which variables to include—and which variables to consider as being extraneous and therefore not to include, or perhaps to control. As considered briefly earlier, when we considered the justification of scientific statements, this choice is often by no means clear. In any case, considerable theoretical inference is often a precondition for the effective design of models and experiments that relate independent to dependent variables. We will return to such issues in Chap. 6, "Experimentation".

5.2 Criteria in Measurement

Three criteria are in focus with respect to the measurement process: *validity*, *reliability* and *replicability*. Although there is no general consensus as to the relevance of these criteria or their meaning, they are very frequently referred to in the literature when characterizing the quality of experiments and measurements. In the present chapter we will only briefly consider their application to measurement processes, while in the following chapter we will discuss their application to experimentation.

---

[2]For example, a large number of different measures have been designed to capture the concept "well-being". One of the most used such measures is the Personal Well-being Index (Wills 2009).

[3]Since 1974 the journal *Social Indicators Research* has presented research results dealing with indicators that can be used to measure the concept "quality of life". Another such journal is *Applied Research in Quality of Life*, the journal of the International Society for Quality-of-Life Studies (http://www.isqols.org/).

5.2.1   Validity

This criterion is perhaps the most general of the three and is used in many different ways in scientific literature, quite often very loosely, where the term often indicates that what is being referred to is trustworthy. Once again, let us try to make the terms we use in a scientific context more precise. Here, in connection with measurement, I will refer to a characterization of validity in terms of so-called "internal validity" while "external validity" will be treated in connection with experimentation.

Internal validity refers to whether a measurement process actually measures what it is intended to measure, i.e. whether a measurement lives up to its claims. By a measurement process here I refer to how the variables that are used to operationalize concepts are in fact measured, as well as to how instruments are actually used to perform the measurement. So validity here focuses on the soundness of the way that the concepts we employ in our models are operationally defined, and on how well grounded our claims are regarding the measurements that have been performed.

Considering first the operationalization of the concepts employed in the measurement process, let us refer once again to the notions of intelligence and the indicator, IQ (which is an abbreviation of "intelligence quotient"). Just how valid IQ is for measuring intelligence depends on how one chooses to define intelligence. And how one makes such a choice depends ultimately on the purpose involved in making the definition and the measurements. This became quite clear in 1995 when the American psychologist and author Daniel Goleman published the book: *Emotional Intelligence*. In that best-selling and highly influential book he demonstrated that there are certain aspects of "intelligence" that are far removed from the cognitive/intellectual perspectives that traditional measures such as IQ focus on. Emotional intelligence included measurements of such concepts as self-awareness, self-discipline and empathy. The empirical evidence provided by Goleman indicated that in many situations people with lower IQs outperform people with higher IQs and that this is due to their emotional intelligence, their EQ. So either must concepts of intelligence henceforth be expanded to include not only IQ but also EQ when considering the performance of people at work, or else one must be prepared to offer more selective definitions of intelligence so that this term is broken down into various kinds of intelligence, each with its own measures.[4]

The message here clearly applies in general: improvements in the operationalization of a concept can be obtained by including better and/or additional indicators/

---

[4]More recently, the concept of "spiritual intelligence" (SQ) has been introduced to further expand more traditional notions of intelligence, particularly in the fields of psychology and management; see e.g. Zohar and Marchall (2000) *SQ: Connecting with our Spiritual Intelligence*, New York, NY, USA: Bloomsbury Publishing. I note that the concept of spiritual intelligence has been discredited as being pseudoscientific due to its use in more popular "new age" discourses and difficulties in its operationalization.

measures, or by splitting the concept into a number of components, each of which is to be operationalized with its own indicators/measures.

The other aspect of the internal validity of measurements that we consider deals with the *precision* of the measuring instruments used and the resulting *accuracy* of the measurements performed. Both the precision and accuracy must be considered, since even though a measuring instrument can enable very precise measurements, the resulting accuracy of the measurements depends upon how well the researcher has calibrated the instrument and the way that he or she actually carries out the measurements and records them. A trivial example: when measuring the distance between two points on a map using a ruler which is divided into tenths of a centimetre, one's readings can be accurate to within at least ±0.05 cm. However, if the ruler is not placed correctly, the measurement will be less accurate than it could have been and the measurement error will be greater than ±0.05 cm. We will reflect on measurement errors in greater detail in Sect. 4.3.

We have earlier considered how, for example, the experience of the researcher can play a role in the accuracy of the measurements. In Chap. 2 we considered how different researchers can arrive at different conclusions due to the way they use microscopes or X-ray equipment to make their observations. We argued that perceptual experiences in the act of seeing are not uniquely determined by the images on one's retina but depend on the experiences and expectations of the observer, and that "facts" therefore may not be independent of who you are and what you know in advance. Researchers who have been through the experience of having to learn to see through a microscope or to work with X-ray pictures will need no convincing of this. Repeating a point we made earlier: One has to learn to be a competent observer in science.

Both the precision *and* the accuracy aspects of measurement must be seriously considered by the researcher.

5.2.2   Reliability

This criterion generally refers to the consistency of a measurement, that is, whether repeated measurements of the same object/phenomenon provide consistent, stable results. Consistency is a necessary condition for internal validity, but it is not a sufficient condition. For example, one may make consistently biased measurements (for example, due to an improper calibration of one's measurement instrument) but this certainly will lead to inaccurate measurements and therefore, perhaps, to invalid conclusions as well.

Again, let us use IQ as a frame of reference since this can be understood by researchers no matter what their field of study. Suppose an IQ test is performed on each of 30 children in a class and that this is repeated a year later. Suppose too that when the results of the two sets of tests are compared, statistical analyses indicate that the "populations are different" (i.e. the null hypothesis that the children's IQs have not changed over time is rejected or, what is equivalent here, the test considers them to be two different groups). Then one could conclude that the IQ test is an unreliable measuring instrument, since "intelligence" is presumably something that is considered to be very stable and neither affected by time or by e.g. schooling—in contrast to specific knowledge, say as to one's field of study. Note that a conclusion regarding the unreliability (the instability) of the IQ test might be incorrect if the variability of the results is due to differences in the way that the tests were given at the two different points of time. So once again, attention must be given not only to the resulting measurement, but also to the measurement process, including the choice of measuring instruments and how they are used.

To indicate the relation of the broad concepts of reliability and validity to each other, let us now consider the situation where a scanner takes a number of pictures of a brain tumour (for example, to make a diagnosis or to evaluate the effectiveness of a new drug). Consider four situations characterized by the scanner having either high or low reliability and high or low validity.

1. Suppose that all the pictures taken by the scanner measure the size of the tumour correctly (the correctness being controlled by some super-scanner or by comparing the measurements with those performed after a patient has been operated on and had the tumour removed). Then the scanner is characterized by both high validity (it correctly measures what it should, the dimensions and mass of the tumour) and high reliability (all the pictures indicate the same size).
2. Suppose that the scanner consistently indicates that the tumour mass is 20 % smaller than it really is. Then although the scanner has a high reliability, it has a low validity; it is biased.
3. Suppose that though the scanner measures quite erratically, sometimes measuring the tumour to be larger and at other times to be smaller than it actually is. Suppose too that the errors balance out so that the average of the measurements is correct. In this case the scanner appears to be characterized by a reasonable validity (on average) but low reliability since the measurements are not free of random error.
4. Finally, suppose that the measurements are both erratic and are not centred around the actual size of the tumour. Then the scanner has low validity and low reliability; it is both inaccurate/biased and unstable.

Decisions by researchers as to how to use the scanner for diagnostic purposes or for research purposes will require an evaluation of the internal validity and reliability of its measurements—as well as of the possible advantages and disadvantages of using the scanner compared to other alternative means of making diagnoses or of evaluating the effect of the new drug on the size of brain tumours.

Over and above stability, another perspective on reliability is offered by the differences between measurements performed at the same point of time but by different researchers. This aspect of reliability is sometimes referred to as "equivalence" . For example, consider the case where the effect of a new drug on brain tumours is to be evaluated by several scientists in a research group. Each of them looks at the same set of photographs obtained via scanning a number of patients immediately before they used the drug and then at certain regular intervals after they used the drug. If the experts come to different overall evaluations based on the many aspects of what they see on the photographs, then this way of performing an evaluation would be rejected since the evaluations cannot be said to be equivalent—even though the photographs themselves are characterized by a high degree of accuracy regarding the size of tumours. This could for example be the result if the scanner is not able to provide good representations of other criteria than size, such as symmetry, colour, smoothness, or whatever, all of which can influence the visual judgements as to the development in a tumour. In this case, supplementary measurements would be required in order to achieve an acceptable level of reliability—or perhaps the insights as to the inability of the scanner to measure certain aspects of tumours could lead to the development of new scanning technology.

5.2.3 Reproducibility/Replicability

This criterion as to the measurement process refers to the ability of others to replicate one's findings. This is an extremely important criterion in empirical science since unless other scientists are able to repeat one's measurements and obtain similar results, the results will not be accepted by the scientific community. One possible reason for a lack of similar results could be that the original results were due to some extraordinary conditions or events. For example, imagine a situation where the apparent positive effects of a new drug are not found by other researchers and that this is due to the fact that the original tests were performed by highly skilled and empathetic researchers whose behaviour had a positive effect on their patients, over and above that due to the drug itself. Another possible reason for a lack of replicability of results could be that the researcher who communicated his results simply did not tell the truth. And yet another reason could be that he made errors in his experiment or that his measuring instruments were biased. And so on. It is exactly such reasons that make it essential for good scientific behaviour to include not just good observations and experiments, but also good descriptions of one's observations and experiments, i.e. in a way that enables others to replicate your experiment to see whether they obtain results that support your conclusions. Which again leads back to earlier comments on the importance of keeping detailed and accurate records of one's research activities.

5.3 Measurement Errors

Since measurement is fundamental to scientific investigation, all reflective practitioners of science, in particular students preparing to do research for a Ph.D. should have at least a basic knowledge of the uncertainties and possible errors associated with measurement; see for example *An Introduction to Error Analysis: The Study of*

Uncertainties in Physical Measurements (Taylor 1997). Being able to analyse measurement errors is important because this can help the researcher to:

- estimate the size of the errors, and
- reduce the size of the errors.

Attention to both of these is necessary for the effective design and performance of one's observations and laboratory experiments and for the acceptance of the research results by one's peers.

As mentioned at the very beginning of this chapter, any measurement consists of two parts: the magnitude (e.g. 37.903) and the unit (e.g. kilograms). Since uncertainty is inherent to measurement, measurement in science is frequently accompanied by other numbers that provide information on the statistical uncertainty that characterizes the measurement. We will return to the general concept of uncertainty in greater detail in Chap. 8; here we only consider more pragmatic aspects of measurement errors.

5.3.1   Potential Sources of Measurement Error

Measurement errors can result due to several different (but often related) factors. These can include the (1) measurement process, (2) inadequacies of the technology employed and (3) the definition of what is actually being measured.

For example, consider the apparently simple task of measuring the area of a window (a simple task in comparison to measuring say the area of an irregular object such as a leaf of a plant or of an irregular tumour). Here, one would clearly think that all that is necessary is to know the height and width of the window. However, as with the leaf and tumour, depending on the accuracy required, this may be more complex than we first think. For example, if we only want to know the area for the practical matter of roughly determining the size of a piece of window glass to be installed, then rather rough measurements may suffice. On the other hand, if we want to know the area of a lens in connection with the design of a high power telescope, where we want to make a very accurate determination of the amount of light radiation that can be transmitted through the lens, a pre-consideration of potential sources of measurement error can be most important.

Simple reflection tells us that there are a number of potential sources of measurement error when determining the area of the window. For the sake of simplicity here, assume that only the measurement of the width is to be made and that this will be done using a high quality tape measure; clearly, the list of possible error sources presented below applies as well to the height measurement.[5] I note that the list is

---

[5]Note that since the area to be measured is the product of two quantities, width and length, it will be required to estimate the resulting uncertainty in the product of the two measurements, each characterized by its uncertainty; see (Taylor 1997; 31–34, 51–53) for the derivation of rules as to how to determine uncertainties resulting from products and quotients.

not intended to be inclusive; it is only indicative of the potential sources of error—
and clearly it is in principle relevant no matter whether a tape measure is being used
or some other measurement technology, e.g. laser technology.

(a) Supposing that the tape measure is graduated in one tenth of a centimetre, an
     error occurs when a measurement does not coincide precisely with one of the
     marks on the tape measure (and I note that these have a non-zero width) and
     interpolation is required. Presumably this would lead to an uncertainty of order
     ±0.05 cm.
(b) You may simply read the tape (or whatever instrument you are using)
     incorrectly.
(c) The tape may not be placed at exactly a right angle to the frame.
(d) The tape may not be held firmly so that it is not perfectly flat, leading to a
     biased reading whereby a measurement will be larger than if the tape had been
     held firmly.
(e) The tape itself, even though said to be of high quality, may not provide precise
     measurements. For example this may be a result of how it was produced, or it
     might have been stretched due to incorrect usage, leading to ed readings.
(f) The light conditions may adversely affect your ability to read the tape measure.
(g) If the wood that was used to build the window frame was not completely dry,
     the temperature and humidity might affect the dimensions of the frame;
     therefore the measurements might not be reliable/stable due to the fact that
     what is being measured is itself not stable.
(h) If the frame has not been meticulously cleaned prior to the measurement, then
     the possible existence of dust or loose paint introduces another potential source
     of error.
(i) If the window is not perfectly rectangular, such that the width is not constant
     independent of where you measure it, then the "width" cannot be determined
     from a single measurement and a number of readings will be required in order
     to determine the average width and the distribution of deviations from this
     average.

   Certainly, depending on the purpose for which the measurement is being made,
several of these potential sources of error might be considered to be insignificant.
Furthermore, depending on the resources available, finer measurement technologies
than a tape measure might be employed and this could remove or reduce the
potential impact of several of the potential error sources listed above. The example
was provided simply to demonstrate the type of reflections that would have been
required in order to reduce/minimize the measurement errors.
   This section on "Measurement errors" was introduced with the comment that
"uncertainty is inherent to measurement", meaning that no measurement is perfect;
all measurements have some degree of error compared to the true dimensions of the
object/phenomenon being measured. Therefore, when performing analyses based
on measurements and reporting the results, the precision of the measurements
should be correctly and clearly utilized and communicated. Appendix B presents
guidelines with respect to two important aspects of such analyses: "significant

figures" (or "significant digits"; the two terms are used synonymously in scientific literature) and "rounding".

To conclude this sub-section, the following two generalizations are always relevant when performing measurements in connection with scientific research:

1. Serious reflection should be made as to the potential sources of measurement error. This should be done *prior* to the actual design of the measurement process—and is not a trivial matter, involving as it does considerations of the technological, economic and human aspects of measurement.

2. If one only focuses on potential errors arising from the measuring instrument (its precision), and not on how the measurement is actually performed (its accuracy), the total uncertainty can be significantly underestimated. In the case of an inexperienced researcher, only considering the measuring instrument's scales and not the measurement process often leads to underestimation of uncertainties by a factor of 10 or more (Ibid.; 47).

5.3.2 Random and Systematic Errors

I mentioned earlier that, when practically possible, it is generally considered to be good practice to repeat one's measurements since this permits evaluation of the measurement's reliability. However, since multiple measurements of the same phenomenon/object may lead to different results, consideration is required not just of how to make the best estimate of the actual quantity being measured, but also of how to provide good estimates of the resulting uncertainties. Here we will only briefly focus on the concepts of random and systematic errors in experiments while consideration of the statistical confidence one can have with respect to the estimate will be delayed until Chapt. 8.[6]

The errors or uncertainties that can be estimated via the repetition of a measurement are the so-called *random errors*. Returning to our trivial example of measuring the dimensions of a window with a tape measure, since there are only a finite number of scale markings, say at every tenth of a centimetre, then each time a measurement is made there will be a need to interpolate between the scale markings. If we reasonably assume that when so doing we will be equally likely to underestimate a reading as to overestimate it, then the measurements will be characterized by random (unbiased) errors. An even more subtle (and most likely far less significant) source of random errors here could be that when reading the tape measure, just how you position your head might influence the reading and the interpolations you make. No matter how careful you may be, you will not be able to

---

[6]Pagels (1982; 102–110) provides concise and fascinating reflections on the concept of "randomness" in science, including the observation that "Being able to say precisely what randomness is denies the very nature of randomness...". (Ibid.; 105).

position your head in exactly the same way each time you make a reading. As a result the readings will probably be characterized by inconsistency. If we assume that the way you position your head with respect to the tape is unbiased so that roughly half the time you make a reading that is slightly too large and half the time a reading that is too small, the measurements will be characterized by additional random errors. Both such random errors can be analysed statistically and procedures can be developed for reducing them. Such random errors can be revealed and their impact reduced via the repetition of the measurements; the average of such errors will tend towards zero as the number of measurements increases.

In contrast, so-called *systematic errors* cannot be revealed and reduced in this manner. In our simple example, such errors can be the result of a number of factors. For example, the tape measure (or whatever measuring instrument is being employed) may be incorrectly calibrated; as mentioned earlier, the tape measure may have a built in bias due to the fact that the tape has been stretched—in which case we will systematically underestimate the dimension being measured. Such systematic errors are usually difficult to detect and evaluate but can significantly affect the results of a measurement. Therefore, when performing observations/ experiments, it is essential to anticipate potential systematic errors and to eliminate or reduce them so that they do not seriously affect the correctness of the measurements. At a minimum, this typically involves following accepted procedures for calibrating one's instruments.

5.3.3 Whether to Discard a Measurement

Several times earlier we considered whether or not to reject a scientific statement, for example whether to accept or reject a hypothesis depending on the empirical evidence resulting from one's observations/experiments. In particular, we went into some detail when considering the criterion of falsification proposed by Sir Karl Popper. We saw that in some cases where the evidence appears to falsify a theory, there may still be good reasons for *not* rejecting it. To conclude this brief treatment of the analysis of measurement errors, let us now consider a related—and often difficult-question: whether or not to discard from one's data a measurement that seems so highly improbable that one suspects it is the result of a mistake. In other words, whether to decide that an unexpected and apparently highly unlikely measurement is *not* a random result of the phenomenon one is investigating, but is instead the result of a measuring error, and therefore should be discarded.

There are mixed opinions on this matter, and some scientists argue that data should never be rejected unless there is strong external evidence that the measurement in question is incorrect, i.e. the result of a measurement error. Such evidence could, for example, be in the form of information that the measuring instrument malfunctioned—or that when a new measurement was carried out, a different technology was employed than was used with previous measurements—or that a number was incorrectly copied from one document to another.

If you decide to remove data that appear to be highly improbable, but where there is no clear evidence that this is due to a measurement error, you risk being accused of "fixing" the data, that is, of removing data that appear to challenge your hypotheses (of course, you may also be challenged if you do *not* remove such anomalous data in a situation where they appear to support your hypotheses).

Here too it should be noted, as we did earlier when considering the concept of falsification, that sometimes important scientific insights and discoveries result from what appear to be anomalous observations; well-known examples include the discovery of radioactivity and penicillin. Thus, in discarding such improbable data one might be removing the most interesting part of the evidence collected.

The question naturally arises as to what to do in such situations. One common sense approach would be to carry out the statistical analyses two times; once with all the data included, and another time with the apparently anomalous data removed. If the conclusions that are a result of both analyses are the same (in other words no matter whether the suspected data have been included or not), then the data can/should be included. In this case, it would be appropriate to include a note that describes the situation. If, on the other hand, the analyses show that one's conclusions will be affected dependent on whether the apparently anomalous data is included or removed, a subjective choice must be made—and discussed in detail in your reporting.

Another common sense approach could simply be to repeat the measurements not just two times but many times. According to (Taylor 1997; 166), "If the anomaly shows up again, we will presumably be able to trace its cause, either as a mistake or a real physical effect. If it does not occur, then by the time we have made, say, 100 measurements, there will be no significant difference in our final answer whether we include the anomaly or not. Nevertheless, repeating a mea- surement 100 times every time a result seems suspect and s frequently impractical."

There exist formal criteria for identifying observations that appear to be so highly improbable that we may suspect that they are the result of an error. For example, based on the assumption that the data are the result of a process that leads to a Normal/Gaussian distribution of the output, calculations can be made as to the probability of obtaining a result that differs as much from the mean as the suspected observation. Such matters will be in focus in Chap. 8 when we consider the sta- tistical significance of observations.

Let us conclude these reflections on measurement in science by noting that whenever we make measurements and provide descriptions there are *always*, even if only at a subconscious level, underlying hypotheses. Otherwise we would not be able to make basic decisions dealing with: What to measure? Which technologies to use? What precision is required? How to describe the measurements? The challenge to a reflective scientist is to design and perform one's measurements in such a manner that one's level of awareness, reflection and insight is heightened—so that what was otherwise subconscious/implicit becomes conscious/explicit. In other words, so that we do not only describe what we have done but also *why* we did it.

Appendix A: Units of Measure

This appendix introduces the International System of Units, abbreviated SI. I do not, in general, elaborate on either why the various units of measure are defined as they are, or on how they are used in the various branches of science. For additional reading, see e.g. (http://physics.nist.gov/cuu/Units/), (www.unc.edu/~rowlett/units/index.html) and (Lee 2000; 147–158).

In SI there is only one unit for any physical quantity. Units are divided into two classes: (1) the seven base units, and (2) the derived units formed by combining base units.

1. SI base units/quantities

Quantity measured	Unit	Symbol
Length	metre	m
Mass	kilogram	kg
Time	second	s
Electric current	ampere	A
Thermodynamic temperature	kelvin	K
Amount of substance	mole	mol
Luminous intensity	candela	cd

Their official definitions are:

Meter: The length of path travelled by light in vacuum during a time interval of 1/299,792,458 of a second. Originally (in 1791) a metre was defined as 10^{-7} of the distance from the equator to the North Pole as measured along the meridian passing through Paris. In 1795 a provisional bar of brass was constructed to be used as the standard reference; in 1799 it was replaced by one of platinum; in 1889 an even more stable bar (alloy of platinum and iridium) was constructed to be used as the standard reference for the metre. Subsequent measurements have shown that this distance from the equator to the North Pole is closer to 10,002,290 m. Therefore, since 1960 a metal bar is no longer the standard reference, and the definition of a metre has been changed to that given here.

Kilogram: The unit of mass equal to the mass of the international prototype kilogram (still defined as the mass of water in a cube one-tenth of a metre on a side; a reference cube was made of platinum and iridium). This is the only basic unit still defined by a physical object. All other weight units, including those used in the UK and USA and which are not ordinarily expressed in terms of the metric system, are weighted against the standard kilogram.

Second: The duration of 9 192,631,770 periods of the radiation corresponding to the transition between the two hyperfine levels of the ground state of caesium-133 atom. A second was originally defined in terms of the Earth's daily rotation; s = [1/(24 × 60 × 60)] × (the length of a day), but later it was shown that the speed of rotation of the earth is not constant. Then in 1956 a second was redefined in terms of the Earth's complete rotation around the sun in a year. In 1967 it was decided that also this modified standard was not sufficiently stable/reliable compared to the current definition based on atomic physics. There is considerable activity at present to make certain that the three atomic watches that control each other in Paris are precise; this is to some extent due to the need of GPS-systems for more precise measurements.

Ampere: That constant current which, if maintained in two straight parallel conductors of infinite length, and of negligible circular cross section, and placed 1 m apart in a vacuum, would produce between these conductors a force equal to 2×10^{-7} newton per metre of length.

Kelvin: The unit of thermodynamic temperature; the zero point for Kelvin is absolute zero, or the lowest temperature theoretically possible. $0\ °C = 273.15\ K$

Mole: The amount of substance of a system which contains as many elementary entities (atoms, electrons, ions, molecules, etc. or specified groups of such particles) as there are atoms in 0.012 kg of carbon 12. The number is not precisely known (it is roughly 6.02×10^{23}), though the mass of a "thing" relative to that of the carbon-12 atom can be determined.

Candela: The luminous intensity, in a given direction, of a source that emits monochromatic radiation of frequency 540×10^{12} Hz and that has a radiant intensity in that direction of (1/673) watt per steradian. Originally, luminous intensity was defined in terms of amount of light given off by candles.

2. Derived units (created using the base units):

   The following is a list of some of the derived units. Some are named from the base units used to define them (e.g. metres/second) while others have been given special names (e.g. newton for force; pascal for pressure; note though that SI base units are used to determine all of them).

Derived quantity measured	Unit	Symbol
Area	Square metre	m^2
Volume	Cubic metre	m^3
Speed, velocity	Metre per second	$m\ s^{-1}$
Acceleration	Metre/second squared	$m\ s^{-2}$
Mass density	Kg/metre cubed	$kg\ m^{-3}$
Specific volume	Metre cubed per kilogram	$m^3\ kg^{-1}$
Plane angle	Radian	rad
Solid angle	Steradian	sr
Angular velocity	Radian per second	$rad\ s^{-1}$
Frequency	Hertz	Hz
Force	Newton	N
Pressure, stress	Pascal	Pa
Energy, work	Joule	J
Power	Watt	W
Celsius temperature	Degree Celsius	$°C$

(continued)

(continued)

Derived quantity measured	Unit	Symbol
Current density	Ampere per metre squared	A/m^2
Magnetic field strength	Ampere per metre	A/m
Luminance	Candela per metre squared	cd/m^2

Prefixes are used when the sizes of the base units are not convenient for a given purpose, e.g. it is easier to write the distance from Shanghai to Copenhagen in kilometres than in metres, and similarly it is easier to measure distances at the level of the atom in terms of e.g. attometres (10^{-18}), zeptometres (10^{-21}), or yoctometres (10^{-24}) than in metres, centimetres or millimetres.

Note too that there are certain units which are not part of SI but which are accepted since they are used widely. For example, minute: 1 min = 60 s; degree: $1° = (Pi/180 \text{ rad})$; litre: $1 \text{ L} = 10^{-3} \text{ m}^{3}$; hectare: $1 \text{ ha} = 10^4 \text{ m}^2$ ...

Appendix B: Significant Digits/Figures and Rounding

The following guidelines are based on (Lee 2000; 159–164) and (Morgan 2014); many websites provide similar guidelines.

Significant Digits
All measurements have error. The digits that represent the accuracy of actual measurements are called *significant digits* or *significant figures*; their number indicates the accuracy of a measurement.

The following are some widely accepted simple rules for determining the number of significant digits in a measurement.

1. All numbers without zeroes are significant: 1234.56 kg has six significant digits.
2. Zeroes between non-zero digits are significant: 35.09 has four significant digits.
3. Zeroes to the left of the first non-zero digit are not significant (they only show the position of the decimal point): 0.00023 has two significant digits
4. Zeroes to the left of a decimal point and that are in a number greater than or equal to 10 are significant: 10.123 has 5 significant digits
5. Zeroes to the right of a decimal point and that are at the end of the number are significant: 5.60 has three significant digits.

These five rules should be clear. The next rule requires some comments.

6. When a number ends in zeroes that are not to the right of a decimal point, the zeroes are not necessarily significant: 2090 can have three or four significant digits. If it means exactly 2090 it has four significant digits. If it does not mean exactly 2090, it could mean "about 2090", which could mean closer to 2090 than to 2080 or 3000, i.e. 2090 plus or minus 5. In this case it would have three significant digits.

Possible ambiguities in this sixth rule can be avoided by writing numbers in exponential notation. In the example here, depending on whether the number of significant digits is three or four, we could write 2090 as: 2.09×10^3 (three significant digits) or 2.090×10^3 (four significant digits). In this way the number of significant digits is clearly indicated by the number of *numerical figures* in the 'digit' term.

An alternative way of avoiding ambiguities is as follows: If the number of significant digits is four (such that the measurement 2090 is in fact accurate to within ±0.5 due to the precision of the measuring instrument you are using), this could be written: 2090 (±0.5). Otherwise, in the absence of such a clarification, it would be natural to assume the smallest number of significant digits, here three, leading to the understanding that the measurement means 2090 s (±5).

When interpreting measurements provided by *others*, a good rule to follow is to assume the smallest number of significant digits unless there is information that the accuracy is greater. Similarly, when reporting your own measurements, you should make it clear what the number of significant digits is so that the reader does not have to guess/interpret the accuracy you are working with.

Measured quantities always have significant digits. Other numbers that are not the result of a measurement may have perfect accuracy; "Five people were interviewed" means exactly five people, not 4.5 to 5.5 people, and 1 m is by definition exactly 1000 mm.

Rounding

Quantities are rounded to bring them to the proper number of significant digits after either converting units (say from miles to kilometres) or using quantities in mathematical operations (say after multiplying two quantities).

It is improper to report quantities with greater accuracy than is justified by the accuracy of the original measurements; if you present data with more digits than are significant it misleads others into thinking that the measurements are better (more accurate) than they really are, and this affects how the data are interpreted.

When rounding, first determine the appropriate number of significant digits. Then the following general and broadly accepted rules can be followed:

1. If the first insignificant digit is greater than five, the last retained digit is increased by one: 8.472 rounded to two significant digits is 8.5.
2. If the first insignificant digit is less than five, the last retained digit is left unchanged: 13.37 rounded to two significant digits is 13.
3. If the first insignificant digit is 5 or 5 followed by anything other than zeroes, round up: 6.3753 with three significant digits rounds up to 6.38.
4. If the first insignificant digit is 5 or 5 followed only by zeroes, then round up if the last significant digit is odd, and round down if the last significant digit is even: 0.23500 rounded to two significant digits is 0.24 (since the last significant digit, 3, is odd) and 0.24500 rounds down to 0.24 (since the last significant digit is even). The rationale for this rule is to avoid bias in rounding: half of the time one rounds up, half the time one rounds down.

In general, to minimize error, it is best to round numbers *after* a calculation has been done. This can, for example, be important for many mathematical operations in statistics.

Converting units

When converting units a general rule is to round the resulting quantity so that the accuracy is approximately that of the original measurement. Converting 30 ft to metres on a calculator might give 30 ft × 0.3048 m/ft = 9.1440 m. This should be rounded to 9.1 m (two significant digits). The logic is as follows. The accuracy of the original measurement is one-half foot (6 in.) which is roughly comparable to 0.2 m: (6 in. × 0.0254 m/in. = 0.1524 m). So if one rounded down to 9 m (one significant digit, corresponding to ±0.5 m = ±19.69 in. or roughly ±20 in.), the result would be too imprecise compared to the accuracy of the original measurement. Similarly, if one rounded to 9.14 m (three significant digits, corresponding to ±0.005 m = 0.1969 in. or roughly 0.2 in.) the result would be too accurate compared to the accuracy of the original measurement.

Mathematical operations

When doing arithmetic with measurements, significant digits must be considered. The accuracy of a calculated result is limited by the least accurate measurement involved in the calculation.

Addition and subtraction: The result should be rounded so that there are no significant digits further right than the last significant digit of the least accurate measurement. For example:

(a) 1585.236 + 234.76 + 1.2 = 1821.196 which is rounded to 1821.2 because the least accurate measurement, 1.2, has only one decimal place.
(b) 1346.15 − 1218 = 128.15 which is rounded to 128 because the least accurate measurement, 1218, has no decimal places.

Multiplication and division: The result should have the same accuracy (same number of significant digits) as in the measurement with the least number of significant digits. For example:

13.32 (4 significant digits) × 1.08 (3 significant digits) × 2.0 (two significant digits) = 28.7712 which is rounded to 29 (2 significant digits).

Measurement ranges

When we have an estimate of a measurement error, for example when we know details about the precision of the measuring instrument we used, the best way to provide the result of a measurement is to give both the best estimate we can of the quantity being measured *and* of the range within which we can be confident that the quantity lies. In general, as a rule of thumb, one can say that the last significant digit in any reported result should be of the same order of magnitude (in the same decimal position) as the estimate of the uncertainty that is reported.

Thus a statement that the measured volume of a tooth of a fossil is $238{,}651.7924 \pm 40$ mm^3 is misleading. The estimated uncertainty, 40, means that the digit 5 could be as large as 9 and as small as 1. Clearly the five digits to the right of the 5 have no significance and should be rounded. This leads to the more correct statement that the measured volume is $238{,}650 \pm 40$ mm^3. If the estimate of the uncertainty is not 40 but 4, then the rounding should lead to $238{,}651 \pm 4$. And if the estimate of the uncertainty is 0.4, then the rounding rule of thumb here should lead to $238{,}651.8 \pm 0.4$.

Chapter 6
Experimentation

6.1 The Roles and Limitations of Experimentation

The following anecdote in Chalmers (1999; 27–8) provides an excellent and hopefully amusing introduction to this chapter. "When I was young, my brother and I disagreed about how to explain the fact that the grass grows longer among the cow pats in a field than elsewhere in the same field, a fact that I am sure we were not the first to notice. My brother was of the opinion that it was the fertilizing effect of the dung that was responsible, whereas I suspected that it was a mulching effect, the dung trapping moisture beneath it and inhibiting evaporation. I now have a strong suspicion that neither of us was entirely right and that the main explanation is simply that cows are disinclined to eat the grass around their own dung. Presumably all three of these effects play some role, but it is not possible to sort out the relative magnitudes of the effects by observations of the kind made by my brother and me. …To acquire facts relevant for the various processes at work in nature it is, in general, necessary to practically intervene to try to isolate the process under investigation and eliminate the effects of others. In short, it is necessary to do experiments."

Although it is easy to agree with the conclusion to this anecdote, it is not that easy to design a research project and experiments that provide more or less conclusive answers to the question as to why "…the grass grows longer among the cow pats in a field than elsewhere in the same field." Clearly there are a number of possible hypotheses that could provide the starting point for the research; three have been provided above. In addition, there may be other factors that a researcher must consider, the so-called 'extraneous variables', which even though they have not been designated as a hypothesis' independent variables, may in fact play a role. For example, it may be relevant to consider variables characterizing the context. These might include: the type of soil and grass in the field; the type of cows, their density (number per m^2), age, etc.; the climatic conditions; the location of the field (altitude and geography); and so on. Deciding on which variables to include in one's

© Springer International Publishing Switzerland 2016
P. Pruzan, *Research Methodology*,
DOI 10.1007/978-3-319-27167-5_6

investigation and how to include them is an important aspect of any experimental design. It may also be necessary to consider measurement aspects, including the techniques that can be used to measure the rate of growth of grass among the cow pats in order to be able to compare it to the rate of growth of grass where there are no or fewer pats. And so on and so on. And of course there is the whole matter of how to design the relevant experiments so that they will be considered valid and reliable, as well as how to analyse the resultant data and draw conclusions as to one's hypothesis. Think about this—how would you design an experiment to investigate the relationship between grass production and the presence of cow manure? Appendix A to this chapter provides an example of several such considerations in connection with a very simple and very down-to-earth experiment regarding the baking of bread!

In a nut shell, this chapter deals primarily with how experiments can contribute to providing "good" answers to research hypotheses/questions in the natural sciences. As a source of inspiration, the link http://en.wikipedia.org/wiki/List_of_famous_experiments provides a long list of historically important scientific experiments, grouped according to scientific discipline, that demonstrate "something of great scientific interest, typically in an elegant or clever manner". Details of each experiment can be downloaded by going to the website and then clicking on the desired topic.

However, before considering the aims and limitations of experimentation, it is important to underline that research can be carried out without performing experiments. This is typically the case in the social sciences, where active experimentation is rarely used. In fact this is one of the major distinctions between so-called "hard science", characterized by the use of considerable control in experiments and the resultant strong confidence in results, and "soft science", where such controlled experiments are much more difficult to design and perform, with a concomitant reduction in the validity and reliability of the research results. For example, an anthropologist who studies how parents bring up their children in a small isolated village in the Amazon jungle cannot avoid affecting the behaviour of those she observes and thereby of the whole village as well. It is impossible for another anthropologist to replicate the same study. First of all, the village and the families have been changed due to the first investigation. And if a new village is chosen, then we must speak of a new study since the behaviour of parents in such a village may be affected by many factors that are specific to a given local society.

But research can and is also performed in the natural sciences without relying on experiments. In the natural sciences, such non-experimental research activities are often performed in an early stage of an investigation (or for that matter, of a discipline's development), and aim primarily at description, fact-gathering, and answering basic questions. The findings from such predominantly non-experimental and fact-finding (rather than theory-finding) investigations often provide a background for hypothesis testing and theory development later on.

However, as briefly reflected on earlier in Chap. 2, many researchers in the natural sciences may not be concerned with experimental investigations at all. There are sciences, such as astronomy, botany, zoology and geology, where experimentation is either impossible or highly impractical/costly/difficult to perform. Such sciences are therefore primarily oriented towards describing phenomena and answering research questions instead of being oriented towards active experimentation and testing hypotheses. Examples of such questions could be: What is the surface of Venus like? What can we learn about past climatic conditions by analysing ice cores from Greenland? How is the human brain affected by sun-storms? What species of plants live in the eastern Himalaya Mountains? What patterns of nucleotides make up human genes? Is global warming taking place? Does life exist elsewhere in the universe? Such questions are more representative of a descriptive rather than an analytic/causal approach to scientific inquiry.

It should be underlined however that, in general, experiments play a vital role in those domains of the natural sciences which aim at explanation and prediction. Experimentation, in contrast to the more passive observation, can perhaps be most broadly defined as a *manipulative and/or comparative investigation that enables testing hypotheses and the development of causal explanations.*

According to Pagels (1982; 335) there is a symbiotic relation between theory and experiment: "Theory provides the conceptual framework that renders experiment intelligible. Experiment moves the theorists into a new realm of nature that sometimes requires a revision of the very concept of nature itself."

6.1.1 Natural Experiments

As briefly touched upon earlier, in some situations where a researcher would otherwise perform an experiment, the difficulties may be so great as to make it practically impossible to do so. For example, it is (at least at present) impossible to perform experiments that could directly test hypotheses dealing with such matters as the movements of the tectonic plates in order to predict the occurrence of earthquakes,[1] the development of fossils in order to study evolution, or the formation of distant stars in order to study the expansion of the universe.[2]

In such cases it may nevertheless be possible for scientists to resort to so-called *natural experiments*, often referred to as *field experiments* or *quasi-experiments* (Cooper and Schindler 2003; 443).

---

[1]Although plate tectonics, the branch of science dealing with the process by which rigid plates move across molten material, can help to explain global-scale geological phenomena, it has hitherto not been very successful in predicting the occurrence of earthquakes—as the horrendous earthquakes in Indonesia (2004), Kashmir (2005) and Hati (2010), each with more than 100,000 casualties, taught us (http://en.wikipedia.org/wiki/Lists_of_earthquakes#Deadliest_earthquakes_on_record).

[2]It is interesting in this connection to note that in 2005 an "experiment" was performed that only a decade earlier would have been considered science fiction; a space probe called Deep Impact was sent by NASA into a comet, enabling scientists to carry out observations of importance for research on the age of the universe (http://www.nasa.gov/mission_pages/deepimpact/mission/#.U8kh9UBCPos).

In natural experiments one can take advantage of predictable natural changes in simpler systems to measure the effect of such changes on a phenomenon that cannot be manipulated. For example, it would in practice be impossible for astronomers to test the hypothesis that suns are collapsed clouds of hydrogen. Yet the observation of clouds of hydrogen in various stages of collapse as well as of spectral emissions from the light of stars could provide data that can serve as a basis for testing the hypothesis. In this case, no manipulation of variables would take place, no comparisons would be made with control groups; there would simply be a purposeful collection of data that could be analysed to determine whether the observations provide support for or reject a hypothesis (http://en.wikipedia.org/wiki/Experiment).

But also in areas of more applied research, such as often met in economics and epidemiology, where controlled experimentation can be very difficult to perform, natural experiments can be useful. This is particularly the case in situations when a well-defined population has been exposed to a stimulus or been treated for a sickness, so that changes in responses may be plausibly attributed to the exposure or treatments. A famous historical example of this is the investigation of the so-called "Broad Street cholera outbreak", a major outbreak of cholera in a section of London called Soho on 31 August 1854. At the time, even though there had been several major outbreaks of the disease in England during the preceding 20 years, there was no knowledge at all as to the disease and no attempt was made to contain it. During the next three days 127 people near Broad Street died and by the end of the outbreak 616 people died. A medical doctor, John Snow, whose observations and studies had convinced him that cholera was spread by contaminated water, identified what he believed to be the source of the outbreak: the nearest public water pump. He discovered a strong association between the use of the water from this pump and deaths and illnesses due to cholera; nearly all the deaths had taken place within a short distance from the pump. He also had other observations that supported his beliefs; the water company that supplied water to districts with high attack rates obtained the water from the river Thames downstream from where raw sewage was discharged into the river, and, by contrast, districts that were supplied water by another company, which obtained water upstream from the points of sewage discharge, had low attack rates. He convinced the leaders of the parish where the pump was located to remove the pump handle and the spread of cholera fell dramatically (Summers 1989; 113–117). Clearly, no active experimentation was performed (although the removal of the pump handle could be considered a primitive experiment). Nevertheless, there was a purposeful collection of data, the analysis of which enabled Dr. Snow to support his hypothesis that consumption of contaminated water was the cause of cholera (though no hypothesis as such was explicitly formulated).

6.1.2 *Manipulative Experiments*

An activity can be referred to as an experiment without having to make attempts to modify or manipulate variables. For example, in high energy physics one refers to experiments whereby extremely large amounts of data are collected but where manipulation (of independent variables) and comparison (with some control group), are *not* central to the experiment. However, in many areas of investigation, experiments are characterized by the researcher deliberately changing or assigning values to independent variables in a planned manner in order to effectively determine their effect on one or more dependent variables.

In order to ascertain the effects of performing the manipulation of the independent variables, it may, depending on the type of experiment, be necessary for the researcher to also consider the effect of extraneous sources of variations on the reactions/responses of the dependent variable(s). For example, when testing whether a particular form of physical exercise reduces the risk of cardiac arrest, it is the exercise and its effect on heart problems that is in focus, and not such extraneous factors/variables as e.g. the type of shoes the exerciser wears, the person's educational background, the location of the place where the exercise takes place, as well, perhaps, as the weight, gender, and age of the exercisers.[3] Existing theory and common sense will lead the researcher to disregard the first three factors (and the infinite number of other such possible extraneous variables that a priori can be neglected, such as the current rate of inflation or what movie is playing at the local cinema when the exercise is performed). On the other hand, the last three, the weight, gender and age of the exercisers, quite likely will have to be taken into account, though not by being treated as independent variables. For example, this could be done by selecting the participants in the experiment so that they are all overweight (according to the standard Body Mass Index). In this case, we can speak of the direct control of the extraneous variables.

Such direct control of the extraneous variables can be effectively carried out in many types of experiments—often in laboratories under standardized conditions. However, in some cases even though the researcher may believe that extraneous variables may affect the reactions/responses of the dependent variable, due to practical difficulties or limits on the resources available, she may choose not to incorporate the extraneous variables directly into the study design. In these cases, the effects of the extraneous variables can be mitigated by use of randomization in the design, e.g. by a random assignment of patients to groups in an experiment to determine the effectiveness of a new drug. We will consider the use of randomization in greater detail in Sects. 6.3 and 6.6, where we will see that randomization

---

[3]Extraneous variables are often referred to in the literature on experimental design as *control factors* (they are factors to be controlled). In certain situations they are also referred to as *blocking factors* where blocking refers to the allocation of experimental units to groups (blocks) such that the experimental units within a block are relatively homogenous while the major part of the predictable variation among units is associated with the variation among blocks.

can play an important role in the design of experiments, particularly in experiments involving humans, animals and plants.

We can conclude this sub-section by emphasizing that the question of which extraneous variables are to be included in an experiment and how they are to be treated is of major importance; it will provide serious challenges to the reflective researcher who is designing and performing an experiment in the natural sciences.

6.1.3   Comparative Experiments

To demonstrate the likelihood of causality (of an independent variable on a dependent variable), experiments are often designed so as to enable comparing the results that are obtained from an experimental sample with a so-called control sample. The members of this control sample are selected so that it can be said to be as identical as possible to the experimental sample, except for the fact that its members do not receive the hypothesized cause that is being tested.

Although one can meet such experimentation in all domains, it is, once again, perhaps easiest to illustrate with an example from the medical sciences that I presume all readers can follow, no matter what their field of science may be. An experiment is to be designed to determine the efficacy of a new drug with respect to some measurable physical characteristic characterizing people having a certain illness. This is to be done by comparing the effect of the new drug on a sample population of people having the illness to the effect of a so-called placebo on a second sample population of people who also have the illness, the control sample. The members of this second group receive a pill or a tonic or an injection that is known to have no effect whatsoever on the illness (aside from the possible psychological results due to the recipients experiencing that someone is interested in them and their feelings that they may be receiving something that could be beneficial to him and improve their health).

Both the sample populations are selected in such a manner that the two samples can be assumed to be statistically similar with respect to the extraneous variables that have been chosen to characterize their members, for example their weight, height, age, gender, blood pressure, etc. In other words, the sample groups can be said to be probabilistically equivalent in that it would be expected that *if* the members of both groups had received the *same* treatment, the results would have been statistically very similar. This permits concluding that the differences in effect that may be measured between the two samples after they have been treated with the new drug and the placebo respectively, can be ascribed to the treatment alone.

Such testing is generally "double-blind", meaning that neither the members of the two sample groups nor the researcher (nor the staff who administer the drug and placebo, or any others who may directly or indirectly affect the members of the sample groups) know which patients receive the new drug and which patients receive the placebo; this information is not available until the end of the experiment. This is to ensure that: a) the members of the two sample populations do not react due to the knowledge as to how they are being treated (with the real-thing or with placebo), and b) the researcher herself does not unknowingly exert an influence on the test due to her knowing who receives what.

6.2   Experimentation and Research

Having briefly considered the manipulative and comparative aspects of experimentation, that either alone or in some combination characterize experimentation, we can now briefly consider the interdependency of experimentation with other aspects of the research process.

Experimentation is usually a *multi-step, iterative learning process*; it does not in general take place at one particular time and place. Rather, it is often characterized by considerable feedback from and feed-forward to other research activities such as: the formulation of hypotheses/research questions, determining the required accuracy, the design of the experiment, the statistical analyses to be performed, the interpretation of results, the communication of these results to others so that they can evaluate and replicate the experiment, and so on.

Yet another way of characterizing experimentation in relation to research deals with its relationship to existing theory. It can be argued that the propensity of scientists to accept experimental results depends on what they already know—on their more-or-less implicit understanding of existing theories. Sometimes we "see" things because of what we already know/believe and tend to reject other possible observations due to our distrust/disbelief in them should they appear to go against well-established theory.

An example: At the time this chapter is being finalized (fall, 2015), a debate is taking place world-wide among scientists from many fields as to whether or not saturated fat is unhealthy or healthy. Until recently, the overwhelming consensus was that it is unhealthy. Now considerable new evidence has shown that there were some errors in the earlier research and that there is strong statistical evidence, backed by scientific reasoning, that saturated fat is not only not-unhealthy, it can be healthy as well. In spite of this, it is anticipated that it will take years before the deeply ingrained beliefs regarding saturated fat among scientists will change— perhaps due to mechanisms similar to those that Thomas Kuhn referred to in his book on *The Structure of Scientific Revolutions* (see Sect. 2.2.5).

So, as we have also briefly reflected on earlier, scientists, just like ordinary people, are selective in their perceptions. Einstein is reported to have reformulated the well-known American statement: "I wouldn't have believed it if I hadn't seen it" as "I wouldn't have seen it if I hadn't believed it in the first place". So from this perspective, experimentation does not always provide us with clear, undeniable facts, and our observations depend on existent theory. But of course, before scientific statements are broadly accepted as being true by the scientific community, they must be justified, replicated by others (when possible) and published in respected refereed journals. These activities contribute to filtering out a researcher's personal beliefs and biases (and possible mistakes and errors) so that the final results are supported (or at least not falsified) by evidence and scientifically acceptable by one's peers.

Experimentation thus provides new knowledge—and as our knowledge develops, our observations and judgments are also subject to change. Therefore it is also meaningful to consider scientific experimentation as a *dialogue* with both the scientific community and with nature. According to Prigogine[4] and Stengers (1984; 5): "Modern science is based on the discovery of a new and specific form of communication with nature—that is, on the conviction that nature responds to experimental interrogation. ... Experimentation does not mean merely the faithful observation of facts as they occur, nor the mere search for empirical connections between phenomena, but presupposes a systematic interaction between theoretical concepts and observation. In hundreds of different ways, scientists have expressed amazement when, on determining the right questions, they discover that they can see the puzzle fits together. In this sense, science is like a two-partner game in which we have to guess the behaviour of a reality unrelated to our beliefs, our ambitions, or our hopes. Nature cannot be forced to say anything we want it to. Scientific investigation is not a monologue. It is precisely the risk involved that makes this game exciting."

Before proceeding to considerations of how experiments can (and should) be designed, performed and evaluated, let us build upon the above by considering briefly the advantages and disadvantages that experimentation offers in the natural sciences.

Advantages and disadvantages of experimentation

Experimentation has (at least) the following advantages compared to carrying out investigations without experimentation; see e.g. (Cooper and Schindler 2003; 426–7).

Potential Advantages:

1. Experimentation provides the researcher with an improved ability to uncover causal relationships. Better than any primary data collection method, experimentation can establish the probability of one variable being linked to another since the researcher has the ability to manipulate the independent variable(s) in a planned and controlled manner and to observe the changes in the dependent variable(s). In this way it contributes to the effective utilization of investigative resources.

2. Experimentation permits the effective control of extraneous as well as environmental variables (i.e. the physical conditions surrounding an experiment such as the conditions of the laboratory where the experiment is performed) and helps the researcher to isolate the impact of such variables on the outcomes so they do not contaminate the results.

---

[4]Illya Prigogine received the Nobel Prize in Chemistry in 1977 for his work on so-called "dissipative structures" and the thermodynamics of non-equilibrium systems.

3. In some cases experimentation permits the researcher to adjust the independent variables and the environmental conditions so as to evoke extremes that could not otherwise be observed under more ordinary conditions.
4. Experimentation provides the researcher with an ability to repeat tests under different conditions and thereby to discover the statistical relationships between the values of the independent variables and the dependent variables.
5. In this way experimentation also permits either the ruling out of idiosyncratic and isolated results—or the investigation of why such unexpected results occurred.
6. Due to their systematic nature, experiments permit control with respect to sources of bias.
7. Finally, since experimentation is so very much a part of what is considered to be the "scientific method" (see Chap. 7), the analysis and communication of results can be greatly simplified. For example, there are more or less well-structured procedures in the various scientific disciplines for performing and reporting on the results of one's experiments.

Potential Disadvantages:

1. Particularly in the case of the social sciences, but also with many experiments in the natural sciences, in particular those which involve human beings or animals, the artificiality of experiments can be a primary disadvantage. This is particularly true when the sentient beings are aware of the fact that experiments are taking place.
2. Generalization from non-probability samples can pose problems even if the researcher makes serious attempts at randomizing the assignment of units to the samples.[5]
3. Compared to other primary data collection methods, the time and costs required for an experiment can be too demanding given one's research budget.
4. A final potential disadvantage of experiments is the limits to the types of manipulation and control that can arise due to ethical considerations.[6] There are

---

[5]For example, if a controlled experiment is performed to determine the effects of various kinds of fertilizers on the immune system of ticks that live on a particular kind of cow that lives in a particular geographic region, no matter how the cows/ticks/fertilizers are assigned, it can be difficult to generalize the results obtained so that they are applicable to other types of ticks and cows in other geographical environments. Similarly, even though it is easier and less expensive to perform an experiment to evaluate the laboratory skills of Ph.D. students in chemistry at Harvard University rather than of chemists throughout the USA, and even if the students to be observed at Harvard are randomly selected, it is rather doubtful whether results based on the experiment at Harvard can be generalized to American chemists.

[6]Tragically, such limitations have not always been imposed; throughout history dictators have been tempted to perform cruel experiments involving human beings For example, during WWII, due to his horrific experiments on prisoners, particularly Jews, at the huge concentration camp Auschwitz, the Nazi doctor Josef Mengele, was known as the "Angel of Death". Alas, he was not unique; there were many such doctors who performed horrendous experiments, often on fully conscious prisoners, in the concentration camps. Often the patients were murdered after the experiment (http://www.auschwitz.dk/doctors.htm).

many organisations that protest against activities such as animal testing (particularly when performed to test cosmetics), the killing of whales or primates (apes, monkeys etc.) that are to be used for research purposes, and against experiments involving human embryonic stem cells (until recently there were strict laws on this matter in the USA). It can be a very difficult situation for a scientist who faces the dilemma of having to balance the possibilities for the generation of valuable knowledge via an experiment with, for example, the harm that might result from the experiment. This will be a major topic of discussion in Chap. 10 dealing with ethics and responsibility in research.

6.3   Conducting Experiments

The presentation here on the implementation of experiments is structured in a number of steps. Even though there is an apparent logical progression in these steps, it must be underlined that since feedback takes place continually during the learning process that characterizes experiments, these steps should not be considered to represent a strict sequence. Furthermore, in practice there are other activities that tend to characterize the conduct of an experiment, but that I have chosen to ignore here. These include the many practical matters dealing with obtaining funding to carry out the experiment, how to make the best allocation of resources to the various aspects of the experiment, the choice of which technology to employ, how to select experimental and control samples, planning the experiments so as to meet time and budget limits, how the results are to be analysed and reported on, and so on. These may all be important practical aspects of the experimentation process, but are not in focus here, primarily due to space limitations.

Therefore, in the following, only the following six more or less independent stages in the process will be highlighted. The exposition is based on Cooper and Schindler (2003; 427–32) which focuses on research in a field of the social sciences, but where the structure presented is relevant as well for the natural sciences:

1. Specify the relevant variables and operationalize the hypothesis
2. Specify the level(s) of treatment
3. Control the environment
4. Choose the design
5. Select and assign objects to be studied (sampling)
6. Pilot-test, revise, and test

In the discussion of these six stages, reference will frequently be made to the very simple experiment dealing with the baking of bread presented in Appendix A to this chapter.

Regarding 1: Specify the relevant variables and operationalize the hypothesis
This is primarily a matter of selecting variables that are the best operational representations of the original concepts in the hypothesis or research question. For

example, if we consider the discussion in Appendix A on how to perform an experiment to test the hypothesis: "all other things being considered equal, the greater the humidity the fluffier the bread", there are clearly two concepts that have to be operationalized: "fluffiness" and "humidity". The first of these is a term that, while it is easily understood by all of us, most likely requires operationalization, i.e. a definition that permits measurement. When I write "most likely", this is because one could also consider avoiding such operationalization by employing a panel of people—say bakers, cooks, gourmets or ordinary people who consume bread—and have them perform subjective evaluations of the fluffiness of the breads that are baked under various conditions of humidity. This could be done by having them assign a number of points, say from one to ten, or by simple classifications such as "low fluffiness", "medium fluffiness" and "high fluffiness". This is something that many TV-viewers are familiar with in connection with the evaluation by panels of dance or gymnastic competitions, where experts who are members of a panel give a number of points to characterize how well each competitor live up to a number of criteria (e.g. gracefulness, creativity, execution, difficulty of the performance, and so on). Appendix A does not allow for such a method of operationalization, as it states: "The experiment must be based on *objective* quantities—for example 'fluffiness is measured as the total volume of the loaf of bread from one pound of flour'". Although such a demand as to objective quantities is very much in line with traditional scientific thinking, it is suggested here that it is not always necessary to operationalize by defining objective measurement criteria, and that in many cases where the experiment involves variables that readily can be perceived by our senses, objective criteria can be supplemented or even replaced by subjective evaluations of the kind presented above.

The other major concept, humidity, is far more straightforward; hygrometers can provide accurate and reliable measures of this concept. But note too that if (which is not the case here), it is considered important that the humidity in the laboratory where the experiment is being performed is stable and the same everywhere in the laboratory, then the design and measurement process could be more complex.

As we have touched on earlier, specifying the relevant variables in an experiment also includes the matter of determining which variables to include in the experiment and which to ignore. In the simple example of baking bread, only two are specified, one independent variable (humidity) and one dependent variable (fluffiness). But it may not be reasonable to limit consideration of the independent variables to humidity alone; for example, theory may indicate that there is an interaction between room temperature and room humidity in their effects on the dough and therefore it would have to be considered early on in the design of the experiment whether also temperature should be an independent variable, perhaps together with humidity in a non-linear manner, or e.g. considered to be an extraneous variable to be controlled. In general, and particularly in those situations where the number of variables that might influence the dependent variable(s) are rather large, trade-offs must be made when determining which variables to ignore in the design, which variables to include as extraneous variables to be controlled in the design, and which variables should be considered to be independent variables. Such trade-offs will represent a balance between: a) the desire for including many possible causal factors in the experiment, b) the complexity that results from including many variables, and c) the resources required, including time, technology, assistance by others, costs, etc.

Re. 2: Specify the level(s) of treatment

Specification of the levels of treatment will often be more or less determined by the existing body of theory that underlies the experiment, and in particular by the hypothesis or research question that serves as the basis for the experiment.

In the experiment regarding bread, the hypothesis was that bread prepared on days with high humidity will be fluffier than bread prepared on days with low humidity. If the hypothesis is to be tested, then clearly experiments should be performed to determine whether in fact the bread prepared on the high humidity days is fluffier. But this requires clear distinctions between what is meant by "high" and "low" humidity. How to decide upon this probably requires consideration of how the results of the experiment might be used—for example, to help commercial, large-scale bakeries to regulate the humidity of the rooms where they have their ovens and do their baking, or to help their leadership decide whether to purchase ovens where the humidity can be controlled; the hypothesis clearly appears to be oriented towards applied, rather than pure science. And one could very well consider not just waiting to perform the experiment until the humidity changes, but instead to bake the bread to be evaluated in a laboratory where the level of the humidity (together perhaps with a number of extraneous variables) can be closely regulated. This leads up to the next stage.

Regarding 3: Control the environment

We have paid considerable attention to the extraneous variables and argued that such variables must be controlled or eliminated via randomization. Here we focus on what might be considered a special aspect of the control of extraneous variables: environmental control. This aspect of control in experimentation deals with keeping the physical environment constant. In other words, the relevant characteristics of the laboratory/area where the experiment is carried out must remain constant while the experiment is carried out, the instruments used for applying the treatment and for measuring the results must be unchanged throughout the course of the experiment, and so on. Of course by "constant" and "unchanged" I refer only to those aspects of the environment that conceivably can affect the performance and outcome of the experiment. For example, while the room temperature, the time of day, or the amount of daylight may not be expected to affect experiments dealing with radioactive decay, they might have to be controlled when performing experiments dealing with bacterial growth or the reproduction of insects—and as we just mentioned, in the bread baking experiment, it may be necessary to control the room temperature to avoid the possible influence of this factor on the "fluffiness" of the bread.

Regarding 4: Choose the design

Experimental design is a central aspect of empirical research. The design serves as a statistical plan that designates the relationships between the experimental treatments and the experimenter's observations. In determining the "optimal" design, the researcher is faced with making choices that lead to the best balance between: a) living up to constraints as to time, budgets, access to equipment and guidance, etc., b) improving the probability that she will be able to determine whether observed changes in the dependent variable(s) are due to the manipulation of the independent variable(s) and not to other factors, and c) strengthening the generalizability of the results beyond the experimental setting—including here consideration of how the results can be compared with similar experimental results obtained by others.

It should be noted that demands as to the design of experiments differ in the various fields of science. For example, in the social and life sciences, emphasis is often placed on such matters as randomization and the use of control groups, while in the "harder" sciences such as physics and chemistry, less emphasis is placed on such matters; experimentation in the bio-sciences can include elements of both the "softer" and "harder" sciences. In the harder sciences, rather than randomization being in focus when considering validity, it is not uncommon to pay more attention to the repetition of experiments, for example, as we shortly will see in Sect. 6.5 on epistemological strategies in experimentation, by using different instruments/ technology to perform the same measurements. Similarly, rather than focusing on control groups, emphasis in the "harder" sciences often is placed on the control of measuring instruments via their calibration. This can be achieved either by comparing the performance of the instruments when using standard inputs or by comparing the results that have been measured during an experiment to those of identical, reliable experiments that have been performed earlier under the same conditions.

Regarding 5: Select and assign the objects to be studied (sampling)
Although this aspect of design is far more in focus in the social sciences, it can also be important in those experiments in the natural sciences where the objects to be studied must be considered to be representative of the population to which the researcher wishes to generalize. Consider first the use of randomization in experimental design so as to obtain groups that are representative of the populations they represent. Randomization is a simple, impersonal technique for avoiding biases that can result if one uses so-called judgmental or systematic assignment methods. But randomization also plays an important role with respect to extraneous variables. It has been mentioned earlier that in some situations it is not practically possible to control the many extraneous variables that can influence the outcome of an experiment. In such cases, assigning units to an experimental group via randomization can reduce the potential biasing effects that may result when extraneous variables are not controlled. This may e.g. be the case where an experiment is to be performed by comparing the relevant characteristics of an experimental group (of cells, organs, people, animals, plants, etc.) after its members received a treatment with another group (the control group), whose members did not receive the treatment or who received another treatment.

For example, consider the case where there is an interest in measuring the effect on fruit production of spraying a particular kind of fruit tree with a new pesticide.

The evaluation will be done by comparing the average production from a sample of such trees to the average production from a control sample where the control units (trees) have not been sprayed. In such cases the potential effects of extraneous variables (for example, the size and health of the trees) can be mitigated by use of randomization in the design. The assignment of trees to the two groups via randomization may make the groups as comparable as possible, given that the extraneous variables are not directly controlled. This does not provide a guarantee that the groups will be identical (which is of course impossible), but provides the researcher with the assurance that the differences due to the effects of extraneous variables are randomly distributed and therefore are far less likely to be characterized by systematic differences than otherwise.

Another way of obtaining samples that are representative of the population they are chosen from is to employ what is called "matching". This is done via the use of a so-called non-probability quota sampling approach where the goal is to have the members of the experimental and the control groups matched on every characteristic that is considered to be of potential importance in the experiment. To minimize the complications that can arise when the number of independent variables and groups increase, only those variables that are correlated with the treatment and the dependent variable are included. For example, consider the case of a sample of 120 trees chosen from two types of trees, where all the trees are classified as being either large or small and healthy or unhealthy. Three groups are to be formed consisting of (a) trees that are to be treated with a small dosage of a new pesticide, (b) trees that are to be treated with a large dosage of the new pesticide, and (c) trees in a control group that are not to be treated at all and that provide a base level for comparison. Suppose too that the data are as follows:

Frequencies before matching

	Type 1 trees Large trees	Type 1 trees Small trees	Type 2 trees Large trees	Type 2 trees Small trees
Healthy trees	24	9	18	9
Unhealthy trees	18	12	9	21

Suppose now that in order to perform the experiment, a design is made whereby an equal number of trees is to be assigned to each of the three groups. Therefore, to perform the matching, one third of the trees in each category would be assigned to each of the three groups. The composition of the groups would be as follows:

Group composition after matching

Exp. group 1 (large dosage of pesticide)	Exp. group 2 (small dosage of pesticide)	Control group (no treatment)
8 Large healthy type 1	8 Large healthy type 1	8 Large healthy type 1
6 Large unhealthy type 1	6 Large unhealthy type 1	6 Large unhealthy type 1

(continued)

(continued)

Exp. group 1 (large dosage of pesticide)	Exp. group 2 (small dosage of pesticide)	Control group (no treatment)
3 Small healthy type 1	3 Small healthy type 1	3 Small healthy type 1
4 Small unhealthy type 1	4 Small unhealthy type 1	4 Small unhealthy type 1
6 Large healthy type 2	6 Large healthy type 2	6 Large healthy type 2
3 Large unhealthy type 2	3 Large unhealthy type 2	3 Large unhealthy type 2
3 Small healthy type 2	3 Small healthy type 2	3 Small healthy type 2
7 Small unhealthy type 2	7 Small unhealthy type 2	7 Small unhealthy type 2
Total: 40 trees	Total: 40 trees	Total: 40 trees

Note that matching—as well as randomization—is no guarantee that the assignment procedure leads to sample groups that completely equalize, so to speak, the potential effects on the outcomes due to extraneous variables (including variables not considered in the experiment due to the demands on resources and the ensuing complexity—such as for example, the number of branches on the trees, their position with respect to the sun, soil type and condition, and so on). In general, the assignment problem of removing the effect of the influence of extraneous variables can be mitigated by a combination of matching, randomization, and increasing the sample size. This last factor, ceteris paribus, increases the probability that the objects sampled are representative, but increases the costs as well.

Regarding 6: Pilot-test

Pilot-testing here is similar to its use in other forms of primary data collection and is intended to reveal potential errors due to the design of the experiment and inadequate control of extraneous variables and/or environmental conditions. In this way pilot-testing permits refinement of the experiment before the actual implementation. It provides the researcher with the opportunity to rethink things—to observe problems she might not have thought about with regard to possible laboratory conditions, the instruments to be used, the time and other resources required, external factors that can exert an influence on the results, and so on and so on. These virtues of pilot-testing cannot be emphasized enough—because many researchers, particularly Ph.D. students who find themselves in a time-bind, tend to ignore such testing, with the risk of a concomitant reduction in the quality of their results and of their arguments regarding justification.

In this connection I can refer to the many situations, both in the natural and social sciences, where questionnaires are employed to collect primary data. It is my experience that the vast majority of scientists have very little appreciation of the difficulties associated with designing a questionnaire that provides accurate, reliable information. In addition, they tend to ignore the benefits of pilot-testing such questionnaires, including having people who are knowledgeable as to the field and researchers who have experience in the design of questionnaires evaluate the questionnaire being considered. Such pre-testing can lead to significant improvements in the questionnaire design and thereby to improvements in the validity of the results obtained. It is my experience that a large majority of questionnaires developed by researchers are poorly designed. By this I do not mean that respondents do not provide answers, but that the answers they provide (to poorly designed questionnaires with poorly formulated questions) are not reliable. This is due to the lack of clarity of the questions so that different respondents interpret them in different ways. A result is that in spite of all the analyses performed as to statistical significance etc., the researcher is in fact unaware of the lack of validity of his results.

Let us now consider the concepts of validity and reliability that have just been referred to.

6.4 Validity and Reliability in Experimentation

We can start by reflecting on the question: how can one perform experiments so that the researcher as well as his or her peers can rationally believe in the results. We have previously considered such issues in connection with our discussions of both theory justification and measurement. Here we will commence by briefly considering the criteria of validity and reliability as they are specifically applied to experimentation. This will be followed up by consideration of a series of practical strategies that can be used to support rational belief in the results of experiments.

6.4.1 Validity

We have referred to the concept "validity" a number of times. It is a word that is frequently used in connection with scientific statements. In general, it is a broad criterion dealing with the trustworthiness of one's results. In Chap. 5, where we considered measurement, we specifically focused on validity via the so-called "internal validity" of measurements—whether a measurement process actually measures what it is intended to measure. In the following, we consider both the internal and external aspects of validity as the term refers to experiments. But let me underscore that what is important here, as elsewhere in this book, is not definitions. Rather, what is in focus here is understanding what these terms really mean regarding the quality of your research and the ability to communicate regarding that quality so others may have rational belief in, may trust, your results.

Internal validity
While a *measurement's* internal validity is determined by whether it measures what it is intended to measure, an *experiment's* internal validity is determined by whether the conclusions that are drawn as to a relationship between the independent and the dependent variables really imply a causal relationship. A statement that an

experiment is internally valid refers essentially to two types of validity: *statistical conclusion validity* and *causal validity*.

Statistical conclusion validity in experimentation deals with whether statements regarding the relationship between variations in the independent variable(s) that have been manipulated, and the dependent variable(s) are justified as seen from a statistical perspective. In checking for statistical validity, statistical measures (e.g. a t-value or a correlation coefficient) are used to quantify relationships between the variables. In other words, in order to demonstrate the statistical validity of the results of an experiment, the researcher must justify his statistical inferences. But most researchers today are ill-equipped to do so! Perhaps the major reason is that there is a strong tendency today for researchers to simply apply standard computer software when making statistical tests without adequate knowledge of the assumptions underlying the technique employed and as to how and why the technique works. In this way they have very little understanding of the statistical justification of their results and will find it difficult and embarrassing if they have to defend such results—say in connection with a Ph.D. defence or in response to a question by a referee from a journal. Therefore it is strongly recommended that Ph.D. students become familiar with both the assumptions and the theory under-lying the statistical methods they employ, as well as with how to interpret the output/results of applying the software to one's data; we will return to this issue later on in Chap. 8 and in particular in Sect. 8.3.

The second type of internal validity, "causal validity", deals with the plausibility of relationships that have been shown to be statistically related—that is, whether the results are supported by and are coherent with existing knowledge. Assuming that a statistical analysis indicates that an independent variable appears to be correlated with variation in the dependent variable, causal validity deals with the plausibility and nature of a causal relation between these variables. Such information is provided by existing theory, similar empirical results—and common sense.

I note that if a relationship appears to be causal, the direction of the causality is not always obvious. In fact in certain cases where there are apparent synergetic relationships between variables, the causality is systemic and "points in both directions" simultaneously. For example, many fungi (so-called mykorrhiza fungi) colonize the roots of plants/trees and form a symbiotic relationship that benefits the fungi and the plants/trees. In general, however, the direction of the causality is usually inferred in advance, where existing theory provides a basis for such an inference.

Furthermore, if there are one or more other variables that also contribute to explaining the apparent statistical relationship between an independent and a dependent variable, then what appears to be a causal relationship may be mis-leading. The existence of so-called *spurious* relations, where variables appear to be related, but in fact only are correlated, provides a threat to internal validity. The following exemplifies a situation where an apparent relationship between two variables is spurious (Bryman 2001). A large scale study in the UK based on a random sample of 9000 individuals found a relationship between diet and smoking. The data consisted of structured interviews, self-administered questionnaires and physiological data collected by nurses. But the question that arose is whether there is in fact a causal, and not just a statistical, relationship between smoking and one's eating habits. An analysis strongly indicated (what our common sense can support) that the association between these two is in fact spurious, and that the apparent relationship is due to a third variable, a so-called *intervening variable*: the commitment (or indifference) of individuals to a "healthy lifestyle". Commitment to a healthy lifestyle, where this concept was operationalized in the study, was seen to affect both one's eating habits and one's propensity to smoke; people who have a healthy life style tend to have both healthier eating habits and to be non-smokers/infrequent smokers. If the study had not considered this possibility, the researchers could have been tempted to infer that there is a direct causal relationship between diet and the propensity to smoke. In this case, and given a goal by the government of reducing the frequency with which people smoke, it could have been logical (but ridiculous) to design a campaign to change smokers' eating habits.

External validity
This concerns whether the apparent causal conclusions resulting from an experiment can be generalized beyond the experimental context. That is, whether an observed relationship between the independent and the dependent variables can be said to hold independent of the context, time or other factors characterizing how and where the experiment took place.

For example, in situations where samples are made (e.g. where an artificial growth hormone is to be evaluated by testing it on a sample of children with growth hormone deficiency or where bore wells are sunk in searches for oil), it is obvious that one must somehow justify the generalizability of one's observations to a wider population (e.g. youth characterized by growth failure and short stature, or the total amount of oil in the whole region where the limited number of bore wells are drilled). Even more fundamentally, external generalizability can be said to deal with the whole question of "moving" from the level of hypothesis to that of theory to that of scientific law, since this is essentially a matter of generalizing one's results so that they become more and more universally applicable.

One aspect of external validity that is often met in the natural sciences deals with how to generalize in situations where some things/phenomena are too large or too small or too far away or too complicated or too costly or too dangerous to study in their real contexts. In such situations there is a need for working with models or surrogates that allow experimenting and learning under artificial conditions (e.g. working with wind tunnels when designing airplane wings; with models of DNA molecules; with animals as surrogates instead of humans in biological/pharmaceutical research; with mathematical models of economic phenomena; and so on).

As was the case with internal validity, a subject of major importance for the establishment of external validity is the statistical analyses that are performed on the results of one's experiments. Since this subject matter is far too comprehensive to treat in detail here I will only provide a brief sketch of the underlying logic that characterizes traditional approaches to establishing the external validity of an experiment and refer the reader to the widely available literature on statistical testing. The presentation below is based on what is generally referred to as a "frequentist" perspective on statistical testing, which is the most widely approach used in the sciences. I note however, that in Chap. 8 we will, in much greater depth, discuss the notion of statistical testing within a broader framework of uncertainty, probability and statistics, and will contrast this frequentist perspective with that provided by the so-called Bayesian perspective.

Most often the reason for carrying out an experiment is to test a hypothesis. Based on the outcome of the experiment, the researcher can run statistical checks to work out how likely it would be for the empirical evidence which seem to support/reject the hypothesis to have come about simply by chance.

Therefore the experiment is set up based upon a so-called "null hypothesis", which is the opposite of the "alternative hypothesis", the hypothesis that is in fact to be tested. The null hypothesis predicts the results which would be obtained if the alternative hypothesis, the one the researcher is really investigating, is *not* true. For example, if the alternative hypothesis is that variable A affects variable B (for example, that a new drug, A, cures prostate cancer, B), the null hypothesis is that B is independent of, i.e. not affected by, A. This can then be tested by manipulating the independent variable A and measuring the corresponding variations in the dependent variable B. If statistical tests do not confirm that there is a relationship between A and B, in other words if the tests indicate with some strength that the null hypothesis should be accepted, then the alternative hypothesis should be rejected. If on the other hand the null hypothesis that B is independent of A does not appear to be supported by the experiment, then more confidence can be placed in the alternative hypothesis that variable A affects variable B.

Thus there are two types of errors that can occur when performing such a test of hypothesis:

Type I error: This error occurs when one rejects a null hypothesis that is in fact true.
Type II error: This error occurs when one accepts a null hypothesis that is in fact false.

Note that a hypothesis is only considered as being true or false.[7] The statistical perspective provided by a test of hypothesis does not permit conclusions such as "maybe true" or "maybe false"; reference is made to Footnote 4 in Chap. 4 where we introduced the concept of "fuzzy logic". While in our daily communication such "fuzzy" expressions are widely used and a person may indicate various degrees of doubt as to whether e.g. his stress has been reduced as a result of meditating, the

---

[7]There is a 'hidden' problem here: How to determine whether in fact a relationship is really "true"—here, whether an independent variable A physically somehow affects or causes a dependent variable B or not. If we are not able to establish this convincingly from either existing theory and/or observable causal mechanisms, then we must rely on our empirical results alone, which, as we considered earlier, can never really determine what is true or equivalently, prove something. Induction and statistical testing can only contribute to our willingness to accept or reject a hypothesis, not to "true" causal knowledge.

evaluation of the results of an experiment via a test of hypothesis is far more demanding; we will consider this in greater detail in Chap. 8.

When designing an experiment, a significant factor to consider is how to minimize the likelihood of both Type I and Type II errors, where the emphasis/weighting is almost always placed on reducing the likelihood of an error that causes the greatest "damage", typically of accepting the alternative hypothesis when in fact it should be rejected (i.e. of rejecting the null hypothesis when it should be accepted). After all, the researcher wants to demonstrate something "new", that say A affects B (for example that a new drug cures prostate cancer), and wants to feel very certain about this before presenting this as a conclusion of her research. If she makes such a statement, and others replicate her experiment and do not arrive at the same conclusion, the merit of her work and her reputation as a scientist will be challenged.

Such weighting is in practice determined by defining an acceptable level of risk that the conclusion might be wrong, and is expressed by assigning a so-called level of significance or confidence level to the test results. There are a number of ways of interpreting such a level and the following are several expressions of this concept. In the following, the hypothesis being tested is that an independent variable A affects a dependent variable B—therefore, the null hypothesis is that A does *not* affect B:

(1) If a researcher reports that the results of the test have a 95 % level of confidence, this is usually interpreted to mean that she is willing to accept that there is at most a 5 % chance that the alternative hypothesis (that A affects B) is *not* correct. In other words, there is at most a 5 % chance that what appears to be data resulting from a causal relationship might instead have been due to the play of chance in the form of random variation.

(2) The analyses indicate that if the experiment is repeated a large number of times, then at most 5 % of the outcomes will *not* support the result obtained that A affects B.

(3) There is at most a 5 % chance of committing a Type I error, i.e. there is at most a 5 % chance of incorrectly rejecting the null hypothesis when it is in fact true.

(4) If the null hypothesis is true (so that we are sampling exclusively from cases for which the null is true, i.e. that A does *not* affect B), then the probability of obtaining observations (that A does affect in fact affect B) as extreme or more extreme than actually observed in the experiment is at most 5 %. I note that this last interpretation most clearly and correctly expresses the meaning of "the results of the test have a 95 % level of confidence". We will consider this in greater detail in Chap. 8.

I strongly emphasize that experiments should be carried out by scientists to *test* hypotheses, not to *confirm* them—no matter how much one may believe in them or want them to be accepted! There is a fundamental difference between these two goals! Nevertheless scientists often forget this and may look upon their goal as getting a hypothesis accepted. For example, a researcher may, given his worldviews and his values, feel that it would be good for the world if his hypothesis receives support. I have observed such a focus on confirming one's hypotheses many times when working with younger researchers, in particular among researchers in the social sciences. In addition, a researcher may feel pressure to get results that support his (alternative) hypothesis since this could lead to publication of his results. "Publish or perish" is an aphorism that is becoming more and more accepted at universities, particularly since research positions today are highly dependent on one's publishing track record. And publication is far easier with results that support one's alternative hypothesis than the opposite, the null hypothesis! It is far easier to get results published that indicate that something new and exciting, say cold fusion, in fact takes place, then results that indicate what almost all scientists agreed to in advance—that cold fusion is not possible.

Finally in this connection it must be underlined that it is simply bad/unacceptable science to alter a hypothesis after performing an analysis, so that it appears that positive results were obtained! We will return to such considerations later on when we consider the role of ethics in research.

6.4.2 Reliability

This criterion refers to the consistency or stability of the results of an experiment. The reliability of an experiment is said to increase if the experiment is repeated and consistent results are obtained—or if the experiment is performed using different types of measurement technology and the same results obtain. In particular, one refers to an experiment as being reliable if *other* researchers replicate the experiment and obtain the same results.

There is always the possibility that unknown errors (say in the measuring instruments) can result in the outcome of an experiment appearing to be unstable, but such errors can be located and corrected before the results are made public. Pilot testing can be of considerable importance here.

Another potential source of inconsistency can be extraneous variables. Experiments in poorly controlled situations require more replication because the uncontrolled extraneous variables can lead to variations in the results.

Yet another potential source of unreliability in experimentation is bias. There are two sources of bias to be considered here: sampling bias and the experimenter's own personal bias.

There are well-known procedures for minimizing sampling bias (we have previously referred to matching and to the randomization of the selection of the objects to be subjected to experimental treatment). And while much is written about minimizing sampling bias, very little is available in the literature dealing with experimentation about reducing one's own biases. This is most likely due to the fact that the natural sciences, with their focus on objectivity, tend to deal with technical/objective matters (such as measurement techniques) rather than with more subjective issues such as personal bias that can be the result of the researcher's beliefs, desires (e.g. that an hypothesis is supported by evidence), ambitions etc.

Nevertheless, the experimenter's own bias is a most serious matter to deal with since we are usually blind with respect to such biases. They can influence virtually all aspects of experimental design (as well as the analysis and the writing up of one's results).

In order to be aware of—and then to eliminate or reduce the bias that can result from a researcher's own subjectivity and values—it is strongly suggested that reflective scientists develop experimental protocols in the form of written plans and detailed guidelines to follow when conducting an experiment, as well as detailed records of what took place and when. Such documentation can also be of considerable value should patent disputes arise or should the researcher be accused of misconduct, e.g. as to plagiarism or misrepresentation of the results. However, by far the major motivation for such record keeping is the reflection that automatically takes place when one's thoughts, plans and deeds are made explicit. In particular, regular note-taking is important. This should cover all stages of the experiment, from conceptualizing during the early stages, to planning, designing, and performing the experiment, to the statistical analyses, and to communicating the results. Ideally another researcher should be able to read one's protocol and notebook and to then repeat the experiment based only upon these documents.

A typical example of a situation where good notes can be important is when a researcher faces a decision regarding whether or not to reject data that appears to be anomalous. As considered in Chap. 5, he may find that some of the data seem extremely unlikely and may decide to ignore them as being unrepresentative of the actual relationship he is investigating. For example, this could be the case if he has good reasons to believe that his equipment may be malfunctioning or that a laboratory assistant has made an error when reading a metre. But the decision to reject the data may also be due to his biases; he simply does not like an apparently anomalous result, which goes against his (often sub-conscious) preferences and ambitions. Therefore such decisions as to the removal of anomalous data should be discussed in the notebook, so that he is ever aware of what he does, when, and why —and can defend his decisions.

The keeping of notes is a skill that is ordinarily ignored in most educational programmes. This is a pity. Continual reflection as to one's experiments and the documentation of one's thoughts and practice, although requiring time and discipline, can be of great benefit to the researcher, both when he attempts to remove any personal biases he may have, as well as when he is to document his results—for example in a dissertation or other scientific publication. Note keeping should be learned early on, supported by one's advisor, and worked on diligently throughout one's scientific career! According to Davies (2010; 27) "In a scientific context, laboratory notebooks provide the best method of avoiding self-deception later in life."

6.4.3   *Epistemological Strategies*

We have considered validity and reliability as criteria for establishing rational belief in the results of experiments. I conclude these reflections by summarizing some of the extensive *pragmatic* advice offered by Franklin (1989) as to strategies to support rational belief in the dependability, the genuine results of one's experiments. Franklin refers to this as "the problem of the epistemology of experiment" (Ibid.; 437) and provides a number of strategies designed to establish or help establish the validity of an experimental result or observation.

1. **Use different experimental apparatus**

   If something can be observed/measured using different instruments (e.g. different types of microscopes), then there is more strength behind a statement as to the correctness of the observations. For example, it would be an amazing coincidence if when studying dense bodies in cells the same patterns are observed when using totally different kinds of microscopes, such as ordinary, polarising, phase-contrast, fluorescence, interference, electron or acoustic microscopes. Thus, it can be argued that a hypothesis receives more confirmation from "different" experiments, where different measurement instruments, based on different underlying theories, are used, than from the repetition of the "same" experiment, where the same apparatus is used.

   I note that this strategy is relevant as well for other methods of observation/measurement than experiments. An example is provided by the attempts to determine Avogadro's number (N, the number of particles, usually atoms or molecules, found in one mole of a substance; see the definition of a mole in Chap. 5, Appendix A). According to Pais (1982; 95): "... the issue was settled once and for all because of the extraordinary agreement in the values of N obtained by many different methods. ... From subjects as diverse as radioactivity, Brownian motion, and the blue in the sky, it was possible to state by 1909, that a dozen independent ways of measuring N yielded results all of which lay between 6 and 9×10^{23}." I note as well that these results provided powerful evidence as to the existence of molecules.

2. **Indirect validation**

   This is a variant of the use of different experimental apparatus. Suppose that we have an observation that can only be made using one kind of apparatus, call it A. Suppose too that the apparatus A can also produce other, but similar, observations which can be corroborated by various other techniques. Agreement between the results provided by these other techniques and by apparatus A provides confidence in the ability of apparatus A to produce valid observations in general. This then provides confidence in an observation that has been made with apparatus A, even though the observation cannot be checked out by using other instruments.

3. **Properties of phenomena as validation**

   The gist of the argument here is that even without support of existing theory the consistency of observed data provides support that the observations are valid, in opposition to an interpretation of the results as artifacts of the apparatus. For example, in the 1913 report of his famous oil-drop experiment, Robert Millikan supported his observation of the quantization of electric charge by noting that in every single case the charge on tiny charged droplets of oil suspended between two metal electrodes was an exact multitude of the smallest charge that was measured (this was proposed to be the charge of an electron). In other words, the constancy of the data argued for their validity; no remotely plausible malfunction of the equipment could have produced such a consistent result.

4. **Theory of the phenomena**

   Here the focus is on accepted theory of the phenomena being observed such that theoretical explanations of observations can strengthen our belief in the observations. This argument may seem strange, since usually we refer to observations as establishing a basis for developing and believing in theory, and not vice versa. The point is that there is often a complex interaction between observation and theoretical explanation. Referring to the discovery that the microwave radiation emitted from Jupiter was due to synchrotron radiation, Franklin (Ibid.; 446) states: "This is not simply a case where a theoretical explanation provided additional confidence in a set of measurements. The observations helped to decide between the competing explanations, and the chosen explanation helped to validate what had been a confusing, and perhaps somewhat doubtful, set of observations."

5. **Elimination of alternative explanations**

   This strategy entails the elimination of all plausible sources of error and all alternative explanations. An example is provided by the observation of electric discharges in the rings of Saturn. These were recorded by the spacecraft Voyagers 1 and 2, both launched by NASA in 1977, when they flew by Saturn three years later. A number of alternative explanations as to the discharges were hypothesized. These included: poor data quality in the telemetry link between the spacecraft and earth; discharges were generated near the spacecraft through certain environmental phenomena; dust-like particles could possibly have interacted with the spacecraft and caused the discharges; etc. etc. When all of the plausible sources of error were considered and all the other possible explanation were eliminated, the researchers concluded that the most probable source of the discharges was in fact the rings of Saturn and that the observations were valid.

6. **Calibration and experimental checks**

   This is perhaps the most widely used strategy for validating the results of experiments and observations. The argument is that the ability of an apparatus to reproduce already known results provides support for its proper operation and for the results obtained when using it in general. In addition, calibration not only provides a check on the operation of the apparatus, it also provides a numerical scale for the measurement of the quantity involved. Therefore, in any well designed experiment there will be checks to ensure uniform, if not proper, operation of the apparatus.

7. **Prediction of lack of phenomenon**
   The argument here is that if an effect disappears when you predict it will, then there is reason to consider the functioning of the experimental apparatus as well as the observation to be valid. This is a special case of the strategy considered earlier (number 2: indirect validation) where the observation of predicted behaviour helps to validate a result. Logic indicates that the opposite also holds true here: If an effect does not disappear when it is predicted to, then one can doubt the observations as well as whether the apparatus has been operated properly. So if the effect does in fact disappear, this provides support for the correctness of both the observation and the apparatus.

8. **Statistical validation**
   We have already touched on this in Sect. 5.4.1 on "External validity". It is logical that statistical arguments can contribute to the validation of a result or in establishing that a particular effect was observed. For example, while searching for new particles using bubble chamber techniques in the 1960s, researchers performing experiments in high energy physics used as an informal criterion when looking for a new particle that it gave a three standard deviation effect above the background (the usual technique was to plot the number of events as a function of invariant mass and look for bumps above a smooth background). This corresponded to a probability of 0.27 %, which was extremely unlikely in any single experiment. However, it was shown that if a very large number of experiments were performed each year, then one could expect to observe a large number of such 3-sigma results. This led the informal criterion being changed to 4-sigma, which has a probability of 0.0064 % and later to 5-sigma (corresponding to a probability of roughly 3×10^{-7} or about 1 in 3.5 million). In fact examination of the literature from the 1960s reveals that using the earlier 3-sigma criterion several particles were reported that were not confirmed by later observations and are no longer mentioned in the scientific literature.

Summing up, the strategies listed above, based on Franklin (1989), can be said to provide additional general grounds for rational belief in experimental results— grounds that are more general and inclusive than most common accepted practices. There is no suggestion that these strategies are exclusive or exhaustive, or that any subset of them provide necessary and sufficient conditions for having rational belief in one's experimental results. Nevertheless, they appear to constitute a practical and rational checklist that extends beyond most of the guidance Ph.D. students receive as to the "epistemology of experiment".

6.5   Design of Experiments

Earlier, we broadly characterized an experiment as an active (manipulative and/or comparative) intervention so as to enable testing hypotheses and the development of causal explanations. In order to assure that the manipulative and comparative aspects of experimentation are valid requires an effective design, effective implementation, and effective analysis. Here design is in focus.

A good design improves our ability to test hypotheses and to answer specific research questions by permitting tests, measurements and comparisons to be performed as objectively as possible. In particular, when the experimentation involves the systematic alteration of independent variables so as to enable measuring the resultant effects on dependent variables, a good design improves the probability that the changes that are observed in the dependent variables were a result of the manipulations in the independent variables, and not due to other (extraneous) factors. This is the internal validity aspect of the design. In addition, a good design strengthens the experiment's external validity by enabling its results to be generalized beyond the specific experimental setting.

Over and above the relevance of design for determining the validity of experimental results, a good design also provides the researcher with a tool for planning the investigation and for keeping track of the use of time and other resources. As we will see later on, such planning and control of research resources often presents a great challenge to researchers who work under tight time and financial budgets, in particular to Ph.D. students.

Based on the seminal work on experimental design by the mathematician R.A. Fisher, (Darius and Portier 1999; 74) present the following main features of a design that provides as fair comparisons as possible. I note that it is implicitly assumed here that the experiment includes the comparison of a test group with that of a control group, which may not be the case, e.g. in high energy physics, as we have several times noted earlier.

- "The experimental design should allow for unbiased estimation and comparison of the treatment effects. In this way the key study questions can be answered.
- The experiment must facilitate meaningful estimation of the underlying variability (experimental error) in experimental units, assuming treatment factor and control factor effects have been accounted for.
- An adequate number of replicates (application of the same treatment to different experimental units) should be used in order that effects can be estimated with sufficient precision to detect the smallest differences of practical importance.
- Experimental units should be grouped or blocked[8] in order to control for or balance out [9] known sources of extraneous variation. ...

---

[8]Blocking refers to the allocation of experimental units to groups (blocks) such that the experimental units within a block are relatively homogenous while the major part of the predictable variation among units is now associated with the variation among blocks. In many cases extraneous variables are used as a guide to how the experimental units are grouped.

[9]Balancing refers to the assignment of experimental units in such a way that comparison of factor levels is performed with equal precision—e.g. by allocating an equal number of units to each block, and by assigning treatment such that an equal number of units receive each treatment. We have seen such balanced assignments earlier in the illustrative example based on testing a pesticide on two types of trees.

- Treatments should be randomized to experimental units so that the treatment effects are all equally affected by the uncontrolled sources of extraneous variation.
- The simplest design possible to accomplish the objectives of the study should be used.
- The experiment should make efficient and effective use of available resources."

Summing up, we might say that a good design enables the researcher to balance the requirements and limitations (e.g. as to time, financing, available expertise and facilities/technology) of the research so that the experiment can provide the best conclusion about the hypothesis being tested.

In the "hard" sciences, experiments are often designed to precisely measure certain specific quantities. In such cases, comparison, which will be a major consideration in the sequel, is not in focus. Instead, the values for the specified quantities are theoretically predicted from the hypothesis (or hypotheses) underlying the experiment and these predicted values have to be confronted with experimental measurements. Any major disagreements will necessitate a revision of the hypothesis, or possibly its rejection altogether, both of which are in line with the general approach of falsification. Of course, prior to such a revision/rejection, detailed checks will be performed to make certain that the measuring instruments functioned correctly and that errors have not been made in the design, execution of the experiment, measurements etc.

I note too that in the "hard" sciences it is, generally speaking, relatively easy to meet the ideal as to the objectivity of measurements and the control of both extraneous variables and the experimental environment. However, in some domains, such as some fields of biology as well as in medical research, it is often more difficult to meet these requirements. And when we consider the behavioural and social sciences where human beings and groups of humans (such as organisations, institutions, societies) are the direct or indirect objects of investigation (as in sociology, anthropology, economics, and management) it becomes in general very difficult to design and perform experiments that meet a goal of the objective observation and measurement of relationships between independent and dependent variables and where the experiment can be replicated under constant/controlled environmental conditions. In fact experimentation is relatively rarely used in those branches of science where it may be counter-productive to attempt to control the variables (often some aspect of human behaviour) and the experimental environment (social settings are not controllable as are laboratory conditions). It is primarily for this reason that I have referred to such domains of science as physics and chemistry as "hard" sciences, while I have referred to e.g. some areas of the bio-sciences, the behavioural and social sciences, as "soft" sciences. Objective and controlled measurements are, relatively speaking, far more realistic goals in the former than in the latter.

The statistical aspects of the design of experiments, i.e. how to design a given experiment so as to provide the most reliable statistical results as well as how to perform the relevant statistical tests, dominate the literature dealing with the design of experiments. The focus is often on applications in industrial processes. In contrast, the presentation here will provide a rather brief treatment that indicates the nature of the statistical considerations involved when making a design. The reader is referred to many available books for a more thorough treatment. One might e.g. consider the following publications, listed in alphabetical order, which I identified in 2014 by searching for literature on the design of experiments on the Internet and therefore should *not* be considered as recommended reading:

Antony, J. (2003) *Design of Experiments for Engineers and Scientists*, Oxford, UK: Elsevier.
Bailey, R.A. (2008) *Design of Comparative Experiments*, Cambridge, UK: Cambridge University Press.
Barrentine, L.B. (1999) *An Introduction to Design of Experiments: A Simplified Approach*, Milwaukee, WI, USA: ASQ Quality Press.
Hoshmand, A.R. (2006) *Design of Experiments for Agriculture and the Natural Sciences*, 2nd edn., Boca Raton, FL, USA: Chapman & Hall/CRC Press.
Mason, R.L., R.F. Gunst and J.L. Hess (2003) *Statistical Design and Analysis of Experiments With Applications to Engineering and Science*, 2nd edn., Hoboken, NJ, USA: Wiley-Interscience.
Montgomery, D.C. (2000) *Design and Analysis of Experiments*, 5th edn. Hoboken, NJ, USA: Wiley-Interscience.
Roy, R.K. (2001) *Design of Experiments Using the Taguchi Approach: 16 Steps to Product and Process Improvement*, Hoboken, NJ, USA: Wiley-Interscience.
Ruxton, G.D. and N. Colegrave (2010) *Experimental Design for the Life Sciences*, Oxford, UK: Oxford University Press
Stamatis, D.H. (2002) *Six Sigma and Beyond: Design of Experiments*, Boca Raton, Fl, USA: St. Lucie Press.

So rather than focusing on the statistical aspects of the design of experiments, the presentation below focuses on the overall concept of design and provides short sketches of prototype designs as well as of the nature of the statistical considerations involved when making a design. The presentation is based on Cooper and Schindler (2003; Chap. 14) and follows the structure they provide. I note that although that book is not about the natural sciences, the exposition is well-suited to our purpose, where the intention is simply to stimulate thinking about how to evaluate and choose a design.

I note too that there is no "one size fits all" when designing an experiment. In particular, as introduced earlier, in the softer sciences, but also in the medical sciences with their extensive use of clinical testing, as well as in some fields of the bio-sciences, the results of "treatments" of the members of an experimental group are compared to the results in a control group whose members do not receive the "treatment". In other words, emphasis is placed on comparing the results of manipulating the independent variables on the dependent variable(s) in the experimental group with the results in the control group, where the independent variables are not manipulated.

As we have noted, this focus on control groups is less frequently encountered in the harder sciences. However, this does not mean that those designs considered below which deal with control groups are irrelevant for reflective practitioners of the harder sciences. In particular, as will be discussed, assuming that the objects to be studied are randomly assigned to a test group and a control group is often an oversimplification; the experiment can simply assign different levels of treatment to any number of groups, in which case the distinction between an experimental group and a control group no longer is meaningful!

Similarly, the designs considered below often refer to the concept of randomization as a basic way to achieve equivalence among treatment groups. As also briefly considered earlier, in some of the hard sciences, particularly physics, such an approach is less frequently used. Rather, experiments there can involve such large samples that the variance characterizing the results is extremely small; in such cases it is not meaningful to speak of using randomization (or other means) to obtain equivalence between groups. Instead, experiments are simply performed by measuring the results of various levels of the independent variables on the dependent variables. This should be kept in mind when reading about the designs below. In particular, the symbol R, indicating the random assignment of test units to a group, can simply be ignored in experiments characterized by very large samples and very small variances. The designs are still relevant, but one simply ignores the concept of randomization.[10]

With the above reservations in mind, and since experimental designs differ in their power to control extraneous variables, the following are sketches of designs that are based on the concept of control. Following the structure in Cooper and Schindler (2003; 435–444) and using their design symbols, the designs are categorized into two groups: (a) pre-experiments and (b) true experiments.

The following are the design symbols to be employed:

IV, DV	Independent variable, Dependent variable
X	An experimental treatment or manipulation of an IV
O	A measurement/observation is performed
R	Indicates that the group members have been randomly assigned

---

[10]However, even in the case of experiments where the number of units in the 'sample' is very large, cases do arise where it may be necessary to justify the "representativeness" of one's results. This may be the case e.g. in experiments in high-energy/particle physics where a huge amount of data results, but where a large number of individual measurements are to be made by a number of researchers, each of whom performs cuts on the original data to suit the needs of the particular hypothesis being considered. If such cutting reduces the number of observations in a cut to a relatively small size, the researcher may face the problem of having to justify that the sample he is working with is in fact random (i.e., that the variance is still, relatively speaking, small). Here frequently one resorts to comparing the actual data with so-called "Monte Carlo simulations" as well as the results of a host of other tests to make sure there are no biases.

The X's and O's in a diagram are read from left to right, indicating the order of sequence. For example, O1 X1 O2 X2 O3 means that the sequence is: Observation 1, Manipulation 1, Observation 2, Manipulation 2, Observation 3.

X's and O's placed vertical to each other indicate that the treatment and/or observation take place simultaneously.

Rows that are not separated by dashed lines indicate that the objects in groups to be compared have been randomly assigned to the groups. The objects in comparison groups separated by a dashed line have not been randomly assigned to the groups. This may appear unclear here but will be clear once we start considering specific designs.

6.5.1   Pre-experiments

Such designs are weak in scientific power since they fail to adequately control threats to internal validity (e.g. due to spurious relationships or correlations that do not reflect causal relationships). In particular, they fail to provide comparison groups that are truly equivalent. Several versions of this type of design are:

One-shot case study (where an IV is manipulated and the effect on the DV is observed:

$$X \quad O$$

This design is very weak in its ability to meet threats to internal validity; the lack of a pre-test (i.e. a measurement of the DV's before the experimental treatment) and of a control group makes it inadequate for establishing causality.

One-group Pre-test—Post-test:

$$O1 \quad X \quad O2$$

This is stronger than the One-shot design as it includes a measurement of the dependent variables both prior to and after the manipulation of the IV/experimental treatment. An example of such a design is when a characteristic (for example height) of a number of plants of a particular type is measured (O1). Then the plants are provided with some fertilizer according to a specified plan (X), and after a period of time they are measured again (O2). This is better than the one-shot case study, but is still a weak design. Between O1 and O2 a number of events could have occurred that could confound the effects of the fertilizers, e.g. changes in temperature, exposure to daylight, humidity, attacks by insects. These could of course be controlled by performing the experiment in a closely regulated laboratory (thereby controlling what we earlier referred to as environmental variables). Furthermore, without a control group it is impossible to say to what extent observed changes in height were simply a result of the time that elapsed between O1 and O2, and which growth was due to the manipulation (treatment with fertilizer).

Static group comparison:

$$\_\_X\_\_ \qquad \_\_O1\_\_$$
$$O2$$

This design is characterized by two groups, one of which receives the experimental treatment while the other is a control group that does not receive the treatment; the objects have not been randomly assigned to the groups. It represents a considerable improvement due to the existence of the experimental group and the control group (e.g. plants receiving the fertilizer and plants that do not receive the fertilizer). However, there is no way to be reasonably certain that the two groups are equivalent prior to the experimental treatment of one of the groups. That is, the design does not enable us to determine whether the two groups can be considered to be so similar that it would make no difference to the outcome which of the two groups was selected to receive the treatment and which was chosen to be the control group.

6.5.2 True Experiments

These designs overcome the major deficiency of pre-experimental designs by establishing comparison groups that are equivalent to the experimental group. This can be achieved with matching (the use of non-probability quota sampling; see Sect. 6.3 of this chapter). But still better is equivalence via the random assignment of objects to the groups since this permits employing statistical tests of significance as to the observed differences in the experimental and control groups.

Pre-test—Control Group:

$$R\ O1 \quad X \quad O2$$
$$R\ O3 \qquad O4$$

Here the objects to be studied are randomly assigned to either of the two groups. Pre-testing (observation prior to the manipulation of the independent variables) is carried out to provide the basis for the comparisons to be made after the treatment/manipulation of the independent variable(s) is performed in the first group. Note that this may be an oversimplification in that it implies that *no* treatment occurs in the second group. While this may certainly be the case, it is not uncommon for the control group to experience a different level of the independent variable than the experimental group (in fact experiments can simply assign different levels of treatment to any number of groups, in which case the distinction between experimental and control groups no longer applies; we will consider this case shortly when we introduce the concept of "factors").

The effect of the independent variable here can be determined as the difference between (O2-O1) and (O4-O3). The significance of the difference can be determined using appropriate statistical methods.

Post-test-only—Control group:

$$
\begin{array}{lll}
R & X & O1 \\
R &   & O2
\end{array}
$$

Compared to the above design, this design is simpler since the pre-test measurements are omitted. Although pre-testing is a well-established means for carrying out comparisons, it may not be required when it is possible to employ randomization to select the members of the two groups, particularly if the number of objects is reasonably "large" and/or a "large" number of identical tests is to be carried out, each with a new set of randomly assigned groups. In the design shown here, the experimental effect of the treatment/manipulation is determined by measuring the difference between O1 and O2 after the treatment/manipulation.

Completely randomized design:

This is the basic form of true experimental design. To illustrate its use, consider the following investigation.[11] An article in the journal *Hypertension* published by the American Heart Association indicates that the consumption of flavanols-rich dark chocolate decreases the levels of blood pressure, LDL-cholesterol and insulin resistance in healthy subjects (Grassi et al. 2005). The purpose of the present (make-believe) investigation is to determine whether the positive effect on LDL-cholesterol also applies for people with border-line hypertension (i.e. with blood pressure levels higher than normal yet not so high so that they are considered as hypertension, typically defined as blood pressure in the range 130/80–140/90 mm Hg)—and if so, the ideal amount of dark chocolate to be consumed daily. Here we will only focus on the relationship between the amount of chocolate consumed and the physical measurements of the LLD-cholesterol level for people having borderline hypertension. In addition, such matters as the costs of the chocolate, its brand, the potential increases in weight it might lead to, the potential effects on other measures of health, are not of interest here, although they certainly could be in practice.

The design of this make-believe experiment is as follows: 90 groups of 10 adults each are randomly selected in 90 different towns (one group in each town) from a list of the adults in each town who have border-line hypertension. The decision to choose the groups from many different towns enables taking into account that any one town may be characterized by a number of extraneous variables (for example, temperature, air pollution, income level) that could possibly, directly or indirectly, influence the results.

---

[11]This example is modelled after an example with similar structure, but with a very different content than in (Cooper and Schindler 2003; 439–442).

Three treatment groups are formed, each characterized by its treatment level, the amount of the flavanols-rich dark chocolate that its members are to consume daily. 30 of the town-groups are randomly assigned to each treatment group so that each treatment group consists of 300 adults with border-line hypertension, 10 adults from each of 30 towns.

All the 900 individual members are to maintain their normal diets, styles of life, working habits etc., aside from the consumption of the chocolate, for a period of two weeks. Note that since flavanols are also present in other foods than chocolate, an alternative design could have regulated the participants' diets so as to increase the probability that the experiment's results in the form of differing levels of LLD-cholesterol are due to the differing levels of consumption of the dark chocolate. Similar modifications in the design could also be made so as to more directly eliminate the possible influence on the results of extraneous variables (such as temperature, air pollution, income level, etc.). In the present, less complex design this has not been done and the random selection of the relatively large number of groups and of the participants assigned to the groups is assumed to assure equivalence among the treatment groups. A reflective scientist would consider such matters when designing the experiment and when reporting on it.

At the beginning and at the end of the period, the level of LDL-cholesterol is measured for each group. The design is as follows:

$$R \quad O1 \quad X1 \quad O2$$
$$R \quad O3 \quad X2 \quad O4$$
$$R \quad O5 \quad X3 \quad O6$$

01, 03 and 05 are the average level of LLD-cholesterol of members of the three town-groups prior to the experiment; X1, X2 and X3 represent the three levels of chocolate consumed daily (say 20, 40 and 60 g respectively) by the members assigned to the three town-groups; and O2, O4 and O6 are the average level of the member's LLD-cholesterol in the three town-groups after the experiment.

The design permits determining the effect on the dependent variable (the level of LLD-cholesterol) of different levels of the independent variable (consuming respectively 20, 40 and 60 g each day of the dark chocolate). Statistical tests can then be made to determine the effect, if any, of the three treatment levels (daily consumption of dark chocolate) on the dependent variable (level of LLD-cholesterol).

Randomized block design:

This design is used when there is a single major extraneous variable that one wants to consider. Although random assignment is still the basic way to achieve equivalence among treatment groups, it may be important to determine whether different treatments lead to different results among different treatment groups that are characterized by such an extraneous variable. For example, building upon the example above, the members of the groups will be characterized not just by their level of consumption of dark chocolate, but also by their income level (the extraneous variable being considered here).

Let us assume that the researcher motivates this design by the conjecture that people from poorer towns may be less sensitive to the potentially positive effects of eating dark chocolate than are people from more wealthy communities. This is thought to be due to differences in living habits including such matters as diet, alcohol consumption, physical exercise, and the like, where well-documented investigations performed by others indicate a strong correlation between these matters and income level. Therefore, and since the researcher at this early stage does not have the ability to make experiments that take into account all such living habits and how they may affect/interact with the physiological effects of eating dark chocolate, he chooses to consider income as an indicator of (a surrogate for) these living habits.

90 towns are now chosen and, based on available statistics, classified according to their relative income levels (High, Medium, Low). It is assumed here that the towns are chosen such that there are 30 towns in each income level block. Within each of these three income level blocks, the towns are then randomly assigned to the three treatment levels. This results in 10 groups in each of the nine possible combinations of income and consumption levels.

In the following when we speak of *factors* we are referring to treatment levels of the IV's, where each such treatment level defines a sub-group. *Active factors* are those variables an experimenter can actively manipulate, e.g. by manipulating an object so that it receives one level or another of the treatment. Here there is one such active factor, the amount of dark chocolate consumed each day (20, 40 and 60 g respectively). *Blocking factors* are those factors the experimenter cannot manipulate or chooses not to manipulate, but can only identify and classify. Here there is one blocking factor, the income level of a town. We then have the following design:

Active factor—grams of dark chocolate/day	Blocking factor—income level of the town-groups			
		High	Med.	Low
20	R	X1	X1	X1
40	R	X2	X2	X2
60	R	X3	X3	X3

In the above table the O's have been omitted and it is assumed that measurements are made of each member's LLD-cholesterol level both before and after the period where the experiment is performed. This design permits the measurement of both the main effect on the level of LDL-cholesterol and the interaction effects. The main effect is the average direct influence that a particular treatment—here the level of consumption of dark chocolat—has on the dependent variable (level of LDL-cholesterol) independent of other factors, here income level. The interaction effect is the influence of one factor on the effect of another—here whether different income levels have an influence on the physical reactions (levels of LDL-cholesterol) to differentials in the consumption of dark chocolate.

Whether this design improves the precision of experimental measurement depends on how well it minimizes the variance within blocks and maximizes the variance between blocks. If the reactions to the consumption of chocolate are similar in the different blocks, nothing has been gained from the more complex design regarding the main question of interest: the effect of consumption of dark chocolate on the level of LLD-cholesterol.

Latin square:

Let us now consider a design when there are two extraneous factors that are considered so potentially significant that the researcher wants to evaluate their effect on relationship between the dependent variable(s) and the independent variable(s). In our example assume that it is desired to consider blocking not only on the income levels of the towns, but also on the average temperature level (classified here as High, Medium and Low) of the towns. The motivation is that it is hypothesized that also temperature affects the interplay between the formation of LDL-cholesterol and the consumption of dark chocolate. It is convenient to consider these two blocking factors, temperature and income, as forming the rows and columns of a table, as shown below. Each of these two factors is thus divided into three levels giving nine groups, each with a unique combination of the two blocking variables: income level and temperature level. Treatments (the amount of chocolate to be consumed daily) are then randomly assigned to cells so a given treatment appears only once in each column and once in each row. For example, treatments can now be assigned using random numbers to set the order of treatment in the first row, resulting in 2, 3 and 1. Following this, the two remaining cells of the first column are similarly filled. Finally, the remaining treatments can be directly assigned (since they must live up to the restriction that there can be no more than one treatment type in each row and column). Such a Latin Square design must clearly have the same number of rows, columns and treatments.

Temperature	Income level		
	High	Med.	Low
High	X2	X3	X1
Medium	X3	X1	X2
Low	X1	X2	X3

Note that once again, as in the table for the 'randomized block design', the O's have been omitted and it is assumed that measurements are made of the LLD-cholesterol level of the members of each of the nine groups, both before and after the period where the chocolate is consumed. When these measurements are gathered, the researcher will be able to determine both the average treatment effect (the main effect of various levels of the consumption of dark chocolate on the levels of LDL-cholesterol) as well as the effect of temperature and income on the these levels of LDL-cholesterol.

A limitation of this design, compared to the previous design, randomized block design, is that one must assume that there is no interaction between treatments and blocking factors. Therefore we cannot determine interrelationships among income level, temperature level, and the quantity of chocolate consumed. This is due to the fact that there is not an exposure of all combinations of these three factors, which would require 27 cells instead of the nine cells used here. If one is not interested in such interactions, the Latin square design is economical. If interaction is of interest, the experiment can be performed three times in order to consider all 27 combinations of chocolate consumption (three levels), income (three levels), and temperature (three levels).

Factorial design:

This design deals with the simultaneous manipulation of more than one treatment/IV. For example, in the chocolate experiment suppose we are also interested in finding the effect of providing the participants with information on their LDL-cholesterol count each day, the idea being that such information may have a psychosomatic effect on the production of the cholesterol.

The following is a 2 × 3 factorial design that includes both the active factors (the two independent variables): daily chocolate consumption level (three levels Y1, Y2, Y3) and information level (two levels X1, X2). Here the design is completely randomized with groups being randomly formed and randomly assigned to one of six treatment combinations. As in the preceding example, the O's have been omitted and once again, measurements are made of the LLD-cholesterol level both before and after the period where the two IVs (chocolate consumption and LLD-cholesterol information) are manipulated.

Cholesterol information	Consumption level		
	High	Medium	Low
Provided	X1Y1	X1Y2	X1Y3
Not provided .	X2Y1	X2Y2	X2Y3

This permits estimating: (a) the main effects of each of the two IV's, and (b) the interactions between them. In other words, it permits answering questions such as:

What are the effects on LDL-cholesterol of the different consumption levels of dark chocolate?

What are the effects on LDL-cholesterol of informing participants of their cholesterol levels?

What are the effects on LDL-cholesterol of the interrelations between consumption levels of dark chocolate and information levels as to cholesterol?

Covariance analysis:

The direct control of extraneous variables via blocking was considered earlier (randomized block design). It is also possible to apply some degree of indirect statistical control on one or more variables through analysis of covariance. For example, assume that the chocolate experiment was carried out with completely randomized design, i.e. with no consideration of the effects of possible extraneous variables such as income level. With covariance analysis it is still possible to do some statistical blocking on average income level even *after* the experiment has been run. I will not provide further information here but simply refer to Cooper and Schindler (2003; 546–553) and the many excellent texts dealing with analysis of variance (ANOVA).

The brief exposition above of alternative designs has been presented exclusively to stimulate your thinking about the many possible ways that experiments can be designed and to indicate the many possible issues that can influence the choice of a design. One such matter that it was not possible to include is that, in practice, traditions in your field will exert considerable influence on the choice of design. Nevertheless, considerations like those highlighted above can hopefully enable you as a reflective scientist to better appreciate the intricacies and importance of experimental design for your research.

Appendix: An Experiment as to Fluffiness of Bread

The following description of a highly simplified experiment has been downloaded (April 2015) from the website http://roebuckclasses.com/ideas/experiment.htm. It is presented here since I have referred to this fictive but easily understood experiment several times—mainly to indicate some important considerations when operationalizing one's concepts. Some final reflections are provided after the following presentation of the experiment:

"*An experiment in baking*

As a simple example, consider that many bakers have noticed that the amount of 'fluffiness' in a loaf of bread seems to be related to how much humidity there is in the air when the dough is being made. This can be formalized as the hypothesis: 'all other things being considered equal, the greater the humidity, the fluffier the bread'.

While this hypothesis might arise naturally from baking many loaves over time, an experiment to determine whether or not this is really true would be to carefully prepare bread dough, as identically as possible, on two types of days: days when the humidity is high, and days when the humidity is low. If the hypothesis is true, then the bread prepared on the high humidity days should be fluffier.

Several features of this experiment hold in general for all experiments:

- We must try to make all other conditions of the process as similar as possible between the trials. For example, the amounts of flour and water added, the temperature of the butter, and the amount of kneading all *may* have an effect on the fluffiness; so the experiment should explicitly attempt to control the other variables which could have an effect on the outcome. This gives us some confidence in the statement 'all other things being equal...'.

- Although 'fluffiness' may seem to be an easily understood idea, one baker's idea of 'fluffy bread' may be different than another baker's. The experiment must be based on *objective* quantities—for example 'fluffiness is measured as the total volume of the loaf of bread from one pound of flour'. This idea, coupled with the exactness of the description of how the experiment is to be performed, is sometimes called the operational aspect of the experiment; the idea that all actions, quantities, and observations can be agreed upon by reasonable people.
- Noting that once, on a humid day, one baked a fluffy loaf is not enough. The experiment should be *repeatable*; given that one performs the experiment exactly as described, one should expect to see the same results, no matter who performs the experiment or how many times it is performed.

Repeatability of an experiment helps to eliminate various types of experimental errors—one may *think* that one has accurately described all of the relevant techniques and measurements in an experiment, but certain other effects (such as the brand of the flour, trace impurities in the water used in the dough, etc.) may actually be contributing to the observed effects. Someone may claim that they have performed an experiment with a particular result, and thereby supported a particular hypothesis. However, until other scientists have performed the same experiment in the same way and gotten the same results, the experiment is usually not considered as providing a 'proven' result.

- Finally, even though one has baked bread a hundred times, occasionally a loaf will completely fail 'because the kitchen gods are unhappy'. It is important to realize that some hypotheses cannot be tested experimentally—since we cannot make a measurement which will tell us whether or not the 'kitchen gods' are 'happy', we cannot perform an experiment which either proves or disproves the hypothesis 'the best bread happens when the kitchen gods are happy'."

Earlier in this chapter (Sect. 6.3) I provided reflections regarding "Specify the relevant variables and operationalize the hypothesis". I provided critical reflection with respect to the demand that "the experiment be based on objective quantities." Here I will provide an additional critical comment, this time regarding the final paragraph above. While it should be clear that I agree with the statement that one cannot test the hypothesis "the best bread happens when the kitchen gods are happy", I am critical with respect to the conclusion that "we cannot perform an experiment which either proves or disproves the hypothesis". As I have argued earlier, it is not possible empirically to prove *any* hypothesis—and that even if one is able empirically to falsify a hypothesis, this does not, in general, "disprove" the hypothesis or theory that is being tested.

A final comment with respect to the last paragraph in the presentation above: The statement that "even though one has baked bread a hundred times, occasionally a loaf will completely fail" is an example of the concept of 'anomalous errors' referred to in the discussion in the preceding chapter's Sect. 5.3 on 'Measurement Errors'.

Chapter 7
Scientific Method and the Design of Research

This chapter commences with an overview of the term "scientific method". The term is not only used by scientists to refer to a systematic procedure for carrying out a scientific investigation. It is also called upon to justify the privileged status of their findings and profession. After reflections as to "scientific method" in general, the second half of the chapter develops a practical template or structure for designing one's own specific investigation/research project.

7.1 *The* Scientific Method?

In the Preface we considered several fundamental questions of interest for a reflective scientist, including the following two closely related questions:

1. Is there a standard approach to performing research that is widely accepted? In other words, are there any general principles or "rules of the game" that a researcher should follow when performing research?
2. If there are such general principles, are they more or less independent of one's major field of science and one's culture? Or does each branch of science, perhaps even each specialization within a branch (e.g. quantum mechanics in physics, genetics in biology, artificial life in computer science, palaeontology in geology, organic chemistry in chemistry, galactic astronomy in astronomy) have its own research methodology?

Let us focus first on the first of these questions: Is there a standard approach to performing research that is widely accepted? In other words, is there in essence one scientific method, *the* scientific method, that is shared by the different natural sciences? This appears to be the case, judging both by the vocabulary used by scientists, where reference is very often made to "scientific method"—without restricting the term to any specific field of science.

So let us start by looking at several representations of *the* scientific method. The first delineation is a step-by-step description of an apparent logical and/or chronological progression in research activities:

© Springer International Publishing Switzerland 2016
P. Pruzan, *Research Methodology*,
DOI 10.1007/978-3-319-27167-5_7

1. Start with observations and reflections on these observations based on existing theory
2. Formulate a hypothesis
3. From the hypothesis and initial conditions, deduce observable predictions
4. Design the investigation (including the definition and operationalization of concepts and measures)
5. Select research objects and sites
6. Perform experimentation or by other means make observations/collect data that can be used to test the hypothesis
7. Process and analyse the data collected
8. Evaluate whether the hypothesis is justified by the observations and draw conclusions
9. Write up the results including implications, if any, to existing theory

A similar delineation, though regarding research in social sciences, is provided by Cooper and Schindler (2003; 15). I provide this as it indicates how the view as to scientific method varies when one moves from 'harder' to 'softer' science and enables several interesting insights as to the differences between them:

"Good research follows the standards of the scientific method. We list several defining characteristics of the scientific method ....

1. Purpose clearly defined
2. Research process detailed
3. Research design thoroughly planned
4. High ethical standards
5. Limitations frankly revealed
6. Adequate analysis for decision maker's needs
7. Findings presented unambiguously
8. Conclusions justified
9. Researcher's experience reflected"

This characterization of the scientific method introduces a focus on some aspects of research not generally met in publications on methodology in the natural sciences. I refer here to the emphasis on ethics and decision making as well as on the role of the researcher.

Although most writers on research in the natural sciences do not explicitly consider ethical considerations as being inherent to the scientific method, as is argued in Chap. 10, such considerations should indeed be fundamental to the design and implementation of good research in the natural sciences. In addition, although it is extremely rare to see reference to "decision making" in traditional presentations of scientific method, it is felt that since so much research today is financed by external sources (e.g. institutions, governmental bodies, the military, corporations) and can have far-reaching effects on other sentient beings and/or the environment, a focus on the decision making aspects of scientific method may be relevant in the natural sciences. For example, much funding often assumes that particular questions of interest to the sponsor are answered, even though the research is not what is typically referred to as applied research. Finally, I note that while in most characterizations of "the" scientific method in the natural sciences it is implicitly assumed that the researcher is a neutral and objective observer/investigator, the above characterization, particularly steps 4 and 9, explicitly present the researcher as a potentially active participant in the research process.

On the other hand, in comparison with descriptions of scientific method in the 'harder' sciences, this nine-step delineation does not focus on the formulation of hypotheses. In the social sciences the emphasis tends to be on 'research questions' and not on "hypotheses" to be tested. This is due to the fact that the social sciences tend not to focus on "truth" and instead emphasize understanding and decision making. Finally here, the above 9-step delineation does not focus on experimentation (in the social sciences experimentation is seldom employed, as the "objects" to be observed are human beings or organisations and these are affected by/react to experiments), and therefore neither does it focus on replicability and generalizability, concepts that are so central to investigations in the natural sciences.

A third description of scientific method, "the scientific knowledge acquisition web" is provided by (Lee 2000; Fig. 9.1, 140). Here the number of steps is increased and a more complex and realistic, interactive picture is presented.

Note that in the depiction shown here in Fig. 7.1 there is no natural starting point (even though this appears to be the case when we see box 1: Identify problem area).

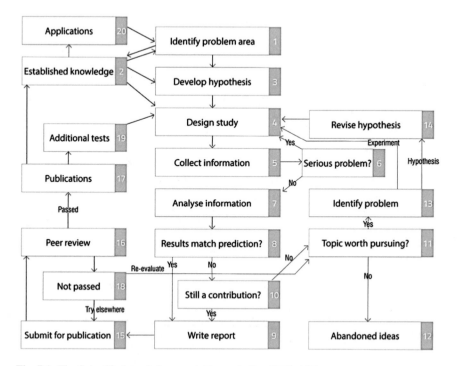

Fig. 7.1 The Scientific knowledge acquisition web (Lee 2000, 140)

This lack of a natural starting point is realistic as in general there is no clearly defined commencement of a research project, which is typically characterized by many reflections, attention to existing theories and results, discussions with colleagues/advisors, and so on. In the structure presented below there is also considerable room for feedback, including room for re-thinking one's hypothesis, the experimental design and the procedures followed. As will be considered in more detail in the concluding sections of this chapter that deal with the design of a research project, such feedback is a far more realistic representation of the process than was provided in the earlier depictions of scientific method.

Perhaps the only aspect of this characterization of the "scientific knowledge acquisition web" that requires elaboration is what happens if in box 8 it is decided that the results do not match the predictions and the study by itself is not a contribution that deserves a write up (box 10), yet the researcher decides that the topic is still worth pursuing (box 11). In this case, the problems with the investigation must be identified (box 13). Here there are two possible ways the experience so far can aid in the preparation of another investigation: (a) either the hypothesis must be modified (box 14), or (b) if the problem was more likely due to the design and/or the implementation of the experiment rather than with the hypothesis, then a modified experiment must be performed. In either case, a new experiment must be designed (box 4 again) and the process continued.

Without going into detail as to what is missing from *all* these three depictions of the progression of steps in scientific method, four aspects will be highlighted; they are all relevant for the ensuing reflections on the design of a specific research project.

First of all, and in particular based on the first two depictions, one gets the impression that the progression is more or less sequential or linear. This is misleading! Research does not develop in such a simple step-by-step manner, and even the notion of "steps" may be misleading, as often the researcher has many, and at times conflicting, thoughts in his or her mind.

Consider for example the earliest stages in the evolution of one's research activity. This often involves qualitative considerations based on one's own reading, on conversations with colleagues, and on reflections on experimental observations carried out by others. But note that even though such more speculative, conceptual and theoretical considerations tend to precede the more active and empirical aspects of research in the natural sciences, similar reflections tend to take place all throughout a research project—even at the very "end"—if there ever really is an end to one's investigation, as the more insights one gains, the more questions and potential new areas of investigation arise.

A second aspect that is missing in the above depictions of scientific method and that is worthy of reflection is the role played by creativity. While the earlier stages of research typically are inspired by a review of the literature and by existing observations and experimental evidence, scientific inquiry can also be inspired by pure curiosity based on theoretical speculation. This must have been the case for many major breakthroughs and discoveries in science, some of which have been referred to earlier: Nicolaus Copernicus' heliocentric model of the solar system

(1543); Galileo's contemplations regarding how objects of different mass must fall at the same speed (1589); Einstein's remarkable theorizing as to relativity (1905, 1916); cosmological theories as to the development of the universe (Big Bang theory by Georges Lemaître in 1927—named as such first in 1949); steady state theory by Fred Hoyle (1948), theories as to "Black Holes" (the original concept goes back to 1783, the modern formulation in 1968 was by my former teacher at Princeton University, John Wheeler); and perhaps also the discovery by Francis Crick and James Watson of DNA (1953).

Continuing this line of thought—some of the world's most famous theoretical developments are the result of intuition and imaginative suppositions. Einstein is famous for his so-called "gedankenexperimente" (the English translation of this often-referred-to German word is "thought experiments"), an example of which is how he at age 16 first asked himself what would happen if he chased and caught up to a light ray, a thought experiment that although not possible to carry out in practice nevertheless played a role later on in his insights as to the relativity of time and to his development of the theory of special relativity. Another well-known thought experiment is 'Schrödinger's cat', often referred to in discussions of quantum mechanics.

In addition, creativity can play a major role not just in theoretical speculation and imaginative suppositions, but also in the design of one's experiments, including the choice or even development of the methods to be employed in one's investigations.

A third aspect of scientific method that is missing in all three depictions of the scientific method is "interpretation". Even in the "acquisition web", matters of interpretation are hidden in steps such as "Results match prediction", "Still a contribution?" and so on. Just as with creativity, interpretation is something that is difficult to teach/learn, but is of major importance, and certainly will play a significant role in the evaluation of the papers a researcher sends to a journal for publication or in his own evaluation of the writings of others. Science is far more than just "results".

A fourth aspect of scientific method that is not considered in the depictions is "uncertainty"; nowhere is there a focus on the many uncertainties characterizing the research process. Even so-called "deterministic systems" may be characterized by considerable complexity and require complex models and large amounts of computation in order to perform statistical analyses of observations—and a whole new set of ideas and approaches is required once we explicitly introduce notions of uncertainty as to causal relationships. This will be the subject of the next chapter.

Before concluding these reflections on the universality of a scientific method in the natural sciences, some final, more general remarks are called for. Consider first the question that introduced this chapter and that dealt with how a researcher can learn the general principles or "rules of the game" that characterize scientific method. Just as is the case with the learning processes of musicians and carpenters, there is no simple, straightforward way to internalize scientific methodology. It is more a matter of "learning by doing" under the guidance of skilled and dedicated mentors whereby the researcher gradually develops an appreciation of the scientific method as well as of the specific techniques and approaches that characterize one's own specialized domain.

But this should not be considered to be a recommendation to passively play the role of an apprentice; the major motivation for writing this book is to stimulate reflection on the many aspects of science and scientific research. It is strongly suggested that a precondition for a truly enlightening, enriching and humanizing research experience, is reflection (and amazement) as to how and why the world is at it appears to be.

In addition, the internalization of the methods of science is also highly dependent on receiving and reacting to feedback from experienced members of the scientific community (e.g.. when one's thesis is evaluated and defended, when one submits articles for publication to reviewed journals, and when one makes presentations at conferences and symposia). In this connection it is important to re-emphasize that one of the important requirements of most scientific research is that a study must be replicable and be shown to produce the same results before it is accepted by the scientific community as a valid and reliable contribution. While this demand is strongest in those domains of science where experimentation is practical, it may not always be reasonable/relevant in other fields, such as astronomy or geology. Note too that in order for this demand as to replicability to be met, others must be able to recreate in a new experiment the same conditions under which one's own experiment was carried out. Therefore these conditions must be recorded precisely, which once again emphasizes the importance of keeping good verifiable records and a diary as part of one's research—something that is omitted in traditional descriptions of "the scientific method".

There is also the need for caution as to the potential "side effects" of this learning process whereby the young researcher gradually becomes accepted as a member of the scientific community. The initiation to the values, language, attitudes and behaviour of his peers may not only lead to the tacit acceptance of well-founded customs and norms; it can also result in tunnel vision where the budding scientist forgets that science is only one of a number of ways of looking at the world. A result can be impatience or even arrogance with others who do not share one's perspectives, particularly as to how to develop and evaluate valid and reliable knowledge. A similar result can be a loss of sensitivity as to the complexity that characterizes scientific investigation. Given their training, many natural scientists tend to implicitly assume that that which is to be studied is independent of one's self. In so doing, and in contrast to the social sciences, objectivity becomes the supreme mantra, even though much of science deals with ideas, interpretation, meaning and creativity—and in contributing to our knowledge of the human condition and to the well-being of sentient creatures and the environment.

We conclude this section on scientific method with brief reflections on the second question that introduced the section: If there is a "standard" approach to performing research that is widely accepted and independent of scientific discipline and one's culture (*the* scientific method), does each scientific domain also have its own research methodology?

It appears to be reasonable to postulate that while specific scientific domains, characterized by their own histories, cultures and technologies implicitly share a broad overall perspective on scientific method, they also tend to look upon scientific method through their domain-specific lenses. For example, according to a report by the American Association for the Advancement of Science (AAAS):

> Scientists share certain beliefs and attitudes about what they do and how they view their work … Fundamentally, the various scientific disciplines are alike in their reliance on evidence, the use of hypotheses and theories, the kinds of logic used, and much more. Nevertheless, scientists differ greatly from one another in what phenomena they investigate and in how they go about their work; in the reliance they place on historical data or on experimental findings and on qualitative or quantitative methods; in their recourse to fundamental principles; and in how much they draw on the findings of other sciences … Organisationally, science can be thought of as the collection of all of the different scientific fields, or content disciplines. … With respect to purpose and philosophy, however, all are equally scientific and together make up the same scientific endeavour (AAAS 1989; 25–26, 29)

In other words, science does not provide a unified epistemology; the unity of the scientific endeavour is characterized solely by its overall methodology and not by the specific methods of inquiry its disciplines employ or by their findings/data. According to (Jeffreys 1983; 7), this perspective on science leads to fundamental demands as to methodology: "There must be a uniform standard of validity for all hypotheses, irrespective of the subject. Different laws may hold in different subjects but they must be tested by the same criteria; otherwise we have no guarantee that our decisions will be those warranted by the data and not merely the result of inadequate analysis or of believing what we want to believe." We will consider such demands on an adequate theory of inductive inference in greater depth in Chap. 8: Uncertainty, Probability and in Research.

7.2 Research Design

The above reflections on scientific method provide a basis for discussing the concept of research design, where the emphasis is not on "principles" or "rules of the game" but on practical guidelines for structuring your thoughts as to *how to design your specific research project*. I note that to simplify the exposition, I will assume throughout this section that the research is to be performed in connection with a Ph.D. project and I will often refer to *your project* and to *you* as its designer. This should in no way restrict the general applicability of the reflections provided.

Let me commence by referring to the following simple diagram (Fig. 7.2) indicating the relationships between five concepts; the diagram and the reflections that follow are inspired by Maxwell (2005). I note that even though Maxwell's book is oriented towards qualitative research its basic ideas are just as relevant for quantitative research in the natural sciences.

Before focusing on the five major components depicted in the model, a few words are called for regarding the overall structure of the model. It can be looked upon as consisting of two closely integrated units. The first is the upper triangle that relates your research questions/hypotheses to your purposes/goals and to your conceptual framework, i.e. what is/can be known about the phenomena you intend to study and

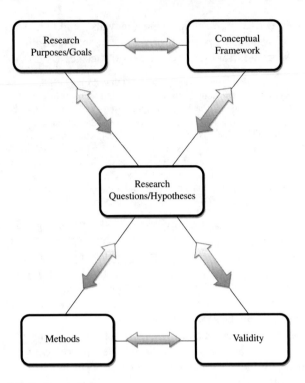

Fig. 7.2 A model of research design

their relationships. This conceptual framework is typically based on your review of the literature, your and your advisor's knowledge and experience, and the theoretical foundations for your study. In addition, the upper triangle underlines that the purposes/goals should be informed by existing theory and knowledge as to the field while decisions as to your conceptual framework (which literature, theories, models and experiences are relevant) depend both on your purposes/goals and how you intend to live up to these via your research questions/hypotheses.

The components of the bottom triangle are also closely interrelated. The methods you will use to collect and analyse your data must be coherent with your research questions/hypotheses and must be effective in meeting challenges to validity threats to your results regarding the questions/hypotheses. Similarly, in order for the questions/hypotheses to be wisely framed, they must consider the feasibility of the methods in your particular context as well as the possibility of validity threats. And how these threats can be dealt with depends on your research questions/hypotheses and the methods you chose to collect and analyse your data.

In both triangles the research questions/hypotheses are the heart of the model, connecting all the other components and informing and being informed by these components.

As Maxwell emphasises (p. 6) there are many other factors that will influence the design, including your research skills, experience and personal goals, your university's rules and traditions and the resources available (funding, technology, you advisor and other sources of guidance, etc.). Although these aspects are not directly included as major components of the design activities presented here, they must all be taken into account by the reflective researcher when designing her or his project.

7.2.1 Components

In order to provide a well-informed discussion of the design process in terms of relationships between the five major components depicted above in Fig. 7.2, the following is a brief presentation of the individual components. The motivation is not to highly structure the process as though there are "rules" to follow. Rather, once again, it is instead to stimulate reflection on what you are doing and why you are doing it when designing a research project.

Research Purposes/Goals
We can fruitfully distinguish here between *Purposes* that refer to the more specific objectives of your study, and *Goals*, that refer to the motives that underlie your research. "Purpose" is more closely related to the Research Questions/Hypotheses than are the "Goals" that are more abstract as well as the more personal. Typically these are in essence answers to questions such as: "Why do you want to carry out the investigation?" "Why is it worth doing—what will it contribute to science? to society? to your own personal and professional development?"

Clearly such reflections should be in focus in a research investigation—but seldom do they receive the explicit attention they deserve in either project proposals or dissertations. It is my experience that there is a tendency to focus almost exclusively on "how to" questions rather than on the more fundamental "why" questions. Perhaps this is an expression of the earlier referred to tendency for many Ph.D. projects to resemble an apprenticeship. During an apprenticeship the individual is trained to be a skilled craftsman in a learning-by-doing process where he performs assigned tasks under the guidance of an experienced craftsman. Similarly, in a typical Ph.D. project in the natural sciences the student works in relative isolation on a well-defined empirical investigation whose goals as well as the tools of investigation often have been more-or-less defined by the advisor (the "experienced craftsman") who supervises his work. In other words, the goals, the "why" aspects of the research design process, tend to be neglected, at least by the young researcher who is eager to satisfy his advisor and may be quite pleased to work on a, more-or-less, pre-defined project.

Conceptual Framework
This refers to the concepts, assumptions, beliefs and experiences that inform your research. They draw upon your background and are founded on your preliminary studies and prior knowledge as to your field and to relevant theory—and evolve via your on-going study of the literature, your interactions with your advisor and

colleagues and via your investigations/experiments. In other words, your conceptual framework can be considered to be a tentative "theory" or "model" of the phenomena and their inter-relationships that you are investigating; the words theory and model were in quotation marks to indicate that they are being used in a very informal manner, as in ordinary language.

It is appropriate here to warn against focusing too narrowly on the literature and on simply summarizing what appear to be relevant publications when developing your framework. Doing this can tempt you to consider your research design as a descriptive task, while it should in fact be primarily considered to be a constructive task. The literature is to be considered as a useful and hopefully inspiring source but not as an authority. In other words, instead of trying to find your conceptual framework "out there", you yourself must construct it.

An effective and challenging way to develop your framework is via the use of a so-called "concept map". Such a map depicts the major concepts you develop and helps you to reflect on them and on the relationships among them. It enables you to think on paper and thereby to clarify your understanding of existing theory, to make visible your own implicit, tentative conjectures as well as to identify holes or contradictions in your tentative plans. The link http://alternativeto.net/software/microsoft-visio/ provides a number of tools that can facilitate the drawing of concept maps, mind maps, flow charts and the like.

The following is an example of such a concept map from a Ph.D. research proposal dealing with outsourcing of R&D (Moitra 2004; 20). I have chosen this topic since, although the focus of the investigation, outsourcing of research and development activities, is not directly within the natural sciences, the map can nevertheless provide inspiration for your own deliberations and is understandable no matter what your specific field is—which may not have been the case had I provided a concept map from a particular field of specialization (Fig. 7.3).

Research Questions/Hypotheses

Earlier I introduced a distinction between Research Questions and Hypotheses. I considered "Research Questions" to be statements of what you want to learn from your research and "Hypotheses" as tentative answers to these questions, where these answers can be tested. While many research projects in the natural sciences, typically characterized by experiments and statistical testing, are formulated using hypotheses, research projects in the "softer" sciences tend to be formulated, at least in their early, more qualitative, phases, using research questions.[1]

As visualized in the model of research design presented Fig. 7.2, these questions/hypotheses are at the very heart of your research. While descriptions of Scientific Method tend to imply that they are formulated at the beginning of one's research and determine the rest of the design, in fact experience shows that this may

---

[1]In those cases where the results of a qualitative investigation lead to quantitative research, hypotheses are formulated *after* the researcher has carried out preliminary empirical investigations rather than being prior conjectures that are simply tested against the data—as is typically the case in the natural sciences (Maxwell 2005; 69).

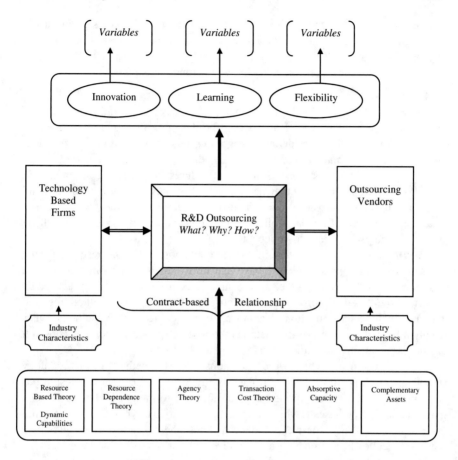

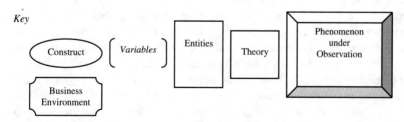

Tentative Concept Map of the Study

Fig. 7.3 Concept Map of Research on R&D Outsourcing (Moitra 2004; 20)

be quite misleading. It does not respect the interactive, dynamic, learning aspects of research. Of course, if prior investigations or demands from a research advisor lead to clear—and hopefully, feasible—questions/hypotheses and thereby bypass some of the major learning aspects of the research, these pre-determined questions/hypotheses will then determine design choices as to conceptual framework and methods. However, in most research this is not the case and locking in on questions/hypotheses too early in the design process, i.e. before sufficient reflection as to the theoretical and methodological options available, can lead to what often is referred to as a "Type III error"— answering the "wrong" question or testing the "wrong" hypothesis.

Methods

While (Maxwell 2005) uses the term "methods" in his exposition, a more correct terminology for a reflective scientist would be the more inclusive term "methodology". Methodological considerations regarding this component of design extend well beyond the normal use of the term "method", which typically refers more narrowly to the use of specific techniques/procedures for collecting and analysing data. "Methodology" on the other hand implies not only focus on such "how to" questions, but also on one's more inclusive reflections on the "why" choices regarding the choice of the data collection methods and technologies and the choice of how to analyse the data so-collected.

Nevertheless, to avoid the terminological confusion that could result if I reformulated Maxwell's presentation, I have chosen to use his more limited and less ambitious term "methods" here. I note in this connection that, in practice, the content of the "Methodology" section in most Ph.D. theses is *not* methodology, i.e. does not include the reflections leading up to the choice of the methods, but is simply a brief description of the methods the researcher will choose/has chosen to use in his or her investigation to collect and analyse data.

However, no matter which word one chooses to use, methods or methodology, serious attention should be given to the more reflective aspects of the design process, including the "why" questions regarding: the formulation of your research questions and hypotheses ("why" choose these questions/hypotheses and not others?), your tests ("why" these experiments and not others?), and your analyses (why these approaches to justification and validity and not others? why these statistical procedures and not others?). Such methodological considerations are of primary importance in the whole design process.

Unfortunately even though as a young researcher you may (hopefully!) reflect on such matters, you may not always have the privilege of being able to choose your methods. For example, in some cases, such as if the project is being funded from external sources, the choice of methods may be pre-specified. In other cases, for example where an advisor looks upon the project as one part of a larger investigation he or she is leading, the researcher may be granted only limited degrees of freedom to consider the relationship between the formulation of the research questions/hypotheses and the data collection methods. In still other cases you may be replicating the research carried out by someone else. Finally, you may be in the situation where you present proposal builds upon research you have carried out earlier such that it is not reasonable to significantly change the choice of methods for data collection and analysis.

Fortunately however, in most cases the choice of method is a major part of one's research design since the choice of methods does not automatically follow from the research questions/hypotheses. There is no straightforward, more-or-less logical way to convert the questions/hypotheses into methods; no compelling, mechanical way to operationalize and convert them into methods.

And the relationship between the choice of research questions/hypotheses and the choice of methods is symmetrical. The availability in practice of data collection and measurement technologies can influence the formulation of one's research questions/hypotheses. For example, it should be self-evident that it is not productive for a Ph.D. student to formulate a hypothesis he could have determined was not, in practice, possible to test. In addition, over and above the availability of suitable technology, other down to earth practical matters regarding the choice of method, such as the traditional approaches that have been used by others who have investigated similar phenomena/relationships, your own experience and skills and those of your advisor, and the availability of resources, can exert a strong influence on the formulation of the research questions/hypotheses.

It is for all the above reasons that in Fig. 7.2 depicting the design process, there is a double-headed arrow connecting "Research Questions/Hypotheses" and "Methods".

Furthermore, an integral part of the Methods component of the design process is also reflection as to how you will analyse your data, once they have been collected. If possible, such reflection should be included in your written research proposal and should clearly delineate the relationships between your choice of methods and your decisions as to how to perform the analyses—instead of simply assuming, as often is done implicitly, that once the data is available, you will be able to perform the analyses. I have a number of times read Ph.D. theses where it was clear that data collection was performed prior to the reflection as to the choice of method for data analysis. In such cases, the lack of reflection on the combined activity: data collection—data analysis often led to a less harmonious and effective project than otherwise might have resulted.

Validity

Earlier, in Chap. 4, we reflected on the concepts of "justification", "verification", "falsification" and "acceptance" while in Chaps. 5 and 6 we considered the concepts of "validity" and "generalizability" with respect to measurement and experimentation. I suggest that you refer again to the relevant sections in these chapters before continuing with the present discourse.

To simplify things here and to make the presentation more in line with that of Maxwell, I will use the term "validity" as an overall expression regarding the integrity of your research—to the soundness of your concepts and to the credibility of your results in the form of measurements, descriptions, explanations, interpretations and conclusions. Another way of expressing this is that reflection on validity is also in essence reflection on how you might be *wrong*, on how trustworthy your measurements, models and claims are, and on how to test your accounts to see if they are *in*compatible with observations. This is in line with the underlying logic in falsification (see Chap. 4, Sect. 4.2).

But note that within the context of research design, validity refers not only to the results of your investigations but also to the strategies you chose to identify so as to eliminate threats to the credibility of your results. For example, as we saw in Chap. 6 within the context of experimental design, such strategies can include the use of random sampling, control groups, statistical control of the extraneous variables, control of the experimental environment, the elimination of spurious relationships, avoiding altering hypotheses after data has been collected and analysed,[2] etc.

I note as well that the challenges to validity when performing research that involves qualitative investigations (for example based on interviews) include considerations seldom recognized in the natural sciences. These include potential validity threats arising from the biases of the researcher due to his or her prior beliefs and attitudes (which can influence the interpretation of the information obtained via interviews etc.), as well as potential validity threats due to the influence of the researcher on the "objects" of the research, e.g. people who are interviewed. This is, in contrast to the natural sciences where, aside perhaps from micro-domains such as quantum mechanics, the researcher's behavior is not, in general, considered as being a potential source of influence on the objects/phenomena being observed. Nevertheless, as emphasized several times earlier, the researcher's own biases can in fact influence many aspects of his investigation (as well as the analysis and the writing up of the results) and yet this seldom is considered in the younger scientist's training. A result is that scientists in general tend to be unaware of their biases and of the potential influence they may have on their research.

Finally, I note that validity is a goal to be sought and that it is relative, not absolute, since, as has been argued a number of times, it must be evaluated with respect to your research goals and purposes as well as the context for your research.

7.2.2 Research Design and Your Proposal

Having introduced the components of a design process, the following attempts to demonstrate how these components can serve as a useful background or framework for developing the structure and content of a research proposal. It is important to once again emphasize that since in practice your design will evolve as you carry out your investigations, your proposal should not be presented as a rigid, static model you commit yourself to live up to throughout your research. Thus, while your responsibility is to develop your proposal as clearly and as comprehensively as possible, at the same time it may be wise to make it clear that there may be a need for future modifications, perhaps with indications as to where you feel that such

---

[2]While in traditional quantitatively based research in the natural sciences it is considered unethical to formulate hypotheses after data has been collected, this may in fact be a perfectly logical and acceptable strategy in qualitative research where your research questions often are formulated based on preliminary observations via a learning process. According to Maxwell (102), "If your methods won't provide you with the data you need to answer your questions, you need to change either your questions or your methods."

modifications may be required. Hopefully this will not only be looked upon with favour by the committee evaluating your proposal who will see that you have the ability to design a coherent and feasible study while being aware of issues that may require further development—it will also lead to your own growth as a reflective practitioner of research.

Purpose of Your Proposal

The structure of your proposal will be closely related to your purpose in developing it. A proposal serves the following three functions (Locke et al. 2007; 3–4): (1) It communicates your research plans to your advisors and evaluators, (2) it provides a step-by-step plan for your investigation, and (3) it serves as a "contract" in the form of a bond of agreement between you and your advisors and evaluators. Maxwell (2005; 118) highlights the communicative aspects, in particular as to explanation and justification: "The purpose of a proposal is to explain and justify your proposed study to an audience of nonexperts on your topic." (Maxwell's emphasis).

Perhaps the reference to "nonexperts" may arouse your curiosity. Certainly if there are "nonexperts" in a committee that is evaluating your research proposal, or if your advisor is not an expert in the specific topic you intend to investigate, then the emphasis on "explain and justify" should be clear. Today research is becoming so highly specialized that although your advisor or a member of your committee may be knowledgeable in the broad field you work in, their expertise in the topic you specifically focus on may be limited.

But even if your advisor is an expert and even if the committee (and you cannot know in advance who will be members) consists of experts in your specific domain, the need for you to clearly explain what you plan to do cannot be over overemphasized. My own experience supports the observation in Locke et al. (2007; 128) that "...advisors and reviewers misunderstand student proposals far more often than they disagree with what is proposed." It is my observation that graduate students have a tendency to implicitly assume that they express themselves clearly—and that the reviewers, advisors and evaluators use the same specific language regarding the subject matter as they do, that they share the same perspectives regarding methodology, and in essence that they come from the same intellectual and scientific culture. These assumptions are often quite naïve. This may particularly be the case in connection with applications for funding from national, and in particular international, sources.

Perhaps even more important than the need for you to clearly explain *what* you plan to do is the need to justify your proposal—to communicate clearly *why* your proposal is worthy of receiving support; once again I emphasize the significance of awareness as to 'why?', and not just 'how to?'

This not only includes motivating why the investigation is important (how it will make a contribution and why that contribution will be important), but also why you propose to do it in the way you describe. In fact according to my experience the single most common act of omission in proposals is the lack of attention to such *why* questions, in particular regarding the rationale for the choice of methods. I have often seen proposals where the student simply states which methods he intends to use to collect and analyse his data, with no attention at all on why these choices are the best—with no consideration of possible alternative procedures that could be used, no arguments as to such matters as resource availability, available technologies, relationships to other projects at one's department, the researchers and the advisor's experience with the various methods, etc.

Finally, and closely related to the above as to motivation and rationale for why you plan to carry out the investigation in a certain way, is the reflection that many students simply write their proposals as a description, emphasizing their knowledge as to the relevant literature and perhaps some technical details as to particular methods. Your proposal should be *your* proposal. A proposal that clearly and concisely presents and motivates your planned investigation does not only enable others to evaluate your proposal and to provide you with guidance and feedback; it also serves the vital purpose of providing you with a road-plan for the next few years of your academic life, a plan that can provide you with the self-confidence that often is not in abundance in the mind-sets of younger researchers.

Proposal Structure

In order to live up to the above, the following reflections on how to structure a proposal draw upon the previously introduced five components of research design and upon (Maxwell 2005; 121–128). In addition, the reflections are inspired by Friedland and Folt (2009) which focuses in particular on proposals to funding agencies such as the National Science Foundation (NSF), Environmental Protection Agency (EPA) and National Institutes of Health (NIH) in the USA, and by Locke et al. (2007), which provides advice on writing proposals for both dissertations and grants. Finally, the reflections draw upon my own years of experience in guiding students who prepare proposals and in evaluating research proposals.

I note that in contrast to proposals as to a dissertation, proposals regarding grants may have to live up to specific guidelines from the relevant funding organisations and that these most likely will differ from the general guidelines for a proposal presented below.

Another factor that tends to distinguish grant proposals from dissertation proposals is that reviewers of the former tend to place far greater weight on the applicant's demonstrated productivity and on the quality of previous work (Friedland and Folt 2009; 92–3). Nevertheless, the broad criteria that reviewers use to evaluate grant proposals are often similar to those employed by a committee that evaluates a dissertation proposal although perhaps with greater focus on the significance of the project and its innovative/creative aspects (Locke et al. 2007; 178–83).

Although brief, the presentation below should provide a reasonable foundation for structuring a proposal, and in any case for structuring your thinking about your proposal. It is in no way intended to represent a firm set of guidelines. First of all, different universities and different funding agencies have different traditions—and this of course applies as well to different cultures/countries. For example, while the structure considered in the sequel is based on the earlier exposition regarding research design, a "typical" NSF grant proposal includes the following: (1) Project summary or abstract, (2) Table of contents, (3) Project description, (4) Reference list, (5) Biographical sketches of the investigator(s), (6) Budget, (7) Current and pending support of the investigators (Friedland and Folt 2009; 31). Przeworski and Salomon (1998; 1) emphasize the role of culture with respect to such grant proposals: "… a proposal writer needs a feel for the unspoken customs, norms, and needs that govern the selection process itself."

Secondly, independent of the type of proposal and the university or funding agency it is aimed at, there is often considerable room for flexibility in how you structure your proposal. What you should focus on in the sequel is not the exact terminology employed or the specific structure referred to but the reflections as to logic and content. The presentation concludes with four appendices in the form of checklists that provide, in some detail, practical and down-to-earth hints as to preparing, writing and defending your proposal: (1) Checklist with respect to the structure of a proposal, (2) Checklist with respect to writing and planning, (3) Checklist with respect to viva/oral defence, and (4) Checklist with respect to precautions. I note that these checklists represent the only real 'how to' content of the book, which has primarily emphasized reflection rather than 'nuts and bolts'.

Before starting the actual presentation as to the structure of a proposal, I feel that it is important to emphasize that one of the most aspects of preparing a proposal has nothing whatsoever to do with its structure or content—but deals with your own frame of mind. Since your proposal will in essence represent a concise synthesis of: (a) what you have accomplished so far in your research, (b) your views as to your proposed project's scientific (and, hopefully, its societal) significance, (c) the specific research questions and hypotheses that you intend to investigate, and (d) how you intend to perform the investigation, the proposal represents *your* intellectual expression of scientific method in practice. If you look upon this as an exciting and challenging opportunity that will enable you to tell others—and not least yourself—what you intend to spend the next few years of your life doing and why this is of importance (for science, but also for you yourself), then you will experience developing the proposal to be a most challenging and rewarding transformative activity. An activity characterized by creative conceptualization, methodological rigor, and personal development. If, on the other hand, you look upon it as a burden, a hurdle, something to be done with so that you can move on towards getting a degree, my experience tells me that you will feel frustration in preparing the proposal and that its quality will suffer. I strongly suggest that you reflect upon this—and draw upon your innate strengths and self-confidence to meet the challenges and reap the rewards of developing your proposal!

Now on to the structure for your proposal:

Acknowledgements
This provides recognition of those who have helped you with your preliminary research, who have supplied data, given feedback, inspiration, etc.

Introduction
Although traditions vary at different universities, I am convinced that the Introduction to your proposal (and this applies as well to your thesis) is extremely

important for those who will be evaluating your proposal/thesis. It is likely to have a powerful effect on an evaluator's impression of you and the project; this can be particularly important in the case of proposals for funding, as the evaluators will most likely be reading many proposals. But the Introduction is not only important for evaluators; developing it is also very important for you, its author. Writing the Introduction gives *you* the opportunity to synthesize all that you really want to say, which is also why the final version of the Introduction is probably the very last thing that you will complete—and is almost certainly the very first thing evaluators will read!

It should provide clear and concise overviews of your study's goals and purposes, its theoretical bases, its research questions/hypotheses, your reflection as to the methods you intend to use and the feasibility of the research design, how you will justify/validate the results you intend to obtain and their potential significance for your field and for the scientific community and society in general. In addition, it can provide an overview of your study's historical foundations, its innovative and ethical aspects, and your own qualifications with respect to the investigation.

According to Friedland and Folt (2009; 91), who for well over a decade have taught a course on proposal development and writing at the Dartmouth College in the USA: "After reading a persuasive introduction, the reader should exclaim, 'Of course! What a great idea for a research project? Why didn't I think of that?'"

Conceptual Framework (or Literature Review)

This section builds directly on the earlier discussion of this component of your design. Choosing "Conceptual Framework" as the title of a section of your proposal/dissertation may surprise you as most students refer to this section as the "Literature Review". As was briefly indicated earlier, the reason that the term "Literature Review" can be misleading is that this section in practice aspires to fulfil functions that go beyond a simple review of the relevant literature. First of all, it should demonstrate how your proposed study is related to existing theory and research and therefore indicate how it will make a contribution to the scientific community—and to your own intellectual/academic development. Second—and unfortunately this is often downplayed in many traditional reviews of the literature —it should also present your reflections on the theoretical framework for the investigation.[3]

In other words, this section aims at more than simply reviewing the literature and briefly summarizing what you have read about that has been done in your field. Although this certainly is a very important element of the conceptual framework, it also aims at grounding the study in what has already been investigated and is accepted as scientific knowledge, including providing a clear sense of the theory that informs your approach. And it aims as well at drawing upon your personal experience and knowledge of the field, including e.g. information on what you may have learned from any pilot studies you have carried out.

---

[3]See for example Fig. 7.3, the "Concept Map" of a Ph.D. proposal; at the bottom of the map, reference is made to the theories that underlie the proposal.

I note however that even though the term "Literature Review" does not really communicate what the section could/should cover, you may feel obliged/compelled to use it depending on the traditions at your department. In such cases, I suggest using the term "Conceptual Framework and Literature Review".

Research Questions/Hypotheses

As indicated earlier in the discussions as to the design of your research, this is the heart of your proposal—and even though the questions/hypotheses will have been briefly introduced in the Introduction, your proposal/thesis will gain from a detailed discussion of them in a separate section. The preceding sections with their presentations of the context for the research, the relevant theory and your personal experience and preliminary research provide a firm basis for the discussion in this section. As indicated earlier, your ability to clearly formulate the questions/hypotheses as a coherent "whole" tends to evolve as your experience increases, so the formulations in your dissertation/report may very well differ from those in the proposal.

It should be emphasized that the section aims at presenting a small number of clearly focused, well-justified questions/hypotheses that *can* be answered within the context of your proposed investigation—which leads up to the section on Methods.

Methods/Methodology

Earlier I reflected on whether this section should be called "Methodology" or whether it should be called "Methods". I noted that in many Ph.D. proposals and dissertations the section is called "Methodology" while in practice the major content of such a section usually is not "methodology" in a broader sense, but simply a description of the methods the researcher will choose to use in his or her investigation to collect and analyse data. So alas, in such situations, "Methods" appears to be a more appropriate title than the standard "Methodology".

I strongly suggest that no matter what title is used, this section of your proposal should include broader methodological reflection. It should not be just descriptive but also reflective, not just frame and answer 'how to' questions but also the more fundamental "why" questions regarding your methods ('why these methods and not others'?) and that are so important in developing your world-view and skills as a scientist.

For example, since the choice of data collection methods does not automatically follow from the research questions/hypotheses, this section of your proposal should not only describe the methods that you will use, but also motivate your choice by relating it to the goals/purposes of the investigation, to the resources available (including skills, time, funding), as well as to the efficacy of the methods with respect to the research questions/hypotheses.[4]

---

[4]Although the great majority of proposals within the natural sciences refer to quantitative methods, many also include some form of qualitative methods and analyses. Locke et al. (2007; Chap. 5) provides a fine introduction to how to include qualitative methods in a proposal.

Similarly, there is no obvious way to analyse the data once they have been collected. Such analysis decisions are closely related to your ability to operationalize the project's goals and to meet potential challenges as to validity. Often this will presume a reasonable knowledge of statistical reasoning—a major topic of the next chapter. Thus, in contrast to what most often is the case, where Ph.D. students simply refer to standard experimental procedures and to the use of standard software packages for statistical analyses of their data, this section should include reflection on *why* the actual choices you have made as to methods are in fact the best choices.

I note here as well that it is becoming increasingly common for proposals (and theses) to include reflection on the ethical aspects of the research, either within the Methods/Methodology section or in a separate section. This may be particularly relevant where methods are chosen that involve experiments that can effect humans and other sentient beings (e.g. in projects focusing on medical research, animal husbandry, or on genetic engineering) or with investigations that can lead to the development of products and processes that can threaten the environment (e.g. in projects within agriculture, mining, or energy production). In many countries there are well established formal procedures for getting approval of experiments and investigations that are characterized by potential ethical implications, in particular in studies involving or affecting humans or animals. From such an ethical perspective, the proposal can be seen as a form of "social contract" for performing research in an ethical manner, including the sensitive mentor-mentee relationship (Locke et al. 2007; 25–40). Chapter 10 provides food for thought on Ethics and Responsibility in Scientific Research including a series of guidelines for ethical practices in research (Sect. 10.2).

Validity

In many proposals, issues regarding validity—the credibility and trustworthiness of your descriptions, measurements, analyses, interpretations and conclusions—are dealt with in the Methods/Methodology section. On the other hand, in most dissertations they are often coupled together both with that section and with a section on analysis and results. In spite of these traditions, there are good reasons for having a separate Validity section in your proposal. Such reasons include your ability to develop an increased focus and clarity that enables you to face validity threats. Attention can be focused on strategies for the design of the experiments you may be proposing and for testing the significance of your results. I note that in connection with the resultant dissertation, focus should also be placed on possible competing explanations as to your results as well as on discrepant data/anomalous observations.

Should you already have carried out pilot tests or other forms of preliminary investigation, the Validity section can also include your preliminary results (alternatively, these can be presented in the Methods/Methodology section and provide background for your reflections on your choice of methods for data collection and analysis). Such preliminary results can also provide a strong foundation for the justification of your study's feasibility. This is the case even if you are relatively new to the specific field of your proposed research since these preliminary results can contribute to establishing your competency for successfully carrying out your proposal.

Concluding sections to your proposal

Such sections will include your list of references and appendices, which may e.g. include a pro-forma table of contents for your dissertation, a rough time plan and budget for the entire proposed study, brief technical descriptions of experiments and technology to be used, and details of preliminary results obtained so far.

Finally, a comment as to the length of your proposal. It is in general easier to write longer rather than shorter proposals. It is always a challenge to be concise and clear. Although at most universities there are no general rules or norms as to the length of a proposal, some rules of thumb probably can be obtained from your department. I note that appendices usually are not included when determining the length of a proposal; this is also the case with dissertations. I note too that proposals for funding generally have to follow tighter guidelines than for dissertations. For example, applications to the NSF are restricted to a maximum of 15 pages with single line spacing.

The following are several links that provide valuable information regarding the design and writing of proposals:

Advice on Writing Proposals to the National Science Foundation (the National Science Foundation, NSF, is the major USA source of funds for research projects in the sciences). The document focuses on writing proposals to NSF but also provides general advice that can be applied to writing proposals, either for grants or dissertations: http://www.cs.cmu.edu/ ~ sfinger/advice/advice.html.
Proposal and Award Policies and Procedures Guide, the NSFs detailed and highly informative guide, effective December, 2014 can be downloaded at http://www.nsf.gov/publications/pub_summ.jsp?ods_key=gpg.
Finally, although the note *The Art of Writing Proposals: Some Candid Suggestions for Applicants to Social Science Research Council Competitions* (Przeworski and Salomon 1998) is primarily aimed at grant proposals within the social sciences, its suggestions are, for the main, equally applicable to the natural sciences. The pamphlet can be downloaded free of charge from http://www.ssrc.org/publications/view/7A9CB4F4-815F-DE11-BD80-001CC477EC70/

Appendix: Checklists to Assist You in the Preparation of Proposals (and Dissertations)

I humbly suggest that the four checklists below can be of value to you in planning and writing your research proposal. Even though there are differences between the structure and content of a proposal and a dissertation—one being prepared more or less prior to your investigations, the other being prepared, or at least finalized, after you have completed your empirical work and analyses—the major content of these checklists is also relevant for planning and writing your dissertation. Nevertheless, for the sake of precision and clarity, the presentation below will specifically refer to a proposal.

1. Checklist structured in accord with a traditional template for a proposal

Title page

Question	Needs work	OK as is
Have you checked the rules about what should be entered here? Does the title clearly and concisely encompass the focus of your proposal? Have you considered how your title can be shortened, clarified, be more precise?		

Acknowledgements

Question	Needs work	OK as is
Do you clearly acknowledge help of people who have (a) Supervised you or given you advice underway? (b) Read earlier drafts and provided you with feedback? (c) Helped you to gain access to laboratory facilities, an organisation, data, etc.? (d) Inspired you, perhaps by providing you with new insights or by challenging your ideas/arguments/methods—even if this inspiration was a result of informal discussions?		

Table of contents

Question	Needs work	OK as is
Do you have a well-structured table of contents indicating chapters, sections/sub-sections, appendices etc. and the corresponding pages numbers?		
Do you want to have a separate list of tables, figures etc. and do you have a well-structured numbering system for referring to them?		

Abstract (Whether you should present an abstract depends on the traditions at your institution).

Question	Needs work	OK as is
Do you provide a clear and concise overview of the proposal including the major goals being addressed, why these are important and interesting and how you will address the resultant research questions/hypotheses?		

Introduction

Question	Needs work	OK as is
1. Statement of the Focus and Purpose of the Study		
Do the opening sentences grab our attention and give a clear indication of the focus of the study and its broader significance?		
Is there a clear and concise statement of purpose indicating why you want to do the study and what you intend to accomplish—of the central, controlling idea in the study? Is this statement of goals/purposes concretized in the form of clearly stated objectives and projected results or outputs?		
2. Research Questions/Hypotheses		
Do you provide a clear and concise statement of your Research Questions and/or Hypotheses?		
Do you avoid constructing too many questions/hypotheses?		
Is there a clear and concise statement of the significance of these research questions/hypotheses for: (a) the scientific community, (b) those directly affected by your research (particularly relevant in the social sciences), and (c) society?		
Do you argue that the proposed research is 'original'—that it will contribute new knowledge to the field?		
Do you briefly discuss how your study has evolved from previous research (e.g. your own earlier research, that of your advisor, and in particular, from your review of the literature)?		
3. Theory and Method/Methodology		
Do you briefly present the theoretical approach or perspective you will be using and justify the choice?		
Do you provide a brief overview of your research strategy, design, and the methods you will employ to answer your research questions/test your hypotheses?		
Do you provide information as to the feasibility of the study with respect to time and resources—and your competences and skills?		
Is there a clear and concise statement of how your research design and process will enable you to meet potential challenges to validity and demonstrate the credibility of your results?		

Conceptual Framework and Literature Review

Question	Needs work	OK as is
Do you provide a clear review of the main ideas, existing theory and research related to your field of investigation and its potential significance?		
Do you provide reflection on your choice of theoretical approach based on both the literature, advice from experts in your field and your own experience (including prior investigations/pilot studies) and competences?		

(continued)

(continued)

Question	Needs work	OK as is
Are you selective in your choice of literature? Have you established priorities as to carrying out your literature search (overviews, journal articles, books etc. including emphasis on recent rather than older references)?		
Do you provide the literature review in such a manner that it clearly supports your formulation of your Research Questions/Hypotheses and demonstrates why your research is important? This may include reference to		
(a) A lack of relevant research on a topic, or gaps in existing knowledge, (b) The existence of competing perspectives on a topic, (c) The value of replicating research carried out on a topic by others, (d) Benchmarks that enable comparing your results with other findings.		
Do you provide comments on the literature you review in relation to the formulation of your research questions/hypotheses? Are you reflective and critical in your presentation and not simply descriptive where you essentially just provide a concise abstract?		
Is there a clear structure to your Conceptual Framework and the Review— do you organise these according to themes or important concepts and principles?		
Do you graphically or otherwise demonstrate the interconnections between the themes/concepts?		
Do you provide summaries of and reflections on the individual publications you review and a conclusion to your overall literature review?		

Methodology/Research Methods

Question	Needs work	OK as is
Do you discuss and justify your research strategy?		
Do you formulate your research questions so they can directly serve as the focal point of your proposal or do you formulate them as hypotheses that can be tested?		
Do you establish the interconnections between these questions/hypotheses and your overall objectives and specific aims, as well as the significance of your study?		
Do you discuss the appropriateness of your research designs—including why you do not propose to use other relevant designs?		
Do you discuss the appropriateness of your specific data *collection* methods —including the choice of technologies and experimental designs as well as why you do not propose to use other relevant methods?		
Do you discuss the appropriateness of the data *analysis* methods you propose to use—including why you do not propose to use other relevant analytical methods?		

(continued)

(continued)

Question	Needs work	OK as is
Do you discuss the feasibility of your methods given restrictions as to time, budgets, your own skills, available support etc.?		
Do you reflect on the external validity—on whether the data and the analyses will permit you to generalise your results—to populations, to theory or to methodology?		
Do you clearly present the delimitations and limitations of your proposed investigation?		
Do you reflect on your own role as researcher—including how your own perspectives and values have played a role in your choice of topic and objectives, methods, design of experiments, analysis and documentation?		
Do you discuss what precautions you will take against procedural bias, including possible bias resulting from your own background, worldviews, and values?		
Do you discuss the possible ethical aspects of your proposed study and how these may have affected your research design?		

Presentation of Results: Note that aside from the checklists below with respect to Appendices and References, the remainder of Checklist 1 is <u>primarily relevant for dissertations and not for proposals</u> since it refers to your results—to work that has been completed!

Question	Needs work	OK as is
Are your results clear and concise? Are they comprehensible: (a) to your peers—in particular to the members of your evaluation committee, (b) to others who may be affected by/interested in your results but who may lack special knowledge or skills?		
Are the concepts, hypotheses and theories that you have developed clearly identified and justified?		
Do you discuss the actual development of your research and the accompanying learning process whereby the concepts, hypotheses and theories that you refer to have evolved (in contrast to giving the impression that everything was done in a completely well-structured and planned manner)?		
Do you make good use of tables, figures etc., including those derived from secondary data sources?		
Have you asked yourself what "story" you want a table, figure etc. to convey and do you try to relate that story to the reader?		
Do you provide a rigorous justification of your results? Do you "validate" your findings—do you demonstrate their credibility/transferability/authenticity/confirmability/reliability, etc.? Do you for example refer to		

(continued)

(continued)

Question	Needs work	OK as is
(a) Theory you may have employed to design your investigation? (b) Evidence provided by others, including the literature you refer to and secondary data sources? (c) Your own empirical results—including salient aspects of tables, graphs or other forms of analysis you present? (d) Re. the social sciences: 'Rich, thick descriptions' resulting from your observations in the field—including the use of representative quotes (as well as dissenting quotes) from interviews? Comments from participants as to the accuracy/credibility of your findings?		
Do you present not just the results but also reflect on their implications and relate these to the research questions/hypotheses that have driven your research?		
Do you also present anomalous, discrepant and unexpected data that appear to run counter to your major hypotheses and results and do you discuss how such data can influence and be interpreted with respect to your findings?		
Do you provide clear, concise discussions of the usefulness of your results, including their possible influence on future work in this area?		

Conclusions

Question	Needs work	OK as is
Do you provide strong, clear statements of your findings—including the concepts, hypotheses, and theories you have developed?		
Do you clearly underline the significance of what you have done—both for the scientific community and for society?		
Do you discuss the implications of your acceptance/rejection of hypotheses for theory development and future research?		
Do you propose areas of further research suggested by your findings?		
Do you, with the benefit of hindsight, draw attention to any limitations of your research and its results?		
Have you written your conclusions in a manner that demonstrates that you have a good command of research methodology?		

Appendices

Question	Needs work	OK as is
Do you place whatever is not essential to the proposal/dissertation in an appendix?		
Primarily for the social sciences: Do you include (in separate appendices): questionnaires, interview guides, coding frames, letters sent to sample members …?		

References

Question	Needs work	OK as is
Do you provide a logical and consistent way of referencing, and a logical and consistent list of all references cited in the proposal/dissertation—and of only those references?		

Other obligations

Question	Needs work	OK as is
Have you supplied a copy or a summary of the dissertation to those you promised a copy? To others you feel indebted to? Primarily for the Social Sciences		
Have you carefully respected the confidentiality of all information that was provided to you under the explicit or implicit assumption of its being kept confidential? If necessary, provide transcripts that are to be kept confidential in an appendix that is only made available to the external examiners upon their request		
Have you respected the explicit or implicit assumption by research participants that they would remain anonymous?		

2. **Checklist with respect to Planning and Writing**

Planning

Question	Needs work	OK as is
Do you develop a provisional table of contents, including annotated titles of chapters/major sections, together with the approximate number of pages for each?		
(And when you are underway with your research project: Do you regularly (say once every month) update your provisional table of contents?)		
Do you develop a provisional time plan—including the time to be spent on major activities such as literature search, pilot		

(continued)

(continued)

Question	Needs work	OK as is
studies/experiments/observations, analyses of data, discussions with peers and advisors, writing, etc.)?		
(And when you are underway with your research project: Do you regularly (say once every month) update your provisional time plan? And do you also keep a rough track of the actual time you have spent on such activities in the past period so that you can compare this to your earlier estimates and be more accurate in your planning?)		
Do you include in your time plan *considerable* "slack" to take into account unexpected events and unforeseen difficulties—as well as the need to re-write and "re-polish" your "gem"?		
Do you keep a "diary" of your activities and progress that you can use to modify current plans as well as to document your research process?		

Writing

Question	Needs work	OK as is
Have you edited and revised drafts so that your writing is clear, concise, consistent, coherent, and credible?		
In particular, do you continually check that you are avoiding "sloppiness"—including grammatical and spelling errors and the indiscriminate use of capital letters; inconsistency in the use of terminology; repeating yourself?		
Do you make use of the facility in your word-processing software for checking spelling and grammar?		
Do you continually ask yourself whether others, especially the external examiners, will readily follow your arguments, presentations etc.? Do you try to look at your writing "through their eyes"—even though you do not know who they are?		
Do you structure the dissertation in an easy-to-follow and systematic way, with a clear and logical structure (chapter and section/sub-section headings, line spacing, typography, footnotes/endnotes, references, quotes, tables, models and schematics, figures, use of 1st or 3rd person, gender or gender neutrality …)?		
Do you start each chapter with a brief introduction to the particular issues that are being examined?		
Do you start each major section with a "topic sentence" that clearly tells the reader what the main message of the section is—and then follow that topic sentence with supporting detail?		
At the conclusion of each chapter and section, do you make clear what your results have shown—and draw out any links that may be made with preceding or following chapters/sections so as to provide continuity?		

(continued)

(continued)

Question	Needs work	OK as is
Do you build 'bridges' between your chapters and major sections such that there is a logical 'flow' in the presentation?		
Do you clearly explain each conclusion that you draw from your research in a manner that a scientifically inclined reader could understand and accept?		
Are you persuasive? Do you provide convincing arguments for why your research is important, and why the findings and results are plausible/reliable and significant?		
Primarily for the social sciences: Have you treated confidential material in a proper manner?		

3. Checklist with respect to your Oral Defence/Viva voce

Structure, Visual Aids & Presentation

Question	Needs work	OK as is
Have you designed your presentation to take into account that a listener's concentration curve results in best memory retention at the start and end of your presentation?		
Does your use of visual aids (e.g. PowerPoint) permit you to gather your thoughts while the reviewer's attention is focused on the screen?		
Do you make sure that your use of slides in fact assists you and does not disturb the flow of your presentation?		
(a) Do you make sure that your slides are not 'packed', that they are easy to read and are not so long that they divert attention from your message?		
(b) Do you avoid cutting and pasting sections of your dissertation which can lead to poor slides and poor presentations?		
(c) Do you make only the minimum number of slides required so that you avoid racing through your presentation and are able to make thoughtful pauses? Rule of thumb: maximum 2 slides/minute; use Arial as font (easier to see from distance)		
(d) Do you avoid putting things on a slide that will not be covered in your presentation?		
(e) Do you minimize the use of upper case letters that are more difficult to read?		
Do you rehearse and pre-test your presentation so that your oral language as well as your 'body language' contribute to the credibility of your presentation and are well-suited to the locality where you will make your presentation—and so that you can have good 'eye-contact' with the examiners and the audience?		

(continued)

(continued)

Question	Needs work	OK as is
Do you avoid looking at the screen instead of focusing on your examiners and audience?		
Do you avoid reading from the screen?		
Do you provide some good 'stories' to liven up your presentation?		
Are you positive and enthusiastic yet calm? And do you speak clearly and firmly?		
Have you prepared yourself for likely questions?		
Have you familiarized yourself with publications that have appeared *after* you submitted your thesis?		
Do you have extra slides ready to deal with questions you can already anticipate?		

4. 4. Checklist of precautions (relevant for both proposals and dissertations)

Question	Needs work	OK as is
Do you avoid the use of "normative" statements (employing terms such as "ought", "should") unless these are backed up by strong evidence?		
Do you avoid subjective statements unless you provide good arguments for them (e.g. when you present your own views as to the importance of your research)?		
Do you avoid the use of words such as "clearly", "obviously", etc. unless your arguments really provide credence to such terms?		
Are you aware of your possible biases and preconceptions?		
Do you avoid speaking of "proof" since in empirical/inductive research one is not able to prove one's hypotheses (only to support them via data and analyses)?		
In case you justify the significance of your proposal/results by referring to a gap in existing knowledge, have you been *very* careful with your review of the literature?		
Do you avoid being too self-critical? Do you avoid being self-laudatory? Do you avoid praising the significance of your proposed investigation/your results?		
Do you avoid providing too many citations and quotes?		
Do you avoid giving the impression of a "linear" process, where everything will proceed (has proceeded in the case of a dissertation) according to your plans?		
Do you wait to finalize your Introduction/Abstract until you are pretty much finished with your writing?		
Do you avoid sloppiness?		
The following precautions are primarily relevant for your dissertation and viva		
Do you avoid presenting too many analyses and results? Do you present only those findings that relate to your questions/hypotheses so that the thread of your argument is not lost?		

(continued)

(continued)

Question	Needs work	OK as is
Do you avoid presenting all your data so as not to burden the reader and to avoid giving the impression that your findings are unstructured? Parsimony is a virtue in science.		
Do you avoid giving the impression that everything has proceeded according to your original proposal? In other words, are you honest and reflective as to your process—including how you have had to modify your ideas, research questions/hypotheses, perspectives and research design as you have gained new insights underway?		
Have you remembered that, in principle, your dissertation serves three major purposes:		
(a) To demonstrate that you have a sufficient command of research methodology and of skill in applying this methodology so that your thesis can be accepted—and that *you* can be accepted as a member of the scientific community (b) To contribute to scientific knowledge in your field (c) To contribute to those affected by your research—including yourself as you develop your competences, the scientific community, and society in general		

Chapter 8
Uncertainty, Probability and Statistics in Research

Having considered the major components of scientific method in the preceding chapter, it is now time to consider one of the important aspects of science that, in general, is more or less ignored or played down in most expositions on scientific method: uncertainty. It is also one of the topics that Ph.D. students tend to demonstrate little interest in—as is often evidenced by their lack of knowledge about or reflection on the standard software routines they rely on for performing the statistical analyses that so much of their conclusions depend on. For these reasons, this chapter is relatively long.

Earlier we touched on this inherent aspect of methodology in a number of different contexts. First of all, in connection with discussions on the relationships between methodology and mathematics, we considered the huge shift that occurred when the reliance of the sciences on mathematics as an axiomatic-deductive system was complemented by the notion of uncertainty. From the time of the ancient Greek philosophers, mathematics was essentially considered as consisting of proofs/deductions based on unproved statements/axioms and science was considered as dealing with causes, not with chance. Then, with the development of probability theory from the mid-17th century, science and chance gradually became reconciled, and probability theory spread from gambling problems to almost all domains of scientific inquiry, developing as its applications developed. In some areas such as physics, it is virtually impossible today to develop theories and models where uncertainty does not play a major role, either via statistical computations or via inherent uncertainties that are represented in terms of probabilities. Although classical systems in physics are in principle deterministic, practical considerations when facing many degrees of freedom may compel the researcher to introduce probability distributions, averages over the distributions, measures of deviations, and so on. And at the quantum level, because of the indeterminacy of dynamical quantities like position and momentum, probability distributions of such variables are inherent to the whole field of study; the laws of physics are statistical, not deterministic. Similarly in biology, it is not possible to discuss evolution, at least within the broadly accepted framework of Darwinism, without including concepts of random change via mutations.

Arguments similar to those used above regarding uncertainty in classical systems also apply to more daily phenomena. For example, (Hawking and Mlodinow 2010;

© Springer International Publishing Switzerland 2016
P. Pruzan, *Research Methodology*,
DOI 10.1007/978-3-319-27167-5_8

73–74) consider the case of a person who throws a dart and aims at the bull's-eye:
"… any given dart can land anywhere, and over time a pattern of holes that reflect
the underlying probabilities will emerge. In everyday life we might reflect that
situation by saying that a dart has a certain probability of landing in various spots;
but if we say that, it is only because our knowledge is incomplete. We could
improve our description if we knew exactly the manner in which the player released
the dart, its angle, spin, velocity, and so forth. In principle then, we could predict
where the dart will land with a precision as great as we desire. Our use of probabilistic terms to describe the outcome of events in everyday life is therefore a
reflection not of the intrinsic nature of the process but only of our ignorance of
certain aspects of it." Such arguments are similar to those considered earlier in
Sect. 2.6.2 of Chap. 2, where we considered the concepts of causality and
determinism.

Another area where we touched on uncertainty was when we considered the
concept of external validity in Chap. 6 dealing with experimentation. We briefly
presented there the traditional "frequentist" perspective on testing of hypotheses, a
perspective that will be contrasted to that of so-called Bayesian analysis later on in
this chapter. In Chap. 6 we also briefly introduced probability in connection with
discussions of experimental design; the concept of randomization was introduced as
a means of avoiding bias in experiments and of indirectly taking extraneous variables into account so that samples can be considered to be equivalent.

There is a complex and intimate relationship between observing, testing, probabilistic reasoning, statistical analyses, and the interpretation of experimental results
in the natural sciences. This is the focus of the present chapter: the relationship
between research methodology and uncertainty.

8.1 Probability

Let us commence by considering the nature and role of *probability*; the theoretical
development here will lead up to the more empirical perspectives on inference
offered by *statistics*. In so doing I will draw upon several classic texts on probability
theory and statistical inference: *Theory of Probability* (Jeffreys 1983; 1st edn.
1939), *The Theory of Probability* (Reichenbach 1949), *An Introduction to
Probability Theory and Its Applications* (Feller 1968, 1st edn. 1950), and *The
Foundations of Scientific Inference* (Salmon 1967). In addition, I will structure
much of the presentation based on (Gauch 2003; Chaps. 6 and 7); the reader is
referred to that splendid book for a clear and more inclusive treatment of a number
of the topics presented here.

Our discussions so far underscore that learning from experience is fundamental
to scientific progress—as it is in our everyday life. And such learning is an exercise
in uncertainty and probabilistic reasoning. Empirical science, based on induction
and generalization, cannot be expected to deliver infallible statements; scientific
investigations are characterized by more or less probable conclusions.

Although common sense is a necessary guide in structuring our thoughts, it often is far from adequate when faced with the need for concrete analyses and decisions in scientific investigation that are characterized by uncertainty. The same applies to some of our everyday activities as well; we continually face practical situations where decisions must be made on the basis of imperfect information on phenomena and relationships, and where it is often not at all clear how best to combine this information with our goals, which are also often unclear, in order to somehow arrive at optimal, or at least good or satisfactory solutions.

Probability theory provides us with such a well-structured and self-consistent basis for dealing with uncertainty in scientific investigation—for drawing inferences from empirical (observational or experimental) data. According to (Feller 1968; 1), "Probability is a mathematical discipline with aims akin to those, for example, of geometry or analytical mechanics. In each field we must carefully distinguish three aspects of the theory: (a) the formal logic, (b) the intuitive background, (c) the applications. The character, the charm, of the whole structure cannot be appreciated without considering all three aspects in their proper relation." The presentation here will unfold by calling upon these three aspects, although not in a sequential manner.

Recalling our earlier emphases on the roles of deduction and induction in scientific reasoning, we will see that probability can be considered to provide a deductive approach while statistics represents an inductive approach to uncertainty. This characterization of probability as being deductive could easily lead us to anticipate that there is complete agreement amongst scientists as to probability theory. This is not the case. As we will see in the sequel, there are two major approaches or schools of thought that continue to challenge and inspire each other: Frequentist and Bayesian. The presentation of probability theory here will to a major extent be structured around contrasting these two perspectives.

Instead of proceeding directly to a theoretical treatment of probabilistic reasoning, inspired by Gauch (2003; 211–215), I have chosen to begin by presenting two examples that can illustrate how commonplace it is to make use of *incorrect* reasoning when dealing with probabilities—and therefore how important it is for reflective scientists to have a firm basis for probabilistic reasoning when analysing and drawing inferences from their empirical data. I will then follow the logic of Feller (1968; 2, 3) and provide reflections starting with simple experiences such as throwing dice or tossing coins where the statements have a rather obvious, intuitive meaning and do not call for definitions and the like. Then, based on the intuitive results and accompanied by a few illustrative examples, a more abstract, theoretical model will be developed that generalizes the results.

Therefore, although the first example make use of concepts ("prior" and "conditional" probabilities) that will be more formally introduced later, it is felt that the pedagogical advantage of starting with a challenging example more than outweighs the minor difficulties that might result from a introducing a few concepts that have not been clearly defined. The example is a modified version of (Stirzaker 1994; 25). It demonstrates a typical error in reasoning that not just ordinary people, but also many scientists and students of science make, and underlines how easy it is to make highly erroneous judgements as to the results of tests. Assume that a blood test can be made for a relatively rare and life-threatening disease which, according to considerable and highly reliable data, occurs by chance in 1 in every 10,000 people. The test appears to be very reliable; if you have the disease it will correctly say so with probability 0.99; if you do not have the disease, the test will wrongly say that you do have it with probability 0.005.

Before reading further, answer the following questions:

1. If the test says you have the disease, what is the probability that this is a correct diagnosis? If you are not able to provide a numerical estimate, at least answer whether the probability is "very high", "high", "50/50", "low" or "very low".
2. Assume that you are the scientific member of a group of experts that is to recommend whether a major hospital should routinely use the test. The test is fairly expensive to apply, and medical treatment of a person who is diagnosed as having the disease is costly, painful and has a number of undesirable side effects. Would you recommend that the test be implemented?

Was your answer to the first question a high or very high probability? Most people, including many of my former students of research methodology, answer that the probability of a correct diagnosis is very high, roughly 99 %. In fact, the correct answer (and we will demonstrate this once we start considering the so-called Bayes' theorem) is only about 0.2 %, which most of us would tend to call "very low"!

This answer paves the way for the answer to a second question. The test, while appearing to be very good, is in fact rather useless; for every real instance of the disease that the test detects, there would be about 50 false positives (if the test was used on the general public, one person out of 10,000 would, with probability 0.99, be correctly shown to have the disease, while for every such correct diagnosis $0.005 \times 10,000 = 50$ people would be wrongly shown to have the disease). Since the use of the test at the hospital would lead to considerable costs, fear amongst patients, and uncalled for and potentially damaging treatment, there are very strong arguments for rejecting a recommendation that the test be used to provide a diagnosis.

Generalizing a bit, you have all had many tests in your careers so far. Now imagine a test that indicated that roughly fifty students were qualified to follow a course of study (or to be a pilot or a doctor, or a …) while in fact only one had the qualifications; such a test would be considered a scandal and hit the headlines of the papers and be on the TV news. What a huge, unforgivable error! What huge costs! What potential damage (who wants to be operated on by an unqualified doctor? who wants to fly in an airplane with an unqualified pilot?)!

So what is wrong with most people's evaluation here? What underlies this huge difference in judgements? What is it that makes the test rather worthless as a diagnostic tool, and therefore, given its costs and its negative impacts on those taking the test, very ill-suited for practical, routine use in hospitals? The answer lies in ignoring the rarity of the illness. The prior probability that one has the disease, i.e. the odds that someone chosen at random from the population has the disease, is

$1:10^4$, while knowledge of a positive test (a test that indicates the person tested has the disease) gives likelihood odds of 99:1. Combining these two, the prior probability and the likelihood, leads to the extremely low probability of a person actually having the disease, even though the apparently very reliable test indicates that the person has the disease. The mistake in reasoning is to base conclusions on this likelihood (if you have the disease, the test will correctly indicate this with a probability of 0.99) while ignoring the information on the rarity of the disease, i.e. the prior probability of one in ten thousand that you in fact have the disease. It is disturbing that such a blunder is so common. We will return to this later on and formalize the analysis.

A second example is based on (Stewart 1996; 172) and considers a simple question dealing with the genders of children in a family. As with the example above, it also demonstrates some faults in the reasoning of many people, including my former M.Phil. and Ph.D. students, when drawing conclusions from data. Let us start by making some simplifying assumptions. We assume that the probability of a new-born baby being a boy or a girl is equal (½) and that the gender of each child in a family is an independent factor, i.e. that if say the first child is a boy, then the probability that the next child born will be a boy/girl is ½ in each case. These assumptions do not completely correspond to available evidence. For example, they ignore the possibility of identical twins that always have the same gender. In addition, large scale statistical analyses do not support the assumption of independence. Nevertheless, the assumptions are reasonable and serve our purposes here; minor changes in the numbers would not lead to any changes in the conclusions we draw. The question is: If you are told that Mrs. and Mr. Gupta have two children and that one of them is a girl, what is the probability that the other child is also a girl? Answer this question before continuing.

A very common response is that the probability that the other child is a boy or a girl is identical, ½. This is not correct. There are four gender possibilities with two children: BB, BG, GB, and GG, where B and G denote boy and girl, and where the order of the letters represents the order of birth. Since we know that the Guptas have at least one girl, we can eliminate BB. This leaves us with three equally likely cases, and only in the last combination GG is the other child also a girl, so the answer to our question is 1/3; i.e., there are twice as many chances that the other child is a boy than a girl! As we shall see, this simple reasoning can easily be developed formally using probability theory and the concept of conditional probability.

8.1.1 Probability Concepts

Before providing a definition, it should be clear that the word "probability" is an everyday, common sense word that, depending on the context, might also be replaced by other words such as "chance", "likelihood" or "belief"—as an examination of dictionary definitions of the term clearly demonstrates. For example, if someone told me "It is probable/likely that it will rain tomorrow", I would understand the statement as a rough expression of that person's belief or confidence as to the event of "rain tomorrow"—without demanding any qualifying remarks as to how she arrived at this conclusion, how heavy the rain might be, when it would occur, exactly where it would occur, with what odds it would occur, etc. If on the other hand, the person had said "there is at most a 20 % chance of rain tomorrow", I would interpret this to mean that she had qualified her belief based either on her own personal experience and/or access to external evidence, such as a weather forecast by the meteorological society. And if she then told me that the TV news has just reported about a rapidly approaching storm and that this news has caused her to significantly modify her assertion as to the likelihood of rain, I would also accept this new statement; it is typical of our learning experiences that we update our beliefs as to the likelihood of an event occurring based upon new evidence.

So we meet common sense statements about probability in many contexts—and even children can with confidence speak about the probability of being invited to a classmate's birthday party or about whether the odds are in favour of team X winning over team Y in a highly publicized soccer match.

To illustrate the earlier reflection that probability can be primarily considered to be a deductive tool, while statistics represents an inductive approach to uncertainty, consider the following examples of the use of the word "probability":

The first illustrative example is based on tosses of a die. Even though there may be minute differences in the probabilities of a toss leading to a "1", a "2"....a "6" due to the different number of spots formed by scooping material out on the surfaces of the die, for practical purposes we can assume there is an equal chance of 1/6 that a toss of the die will result in each of the numbers from one to six (Radin 1997; 136–137). We can then deduce that the chance of two successive tosses both being a '6' is 1/36 and that the chance of two successive tosses adding to 10 is 1/12; a 10 can be a result of the following three pairs of tosses: 5,5 4,6 6,4 where the probability of each such pair is $(1/6) \times (1/6) = (1/36)$. So these examples are of theoretical (deductive) rather than empirical, character. It also illustrates that we often express probabilities in terms of odds (fractions).

A second example, this time of empirical (inductive) rather than theoretical (deductive) character, is inspired by the website of the UK's National Health Service regarding chlamydia, one of the most common sexually transmitted infections in the UK http://www.nhs.uk/Conditions/Chlamydia/Pages/Treatment. aspx. Assume that among 10,000 chlamydia patients in a clinical trial of antibiotic X, 9992 (99.92 %) recovered, whereas among 10,000 other patients treated with antibiotic Y, only 8382 (83.82 %) recovered. It would then be reasonable to hypothesize that new patients will have at least a 10 % higher probability of recovery if treated with antibiotic X than with antibiotic Y. Here based on data as to outcomes (two recovery rates), we derive a conclusion in the form of a probability judgement—which is the opposite "direction" of the first example where, based on given probabilities, we deduced the odds of an outcome.

In other words, while the first example was representative of deductive thinking —we began with a theory/assumption/model as to the probability of an occurrence of an event (a toss resulting in a "six") and derived a conclusion regarding the

Table 8.1 Expressions of probability

Probability expressed as				
Event	Fraction	Odds	Percentage	Decimal
A tossed die gives a "6"	1/6	1:6	16.666	0.1666
The sun will rise tomorrow	1	1:1	100	1.0
I will live 1000 years	0	1:infinity	0	0.0

expected outcome of an event (two "six's"), the second example illustrated inductive thinking that progresses in the opposite direction—we started with observations as to the results of a test and inferred a probability judgement as to the relative effectiveness of two drugs.

The examples of probability statements above indicate that the probability (or chance or odds or likelihood or propensity...) of an event occurring can be meaningfully expressed as a number—as a fraction, a decimal, odds, or as a percent; see Table 8.1. I note however that probability theory does not require any assumptions about the actual values of the probabilities in a specific situation or on how they are measured in practice, just as a theory of geometry does not deal with specific measures of distance or position. So just as carpenters, surveyors, pilots and astronomers use differing procedures when making measurements of distance, so will a scientist, depending on his or her field, use different theoretical reasoning and techniques of observation to develop specific probabilities as to e.g. the life time of a radioactive particle or of a parasite, or the effectiveness of a drug.

The following reflections on Table 8.1 can hopefully contribute to our intuitive and practical understanding of the theory of probability. The probability that "The sun will rise tomorrow" is expressed here as certainty, but that is subject to some, albeit minimal, doubt. In spite of the fact that our experience and the experience of others as far back as history has been recorded as well as our knowledge of astronomy and of the laws of dynamics regarding the movement of heavenly bodies make it virtually certain that "the sun will rise tomorrow", there is no proof of this; proof, in a logical sense, does not exist with respect to induction. In spite of all the experience that indicates that the physical universe as we know it is ordered and follows scientific laws that indicate that the sun will rise tomorrow, we cannot deduce that this will in fact be the case—and we cannot call on induction to justify induction. No matter how many observations we have that the sun rises each day, it is only logically valid to conclude that the sun will rise tomorrow *if* induction works. And whether induction works can only be concluded inductively, and we meet an infinite regress.

A similar comment can be made with respect to the statement: "I will live 1000 years". Is there a maximal age beyond which life is impossible, or is any age conceivable? According to (Feller 1968; 7), "We hesitate to admit that man can grow 1000 years old, and yet current actuarial practice admits no bounds to the possible duration of life. According to formulas on which modern mortality tables are based, the proportion of men surviving 1000 years is of the order of magnitude of one in

$(10^{10})^{36}$—a number with 10^{27} billions of zeroes. This statement does not make sense from a biological or a sociological point of view, but considered exclusively from a statistical standpoint it certainly does not contradict any experience ... such extremely small probabilities are compatible with our notion of impossibility. Their use may appear utterly absurd, but it does no harm and is convenient in simplifying many formulas. Moreover, if we were to seriously discard the possibility of living 1000 years, we should have to accept the existence of maximum age, and the assumption that it should be possible to live x years and impossible to live x years and 2 seconds is as unappealing as the idea of unlimited life."

In his clear and concise treatment of probability concepts (Gauch 2003; 192–196) underlines that any informal or common sense definition of probability leaves a number of important questions unanswered, including: Does a concept of probability concern events or beliefs? Does it pertain to repeatable or singular events? Is probability a function of one argument (such as an event) or of two arguments (such as an event and some evidence)?[1]

The illustrative examples above underscore the need for greater precision in our use of probabilistic terminology and reasoning so that it is both meaningful in ordinary language and yet permits the logic and consistency that characterizes mathematical reasoning. This is particularly important in research since the theory of probability plays a central role within the larger setting of learning and discovering—as we will see when we consider the nature and role of statistics in research methodology. So if we are to develop a useful theory of probability we must build upon our common sense notions of probability, but also develop definitions and concepts that enable scientific reasoning about inductive logic and learning.

According to the classic text (Jeffreys 1983; ix), "… the ordinary common sense notion of probability is capable of precise and consistent treatment once an adequate language is provided for it." We will shortly establish such a meaningful interface between the common sense understandings of the term "probability" and a theory of probability that is applicable by mathematicians (probability theory is an established branch of mathematics) and, indirectly, by scientists engaged in drawing inferences from observational data. I note in this connection that we have several times earlier reflected on the fact that since the natural sciences aim at establishing statements (facts, hypotheses, theories and laws) with a higher degree of precision than our ordinary common sense and language enable us to do, as students and practitioners of science we have an obligation to dig a bit deeper and consider the meaning of the fundamental concepts we use in science. This search for meaning

---

[1]I note regarding the last question that (Jeffreys 1983; 15) argues that human belief regarding uncertain phenomena *is* a function of two arguments and that a coherent theory of probability must account for this: "It is a fact that our degrees of confidence in a proposition habitually change when we make new observations or new evidence is communicated to us .... We must therefore be able to express it. Our fundamental idea will not be simply the probability of a proposition p, but the probability of p on data q." As we will see shortly, this is the fundamental starting point for a so-called Bayseian approach to probability theory.

here is not simply a matter of linguistics; according to Salmon in his text on the foundations of scientific inference, "We are not primarily concerned with the ways a language user, whether the man in the street or the scientist, uses the English word 'probability' and its cognates. We are concerned with the logical structure of science and we need to provide concepts to fulfil various functions within that structure. If the word 'probability' and its synonyms had never occurred in a language, we would still need the concept for the purpose of logical analysis ... if it did not already exist, we would have to invent it." (Salmon 1967; 63)

The presentation of probability theory here will be brief; only those aspects of theory will be included that are necessary for a reasonable understanding of the field as a whole, including the ensuing presentation of inductive reasoning via statistics. In line with the overall theme of the book, the emphasis will be on reflection regarding the development of the theory.

We will follow (Jeffreys 1983) and (Gauch 2003) and progress from consideration of fundamental requirements for a theory of probability that is to serve inductive reasoning, to some general rules, and then to the choice of axioms, that provide the basis for the development of theorems—and thereby of a probability theory that can serve both common sense and scientific applications.

Following this methodological development, the remainder of the chapter will focus on the application of probability theory to inference via statistical/inductive reasoning.

8.1.2 Fundamental Requirements

A theory of probability that can serve as the basis for inductive reasoning must live up to a number of very general rules. Fundamental here are two a priori requirements if it is to contribute to the larger setting of inference and learning via inductive reasoning (Jeffreys 1983; 7).

Generality: An adequate theory of probability "must provide a general method ... if the rules are not general, we shall have different standards of validity in different subjects, or different standards for one's own hypotheses and somebody else's."

Unbiased: Since such a theory is to contribute to learning based on observation, an adequate theory must be unbiased or impartial. It would be counterproductive if it contained presuppositions about the world: "If the rules themselves say anything about the world, they will make empirical statements independently of observational evidence, and thereby limit the scope of what we can find out by observation. If there are such limits, they must be inferred from observation; we must not assert them in advance." According to (Gauch 2003; 198), it is remarkable that these two a priori requirements are "all that need be said to ground a good theory of probability and to exclude bad theories."

General rules:

Prior to the selection of the axioms that serve as the basis for developing the-orems of a theory of probability, Jeffreys (pp. 8–16) presents a series of general rules whose requirements enable us to judge the adequacy of such a theory. The rules are essentially as follows:

1. *Explicitness*: All axioms must be explicitly stated and theorems/theory that build on them must follow from them.
2. *Self-consistency*: A theory of probability must be self-consistent so that it is not possible to derive contradictory conclusions from its axioms and observational data.
3. *Applicable in practice*: The rules for induction must be practical, that is, applicable in practice. For example, estimation of a quantity must not involve an impossible experiment.
4. *Revisable*: "The theory must provide explicitly for the possibility that inferences made by it may turn out to be wrong. … It is a fact that revision of scientific laws has often been found necessary in order to take account of new information … But we do accept inductive inference in some sense; we have a certain amount of confidence that it will be right in any particular case, though this confidence does not amount to logical certainty." (Ibid.; 8–9)
5. *Open, empirically based*: Conclusions that are scientifically based must not deny any empirical proposition a priori. In other words, given relevant evidence, any empirical proposition must be formally capable of being accepted.

These five rules are essential for an adequate theory of probability and induction. In addition, the following three rules serve as useful guides:

6. *Parsimonious*: The number of postulates/axioms should be small, not only for aesthetic reasons but primarily because this will "minimize the number of acts of apparently arbitrary choice." (Ibid.; 9)
7. *Agree with human reason*: An adequate theory of probability and induction must agree with human thought processes. Observation informs us that human behaviour often is a better reflection of inductive processes than rational arguments.
8. *Modest*: Since induction is more general and complex than deduction, which only permits three judgements regarding a proposition: proof, disproof, or ignorance, it is not reasonable to aspire to develop a theory of induction more thoroughly than theories of deduction. In addition, we must be modest and accept that a theory of probability has its limitations and cannot be expected to be perfect—just as we must accept that the data provided by experiments are imperfect.

8.1.3 Probability Axioms

In addition to the two fundamental requirements and the five general rules presented above, all that is needed to develop the most fundamental aspects of a formal theory of probability and induction that is useful for the natural sciences is a series of mathematically coherent and physically sensible axioms. Axioms are presuppositions—they are unproved and cannot be proved—and since the theorems and theory that follow are founded on such axioms, it is important to reflect on the acceptability of the axioms. In other words, if a theory of probability is to serve all ordinary and scientific applications in the same way that a single unified theory of arithmetic does, its axioms must serve as an acceptable and adequate foundation for the theoretical development that follows. According to (Reichenbach 1949; 53) , "… the axiomatic system of the probability calculus assumes a function comparable to that of the axiomatic system of geometry, which, in a similar way, determines implicitly the properties of the basic concepts of geometry, that is, of the concepts 'point', 'line', plane' and so on." I note here that these basic concepts themselves are not defined but once the axioms of Euclidean geometry specify relations among them (e.g. two points determine a straight line) and properties are assumed (e.g. a point on a plane is two real numbers), then geometry becomes interpretable and meaningful and can be applied to real-world problems. The same applies to the probability theory that is briefly and informally developed below; here, probabilities are numbers of the same nature as distances in geometry. Hopefully you will experience that as the exposition of the formal theory develops you will gain increased insights into how it can be applied in practice.

I assume in the remainder of this chapter that the reader is familiar with the most common operators in logic such as: $\sim$ (not), $\wedge$ (and), $\vee$ (or) as well as with elementary notation from set theory such as intersection, union and inclusion.[2]

Letting | be the symbol for "conditional on" or "given", the conditional probability $P(Y|X)$ is defined as the probability of Y given X and is determined as the probability $P(Y \cap X)$ divided by the probability $P(X)$, provided $P(X) \neq 0$: $P(Y|X) = P(Y \cap X)/P(X)$.

For example, suppose a set contains 100 elements (for example balls) of which 38 have property X (for example they have a white dot on them), 24 have property Y (for example they have a black dot on them, 7 have both properties (that is,

---

[2]The members of a set are typically listed in brackets, for example A = {x, y, z} is the set composed of the members or elements x, y and z. A $\cup$ B designates the *union* of A and B, the set containing all members of A and all members of B, and A $\cap$ B designates the *intersection* of A and B, the set containing elements that belong to both A and B and nothing else. For example, if A = {v, x, y, z} and B = {q, r, x, z} then A $\cup$ B = {q, r, v, x, y, z} and A $\cap$ B = {x, z}. A $\subseteq$ B designates *inclusion*—that A is a *subset* of B (every member of A is a member of B); for example, the set A = {x} is a subset of B = {x, y}. $\sim$ A designates the *complement* of A, the set containing everything not belonging to A. Two sets are *mutually exclusive* if they have no members in common; the sets {x, y} and {q, w, z} are mutually exclusive.

Y ∩ X contains 7 elements that have both a white and a black dot), and the remaining 100 − (38 + 24 − 7) = 45 elements have neither property X nor Y. Then the unconditional probability that an element has property X is 38 out of 100; P (X) = 0.38. Similarly P(Y) = 0.24 and P(Y ∩ X) = 0.07. It follows that the conditional probability that an element has property Y given that it has property X is 0.18: P(Y|X) = P(Y ∩ X)/P(X) = 0.07/0.38 = 0.184, which rounds off to 0.18. We can also see that the probability that an element has either property X or property Y but not both (that a ball has either a white dot or a black dot, but not both a white and a black dot) is 0.55: P(X) + P(Y) − P(Y ∩ X) = (38 + 24 − 7)/100 = 0.55.

We can use the concept of conditional probabilities to provide a more formal argument when answering the question at the beginning of this chapter regarding the gender of the Gupta's children. You were told that they have two children and that one of them is a girl, and you were asked to determine the probability that also the other child is a girl. Let Q denote that at least one of two children is a girl and R that the other child also is a girl. Earlier, we informally argued that only the three possibilities GG, BG and GB are possible, given that at least one of the children is a girl and that since these three are equally likely, the probability that both children are girls is 1/3—and not ½ as many of my former students stated. But we can now obtain this result from our definition of conditional probabilities. First of all, the unconditional probability that at least one of two children is a girl P(Q) = P(GG, BG, GB) = ¾ since without the condition that one of the children is a girl, the four possible and equally likely gender sequences are BB, BG, GB, GG. We also obtain P(Q ∩ R) = P({GG, BG, GB} ∩ {GG}) = P({GG}) = ¼. Thus, from the definition of conditional probability we obtain: P(R|Q) = P(Q ∩ R)/P(Q) = (¼)/(¾) = 1/3. Although there is little difference between an informal, common sense argument and this more formal treatment based on the concept of conditional probability, in more complex situations the formal treatment is required to help us avoid blunders such as were evident in the examples that introduced this chapter.

To pave the way for the derivation of probability theorems, consider the following four axioms. Although our primary reference here, (Jeffreys 1983; 15–26), provides similar axioms, his presentation is more complex than required for our purposes. Instead the presentation here is based on (Salmon 1967; 59–61) and (Gauch 2003; 202–207). Salmon notes that the axioms "have been chosen for intuitive clarity and ease of derivation of theorems rather than for mathematical elegance" and Gauch emphasizes that "… the axioms are necessary for any probability statement to have any meaning whatsoever."[3]

Axiom 1.   P(Y|X) is a single-valued number such that $0 \leq P(Y|X) \leq 1$, that is, P(Y| X) ∈ [0, 1]

Axiom 2.   If X is a subset of Y, then P(Y|X) = 1

---

[3]Although for sake of generality the axioms presented are based on conditional probabilities, they also cover simple unconditional probabilities as well if the conditioning event X is taken to be a tautology (a universal truth, such as Z ∨ ∼Z) and then removed.

Axiom 3. If Y and Z are mutually exclusive, i.e. $P(Y \cap Z) = 0$, then $P(Y \cup Z | X) = P(Y|X) + P(Z|X)$

Axiom 4. $P(Y \cap Z|X) = P(Y|X) \times P(Z|X \cap Y)$

Axiom 1 states that a probability is a unique real number in the closed interval zero to one. For example, in a setting where a probability is understood to express the relative frequency of an event occurring, the endpoint 0 represents "never", the endpoint 1 represents "always" and the midpoint 0.5 represents that it is equally likely that the event will occur or not occur. We considered representations corresponding to "never" and "always" in Table 8.1.

Axiom 2 states that the probability of a set X being a member of set Y is 1 if X is a subset of Y.

Axiom 3 tells us, for example, that the probability of drawing a red card (a diamond or a heart) from a standard deck of playing cards equals the probability of drawing a diamond plus the probability of drawing a heart.

Axiom 4 enables us, for example, to determine the probability of drawing two red cards in succession from a standard deck of playing cards (containing 26 red cards and 26 black cards). We assume here that all cards are equally likely to be drawn and that the second draw is made without replacing the first card drawn. Y here is the event that the first draw is red, Z is the event that the second draw is red given that the first draw was red and was not replaced, and X is the original starting conditions with 26 red cards and 26 black cards. The probability of getting a red on the first draw is 26/52 and the probability of getting a red card on the second draw, given that the first draw was red and the card was not replaced is 25/51. Thus P(red on first draw and red on second draw without replacement|26 red and 26 black from start) = P(draw red|26 red and 26 black) × P(draw red|originally 26 red and 26 black and the first draw was red) = 26/52 × 25/51 = 25/102 = 0.245.

These axioms are like some predefined rules of the game; they serve to assure consistency among probability assignments and enable us to derive certain probabilities given others, but do not help us in making probability assignments. This is the job of probability rules. For example, according to (Gauch 2003; 203), the most basic rule is that if n X's have been examined and m of these were Y's, then the probability of the next X examined being a B is m/n. If this rule is to provide a reliable estimate of the probability, then the sample chosen must be representative of the whole population (it is unbiased; the selection was done randomly) and the whole population must be large (far greater say than n). The rule will be formalized in our forthcoming discussions of the so-called frequentist approach to statistics where the probability of the occurrence of an event is defined as the limit of the relative frequency of the occurrence of the event in an infinite sequence of trials.

Although a rule for assigning probabilities may appear logical and straight forward, in a specific situation it may lead to results that are not acceptable. Consider for example the following assignment rule, which is similar to the rule just presented above: "If there are n possible alternatives, where for m of these p is true, then the probability of p is m/n." To illustrate the problem with this rule, consider the following setup: There are two boxes, one contains one white and one black ball, and the other box contains one white ball and two black balls. A box is to be selected at random and then a ball is to be selected at random from the box. According to the rule just provided, the probability that the ball will be white is 2/5 (there are 2 white balls out of a total of 5)—while most of us would arrive at 5/12; the probability of drawing each of the two boxes is 1/2, the probability of drawing a white ball from the first box is 1/2 and it is 2/3 from the second box, leading to the calculation $(1/2)(1/2) + (1/2)(1/3) = 5/12$ (Jeffreys 1983; 371). The lesson here is that probability assignments cannot in general be left to others or to some standard computer software. This is similar to what we remarked at the very beginning of this chapter: common sense may lead us to probability estimates that are incorrect.

8.1.4   Probability Theorems

We are now able to deductively develop probability theorems from the axioms. The theorems express probability relationships that are coherent with, i.e. logically build upon, the axioms and that can address probability contexts that are not covered explicitly by the axioms themselves. To illustrate this, consider the following: Assume that extensive analyses indicate that the probability that a specific type of computer hard disk will 'crash' within 1000 h of use is 0.0123. What is the probability that such a hard disk will *not* fail within 1000 h of use? This simple question cannot be answered directly from the axioms, but, as we will see, follows immediately from Theorem 1 below.

The presentation of the theorems is based on (Salmon 1967) and (Gauch 2003). Five theorems are briefly and informally presented in order to convey the underlying ideas of the development of a theory of probability. In the same spirit, only an informal proof of the first three of these theorems is presented.

Theorem 1 $P(\sim Y|X) = 1 - P(Y|X)$.
  By Axiom 2, $P(Y \cup \sim Y|X) = 1$ since every member of X is a member of either Y or its complement, $\sim Y$, i.e. $X \subseteq (Y \cup \sim Y)$. Then, since Y and $\sim Y$ are mutually exclusive, it follows from Axiom 3 that $P(Y \cup \sim Y|X) = P(Y|X) + P(\sim Y|X) = 1$. Subtracting $P(Y|X)$ from both sides obtains Theorem 1. (Gauch 2003; 204), (Salmon 1967; 60)

We can now apply this theorem to the simple question problem posed above, where Y represents the failure of the hard disk within 1000 h of use, $\sim Y$ represents non-failure, and X represents the given probability of failure, 0.0123. We obtain: P(no failure in the first 1000 h of use |probability of failure in the first 1000 h of use is 0.123) = 1 − P(non-failure in the first 1000 h of use probability of non-failure in the first 1000 h of use is 0.123) = 1 − 0.123 = 0.877, which is obvious.

Theorem 2 If $P(X|F) \geq P(Y|F)$ and $P(Y|F) \geq P(Z|F)$, then $P(X|F) \geq P(Z|F)$.
  This follows from Axiom 1 whereby probabilities are real numbers and from the transitivity property of the real numbers. (Gauch 2003; 204; Salmon 1967; 60)

Theorem 3 $P(X|F) = (P(Y|F) \times P(X|F \cap Y)) + (P(\sim Y|F) \times P(X|F \cap \sim Y))$.

Since $X \equiv (X \cap Y) \cup (X \cap \sim Y)$ it follows that $P(X|F) = P((X \cap Y) \cup (X \cap \sim Y)|F)$. Now since Y and $\sim Y$ are mutually exclusive so are $X \cap Y$ and $X \cap \sim Y$. Therefore, from Axiom 3, $P((X \cap Y) \cup (X \cap \sim Y)|F) = P(X \cap Y|F) + P(X \cap \sim Y)|F)$. And from Axiom 4 we obtain $P(X \cap Y|F) = P(Y|F) \times P(X|F \cap Y)$ and likewise $P(X \cap \sim Y)|F) = P(\sim Y|F) \times P(X|F \cap \sim Y)$. Adding these last two results in Theorem 3. (Gauch 2003; 204–205; Salmon 1967; 60)*

Theorem 4 If X is a subset of Y, then $P(X|Z) \le P(Y|Z)$.

As in the case of Theorem 2, this follows from the transitivity property of the real numbers.

Theorem 5 $P(X \cup Y|Z) = P(X|Z) + P(Y|Z) - P(X \cap Y|Z)$.

A justification is not provided. Note that this theorem generalizes Axiom 3, since it gives the probability of $X \cup Y$ given Z regardless of whether X and Y are mutually exclusive; when they are mutually exclusive, as was assumed in Axiom 3, $P(X \cap Y|Z)$ is 0.

It is to be underlined here that the four axioms and the five theorems presented above by no means cover the field of probability theory, about which there are written a large number of excellent books, several of which are referred to at the beginning of Sect. 8.1. The main idea here has been to convey the flavour of the axioms and theorems that constitute a formal theory of probability and to underline that they form the basis for shared perspectives on rational thinking involving uncertainty—perspectives that will be operationalized in Sect. 8.2 on inductive logic and statistics that follows the presentation of Bayes' theorem below.

8.1.5   Bayes' Theorem

In order to pave the way for methodological reflections on statistics it is necessary to introduce yet another theoretical perspective that, particularly in recent years, has obtained growing attention amongst researchers in both the natural and the social sciences—Bayes' theorem. The theorem was developed in an essay on chance (Bayes 1763) that was sent to the Royal Society of London, two years after Bayes' death, by a colleague who found the essay in Bayes' papers. It is considered by many to be the chief rule involved in the process of learning from experience as to how to decide between hypotheses.

Bayes' theorem is: $P(H|D) = P(D|H) \times P(H)/P(D)$

The theorem can be derived directly from the foregoing four axioms. For sake of clarity however, the informal derivation below follows from the definition of conditional probability.

Informal Derivation:

Although the derivation holds no matter how we interpret D and H, it is easier to follow the derivation if one considers "H" to stand for a hypothesis that is being considered, and "D" to stand for the data or the evidence available. Then Bayes' theorem can be interpreted as solving the inductive problem of how to calculate P (H|D), the probability of a hypothesis being true given some data, from the probability P(D|H), the probability that the data are observed given the hypothesis, and P (D), the probability that the evidence is observed, i.e. when no assumption is made as to the truth of H. We can already note here that this last probability, P(D), is unconditional, it is the same no matter which hypothesis is being considered and therefore does not affect the relative probabilities of different hypotheses.

Based on the definition of conditional probability,

$$P(H|D) = P(H{\cap}D)/P(D) \Rightarrow P(H{\cap}D) = P(H|D) \times P(D).$$

Similarly

$$P(D|H) = P(D{\cap}H)/P(H) \Rightarrow P(D{\cap}H) = P(D|H) \times P(H).$$

Since the sets $H \cap D$ and $D \cap H$ are identical, $P(H \cap D) = P(D \cap H)$. Therefore, the two formulas may be equated, resulting in

$$P(H|D) \times P(D) = P(D|H) \times P(H)$$

and thus we obtain:

$$P(H|D) = P(D|H) \times P(H)/P(D).$$

In the sequel we will ignore P(D), which we can consider to be a normalizing constant; as mentioned above it does not affect the relative probabilities of different hypotheses. In other words, Bayes' theorem shows that the conditional probability that a hypothesis is true given the data/evidence, the so-called posterior probability, is proportional to the numerator. Thus, if we substitute proportionality for equality in the above, we obtain the simplified form of the theorem (where the symbol $\propto$ means "proportional to"):

$$P(H|D) \propto P(D|H) \times P(H), \text{in other words}$$

$$\text{Posterior} \propto \text{Likelihood} \times \text{Prior}$$

This form makes it easier to interpret the theorem. The three terms—prior, likelihood and posterior—are traditionally used when discussing Bayes' theorem. Note that in the development of the theorem presented here, the three terms are referred to as *probabilities*; a probability value is the simplest kind of probability distribution, and this considerably simplifies the presentation and the calculations we will be making here. However, in general the three terms are represented by *probability distributions* and this necessitates the use of calculus to solve induction problems like those we typically meet in statistics see for example (Lyons 2013).

Prior: This is a shorthand term for the "prior probability" of a belief, proposition, hypothesis, or event; for simplicity, we will only refer to "hypothesis" in the sequel. Thus the prior ("previous", "initial" or "old"), P(H), is an estimation of the probability of a hypothesis being true where the estimation is made *prior* to the collection of new evidence, for example from an experiment. In the first of the two illustrative examples we used to introduce this chapter, the prior referred to the information that the rare disease occurs by chance in only one out of every 10,000 people; in other words, the prior probability of an individual having the illness was said to be 10^{-4}. We were not informed about the source of this prior; presumably it could have been estimated via the analysis of a large amount of clinical evidence or as a subjective estimate made by experienced researchers and medical doctors.

Likelihood: The likelihood, P(D|H), expresses the probability that the data would be observed if the hypothesis is in fact true. The credibility of a hypothesis that leads to a low probability of obtaining the data observed would be weakened and vice versa. In other words, this conditional probability summarizes the impact that the new evidence has on the probability that a hypothesis is true; depending on the data actually observed, the credibility of a hypothesis will shift according to how well it fits the observed data.

Posterior: The "posterior probability", P(H|D), expresses the conditional probability that the hypothesis is true given the data/evidence, D. It is what we are interested in. In contrast to "prior", it stands for the "final", "new" or "concluding" probability of a hypothesis being true after (posterior to) obtaining the new information contained in the likelihood.

An example from everyday life can illustrate the underlying idea. A person who is preparing to go to work looks out the window and sees that the sky is overcast. This is nothing new as it has rained quite a bit of late. He turns on the radio and listens to the weather forecast, which speaks of a risk of heavy showers in the afternoon. Even if he hadn't looked out the window or heard the radio, he would have thought that there would be a fairly good chance of rain, as the weather pattern of late has been intermittent showers. So his prior, P(H), is his belief about the probability of it raining during the period from after he leaves for work and until he returns home after work, evaluated prior to the collection of data or evidence, here in the form of visual inspection and a weather forecast he hears on the radio. The likelihood summarizes the impact that this new information has on the probability of his hypothesis as to rain during the day. And the posterior is the conclusion he makes as to the risk (the probability) of rain after combining his past beliefs with the new observations/data (here his visual observations by looking out the window and what he heard on the radio). The posterior might then affect his thoughts about whether or not to take an umbrella with him; we return to this decision making aspect later on. Here we can simply surmise that Bayes' theorem prescribes how probabilities are to be modified in the light of new data or evidence.

Note that it is not uncommon for people to be confused about conditional probabilities. In the case of some other operators, such as $\cup$ (the union of two sets) and $\cap$ (the intersection of two sets), order does not matter because $P(X \cup Y) = P(Y \cup X)$ and $P(X \cap Y) = P(Y \cap X)$. This may tempt one to think that it does not matter for conditional probabilities as well. But this is not the case; $P(X|Y) \neq P(Y|X)$. A simple example illustrates this: Suppose you select a card randomly from a standard deck of cards with 52 cards. The probability of the card being a king if it is a heart is 1/13. But the probability of the card being a heart given that it is a king is 1/4.

This distinction is significant since in general researchers want to know $P(H|D)$ and not $P(D|H)$, i.e. they are interested in the probability that a hypothesis is true given the data, rather than the probability that the actual data will be obtained given the hypothesis, that is, if the hypothesis is true (which is the focus of Type I and Type II errors mentioned earlier).

Reflective scientists, all of whom have a background in analytical thinking and mathematics, should be aware of the distinctions made here as to which conditional probabilities are in focus. They should be completely clear as to the difference between $P(H|D)$, the probability of a hypothesis being true given the data resulting from an experiment, and $P(D|H)$, the probability that the data will be observed if the hypothesis is true.

Bayes' theorem: Multiple hypotheses

Since we will shortly be dealing with the evidential bearing of data on hypotheses, and in particular on the use of statistics in comparing hypotheses, the following is a generalization of the above form of Bayes' theorem so that it applies to an enquiry involving n mutually exclusive and jointly exhaustive hypotheses, H_i, $i = 1, \ldots, n$.

For all j, $j = 1, \ldots, n$,

$$P(H_j|D) = \left(P(D|H_j) / \sum_{i=1}^{n} P(D|H_j) \times P(H_i) \right) \times P(H_j)$$

I note that in practice the condition that the enquiry involves n mutually exclusive and jointly exhaustive hypotheses is not restrictive. In situations where a researcher is considering more than one hypothesis to explain an observation, the hypotheses will typically be mutually exclusive (if any one hypothesis is true, the others will be rejected) and jointly exhaustive (in practice, each individual hypothesis is considered to potentially explain the observation).

To illustrate this as well as to make things easier, let us start by considering the case of $n = 2$ (we have two hypotheses, H_1 and H_2 where one and only one of these is true). From the basic Bayes' theorem we have: $P(H_1|D) = P(D|H_1) \times P(H_1)/P(D)$ and $P(H_2|D) = P(D|H_2) \times P(H_2)/P(D)$. Thus $P(D) \times (P(H_1|D) + P(H_2|D)) = P(D|H_1) \times P(H_1) + P(D|H_2) \times P(H_2)$. Now, since either H_1 is true or H_2 is true, $P(H_1|D) + P(H_2|D) = 1$ so $P(D) = P(D|H_1) \times P(H_1) + P(D|H_2) \times P(H_2)$.

Finally, substituting in Bayes' theorem we obtain the posterior probability:

$$P(H_1|D) = [P(D|H_1) \times P(H_1)] / [\{P(D|H_1) \times P(H_1)\} + \{P(D|H_2) \times P(H_2)\}].$$

Note that since the two hypotheses are mutually exclusive and jointly exhaustive, once we have calculated $P(H_1|D)$ we can directly calculate the other posterior: $P(H_2|D) = 1 - P(H_1|D)$.

To illustrate the use of this formulation of Bayes' theorem (where there are two mutually exclusive and jointly exhaustive hypotheses), as well as the rather common blunder in probability reasoning of ignoring the prior, consider once again the example used to introduce this chapter: You have a blood test for some rare disease which occurs by chance in 1 in every 10,000 people. The test itself is very reliable; if you have the disease it will correctly say so with probability 0.99; if you do not have the disease, the test will wrongly say you do with probability 0.005. We consider again the question: If the test says you have the disease, what is the probability that this is a correct diagnosis?

As discussed earlier, most people answer that the probability of a correct diagnosis is roughly 99 %, while the correct answer is only about 0.2 %. This is due to basing one's conclusions on the likelihood (here the probability 0.99 of the test indicating that you have the disease given that you have it), while ignoring the prior (here the probability 10^{-4} of having the disease). This will be clear from applying the formula above, where:

H_1 is the hypothesis that you have the disease and the prior is $P(H_1) = 10^{-4} = 0.0001$
H_2 is the hypothesis that you do not have the disease and the prior is $P(H_2) = 1 - P(H_1) = 0.9999$.
$P(D|H_1)$, the likelihood, is the probability that the test indicates you have the disease given that you have it $= 0.99$
$P(D|H_2)$ is the probability that the test indicates you have the disease given that you do not have it $= 0.005$

Since the conditions regarding mutually exclusive and jointly exhaustive hypotheses are clearly met here, plugging these data into

$$P(H_1|D) = [P(D|H_1) \times P(H_1)] / [\{P(D|H_1) \times P(H_1)\} + \{P(D|H_2) \times P(H_2)\}]$$

we obtain that $P(H_1|D)$, the probability of your having the disease given the evidence in the form of a test indicating you have the disease, is: $[0.99 \times 0.0001]/ [\{0.99 \times 0.0001\} + \{0.005 \times 0.9999\}] = 0.019417$ which is roughly 0.2 %. This is the result presented in the introduction to this chapter.

Bayes' theorem: Odds

Note that sometimes probabilities are expressed as ratios or odds. For example, consider an investigation with two hypotheses, H_1 and H_2. Saying that the odds that H_1 is true compared to H_2 are 4:1 means that H_1 is four times as likely to be true as H_2. *If* the two hypotheses are mutually exclusive and jointly exhaustive, these odds can be converted into probabilities; 0.80 and 0.20. Otherwise, further information would be required to move from odds to probabilities. For example, consider the case where there are not only two hypotheses but also a third hypothesis, H_3, and where the three hypotheses are mutually exclusive and jointly exhaustive. Assume too that we somehow know that $P(H_3) = 0.7$. Then since $P(H_1) + P(H_2) = 0.3$ and since the odds that H_1 is true compared to H_2 are 4:1, $P(H_1) = 0.24$ and $P(H_2) = 0.06$.

The ratio form of Bayes' theorem for the case of two mutually exclusive and jointly exhaustive hypotheses is:

$$P(H_1|D)/P(H_2|D) = [P(D|H_1)/P(D|H_2)] \times [P(H_1)/P(H_2)]$$
$$\text{'posterior odds'} \quad \text{'likelihood odds'} \quad \text{'prior odds'}$$

which follows directly from the basic theorem. So in the case of odds, Bayes' theorem states that the posterior odds equal the likelihood odds times the prior odds. For example, if the information that we have available before an experiment is performed leads to prior odds of 4:1 for the hypotheses H_1 and H_2, and if the experiment gives them likelihood odds of 1:32, then the posterior odds become 1:8, strongly reversing the initial odds so they now favour H_2.

Before leaving our considerations of Bayesian inference, it should be noted that other options are available than simply calculating the posterior probabilities. For example, one might be interested in performing *sensitivity analyses*, such as determining the range of priors that would, given that we are considering two hypotheses and the data obtained, result in H_1 having a greater posterior probability than H_2. This could be of interest in a situation where there is considerable doubt as to the prior. In other words, such a sensitivity analysis can provide information on the robustness of the posterior probabilities with respect to the prior. I note as well that such analyses can be of considerable importance not just in connection with the assessment of the posterior probabilities but also in connection with decision making. We will return shortly to how Bayesian inference can contribute to making rational decisions in situations characterized by uncertainty.

8.2   Inductive Logic and Statistics

Until now in this treatment of uncertainty we have considered probability theory, which, as we have seen, is essentially a branch of deductive logic. Here we will build upon the discussions of probability theory and discuss the role of statistics. In contrast to the emphasis on deduction in probability theory, from a research methodological perspective one can refer to statistics as "applied inductive logic". We are able to directly build upon the preceding results since probability and statistics have the same axioms. They differ however in reasoning and in the questions they answer.

As an introduction to the treatment of statistics/applied inductive logic, the following is a brief recapitulation of previous discussions as to the relationships between deductive and inductive thinking.

The underlying idea of deduction is very simple: one deduces a statement from other, given statements, the premises. If the premises are true, so are the conclusions of a valid deductive argument. In other words, the conclusion of a deductive argument can be said to be contained in its premises, so deduction is truth preserving. In the discussions above as to probability theory, the premises were the axioms and the deductions were the probability theorems that were derived from the axioms and/or from theorems that had already been established. *Deduction reasons from the general to the specific.*[4]

Induction on the other hand reasons from the specific to the general. It deals with drawing inferences/generalizations based on specific observations. In so doing, it presupposes a uniformity or regularity of nature. It is the primary logical method upon which empirical science is based and its essence is the notion of statistical validity. Applied inductive logic in the form of statistics does not provide us with truth in an absolute sense, but it enables researchers to evaluate their test results and provides them with the opportunity to learn—which is what science is really about.

Let us consider a simple example that clearly illustrates how the distinction between deduction and induction can be encapsulated in the current context of probability and statistics (Gauch 2003; 219). For the sake of simplicity here we define a "fair coin" as one that gives heads with probability 0.5 and tails with probability 0.5 so that if it is tossed n times all of the 2^n possible results have the same probability. Let us here define an "unfair coin" as one that gives heads with probability 0.7 and tails with probability 0.3. Assume too, that we somehow know in advance that a coin we are tossing is either "fair" or else "unfair" as defined here —and that these are the only two possibilities—but do not know whether it is fair or not. Consider now the following two problems:

Problem 1: Given a fair coin, what is the probability that 100 tosses will produce 42 heads and 58 tails?[5]

Problem 2: Given that 100 tosses of a coin produce 42 heads and 58 tails, what is the probability that the coin is a fair coin?

---

[4]For a fascinating, challenging and rather unorthodox introduction to deduction, formal systems, mathematics and their relationship to 'reality', see the first two chapters in (Hofstadter 1989); this book also provides a detailed treatment of Gödel's Theorem that was considered in the section on innate limitations of science in Chap. 2.

[5]Readers with a background in probability and statistics will see that Problem 1 can be solved by calculating the binomial probability: P(42 heads|100 tosses and P(head) = 0.5) = (100!/42!58!) $(0.5^{42})(1 - 0.5)^{58}$. Problem 2, which compares two hypotheses (the probability of a head is either 0.5 or 0.7) based on the data—42 heads and 58 tails—cannot be solved at this point in the exposition, but you should be able to solve it after reading the next section on the Bayesian paradigm. Here it should suffice to note that it is extremely unlikely that the coin is unfair (with probability of a head = 0.7) since only 42 heads resulted.

At first glance, these two may seem to be quite similar problems. But they are fundamentally different. Problem 1 is deductive while Problem 2 is inductive. Problem 1 is a probability problem, whereas Problem 2 is a statistics problem.

In other words, deduction and induction pursue answers to different kinds of questions. Deduction reasons from a mental model (here that we have a fair coin with its characteristics) to observable data (the number of heads/tails), whereas induction reasons from actual data (here the observed number of heads/tails) to a mental model (whether the coin is "fair" or "unfair"). As we have seen in Sect. 7.1 of the previous chapter (Scientific Method and the Design of Research), it is just this interplay between deduction and induction that characterizes many scientific investigations. The iterative process whereby there is a progression from hypotheses as to phenomena/relationships in the physical world, to the deduction of predictions, to facts/observations, to knowledge or explanations (in the form of models, theories, laws) as to the facts/observations, and then to new hypotheses, new predictions, new facts/observations, new knowledge/explanations, and so on, is at the heart of the scientific method. The remainder of this chapter builds on the foundation of the deductive probability theory presented earlier, and focuses on the application of inductive logic: statistics.

Earlier (in Chap. 6, Sect. 6.4.1) when considering the external validity of experiments, we introduced the traditional notion of frequentist statistics. And in our treatment of Bayes' theorem earlier in the present chapter we prepared the way for the consideration of Bayesian statistics. These are the two major approaches to applied inductive logic. The former is by far the most widely known and taught—and most likely the approach that your colleagues and advisors are most familiar with. In order to judge which paradigm can best meet specific research needs as well as to be able to critically reflect on the statistical analyses presented by others, the reflective scientist must understand what distinguishes these approaches from each other—that is, which questions they ask and which data/information they require in order to provide answers. Although having many similarities, they represent two fundamentally different paradigms, for statistics. I note that this use of the term "paradigm" (as meaning "approach" or "way of thinking") is more in line with its far broader use today than with Kuhn's use of it in connection with his work on normal science and scientific revolutions (as presented in Chap. 2, Sect. 2.2.5).

8.2.1 The Bayesian Paradigm

In order to present the fundamental ideas of inference based on a Bayesian approach to probability, we will now consider a simple illustrative example. While being easy to follow, the example also clearly demonstrates the advantages of a well-grounded application of the Bayesian paradigm compared to unaided common sense. Building on the concepts introduced earlier, consider the following simple "experiment" with marbles, where the presentation is based on the structure provided in (Gauch 2003; 226–240) but where the details are different:

Preliminaries

Have someone flip a fair coin without showing you the result so that you do not know which of the two, heads or tails, actually occurred. If the toss results in 'heads', have the person who tossed the coin place one red marble and four green marbles in an opaque bag. If the toss results in 'tails', have the person place four red marbles and one green marble in the bag. This too is done without your being able to see the result.

Hypotheses

H_G: the bag is a Green bag (contains one red marble and four green marbles)
H_R: the bag is a Red bag (contains four red marbles and one green marble)

Aim

To determine which hypothesis H_G or H_R is (probably) true.

Your duties

Without looking at the contents of the bag, shake it so that the marbles are mixed, draw a marble from the bag, observe its colour, and replace it, repeating this procedure as many times as necessary.

Stopping rule

Stop when a hypothesis reaches a posterior probability of 0.999.

Before getting on with a formal analysis, it can be instructive to perform a common sense analysis. Since whether the bag is in fact a Green bag or a Red bag has been determined before the draws start by the secret toss of a coin, for an observer, it is equally likely that the bag is a Green bag or a Red bag. Therefore the prior odds for H_G:H_R are 1:1. Furthermore, if after many draws the yield has primarily been green marbles, it is probable that the bag is a Green bag—and vice versa. However, common sense does not enable us to know how many draws should be performed before we can arrive at a reasonably well-founded conclusion, nor just how probable the resultant conclusion is. Since we are only dealing with marbles, you may think that it really doesn't matter and, since the experiment is very simple, a large number of draws can take place, enabling you to feel rather certain as to a conclusion without incurring serious investments in your time, effort and money. But suppose that while carrying out experiments to determine the effectiveness of a potential new drug each clinical trial costs $5,000,000, takes a month to perform and evaluate, and leads as well to considerable discomfort for the many patients participating in the trial. In this case it would certainly be important for the pharmaceutical company to be able to calculate the strength of a conclusion and to avoid collecting either too much or too little data. The reflective scientist will note that to determine the 'optimal' number of trials in this fictive clinical trial one would have to introduce elements of decision theory as the answer will depend on the total expected costs involved, the discomfort of the patients participating, the potential risks involved, and the like. Note too that in the marble problem the stopping rule was arbitrarily set to 0.999, with no reflection whatsoever on this choice. We deal with both of these issues later in this chapter.

To analyse the marble problem, the ratio (or odds) form of Bayes' theorem is easiest to use. We obtain:

$$P(H_G|D)/P(H_R|D) = [P(D|H_G)/P(D|H_R)] \times [P(H_G)/P(H_R)]$$

$\quad\quad$ 'posterior odds' $\quad\quad$ 'likelihood odds' $\quad\quad$ 'prior odds'

where the posterior odds that are determined after a draw become the new prior odds in the next draw. Table 8.2 below shows the development of the marble experiment. The final column, where we arbitrarily focus on the posterior probability $P(H_G|D)$, which is $1 - P(H_R|D)$, enables us to determine when the experiment should stop according to the stopping rule.

The likelihood odds $P(D|H_G)/P(D|H_R)$ that arise from each possible outcome of drawing a green or a red marble are determined as follows. At the very beginning, given the setup with a flip of a fair coin, we deduce that the prior odds of $H_G:H_R$ are 1:1, thus $P(H_G) = P(H_R) = 0.5$. Given that H_G has four of five marbles green, while H_R has one of five marbles green, a green draw is four times as probable given H_G as it is given H_R. Therefore, because $P(Green|H_G) = 4/5 = 0.80$ and $P(Green|H_R) = 1/5 = 0.20$, a green draw at any time contributes likelihood odds $P(Green|H_G)/P(Green|H_R)$ of 0.80:0.20 or 4:1 for $H_G:H_R$, favouring H_G. By similar reasoning, a red draw contributes likelihood odds of 1:4 against H_G. And since, after remixing the marbles, each draw is an independent event, probability theory shows that the results of the individual draws combine multiplicatively in the experiment; this is just a simplification of Axiom 4 for the special case of independent events. Thus, in our sequential experiment, each green draw will increase the posterior odds for $H_G:H_R$ by 4:1 while each red draw will decrease it by 1:4.

As the sequential experiment progresses, we learn about the relative strengths of the two hypotheses. The posterior odds after a draw, $P(H_G|D)/P(H_R|D)$, become the prior odds at the start of the next draw. To follow the calculations in the table, consider the first two draws. As shown above, we deduced that the odds prior to the first draw, $P(H_G)/P(H_R)$, are 1:1. The result of the ensuing first draw, Green, contributes likelihood odds $P(Green|H_G)/P(Green|H_R) = (4/5)/(1/5) = 4:1$ in favour of H_G. Combining these prior and likelihood odds we get the posterior odds after the first draw: Multiply the likelihood odds of 4:1 by the prior odds of 1:1 which results in posterior odds of 4:1. This increases the posterior probability $P(H_G|D)$ to 0.8; this is shown in the last column of the table.

Continuing to the second draw: The posterior odds after the first draw were 4:1 and this becomes the odds prior to the second draw. Given that this draw is a red marble, the likelihood odds are 1:4 and thus the posterior odds after the second draw are given by $(4/1)(1/4) = 1$. Thus the posterior probability after the second draw is 0.5.

A final example of the reasoning and calculations here: Before the seventh draw the prior odds are 16:1, and the green draw contributes likelihood odds of 4:1, resulting in posterior odds of 64:1 in favour of H_G and hence a posterior probability $P(H_G) = 64/65 = 0.984615$.

Table 8.2 Bayesian analysis for the experiment

Draw	Result	Posterior odds $P(H_G|D)/P(H_R|D)$	Posterior prob. $P(H_G|D)$
	(Prior odds)	1:1	0.500000
1	Green	4:1	0.800000
2	Red	1:1	0.500000
3	Green	4:1	0.800000
4	Green	16:1	0.941176
5	Red	4:1	0.800000
6	Green	16:1	0.941176
7	Green	64:1	0.984615
8	Red	16:1	0.941176
9	Green	64:1	0.984615
10	Green	256:1	0.987805
11	Red	64:1	0.984615
12	Green	256:1	0.996109
13	Green	1024:1	0.999024

Continuing in this manner, after 13 draws the posterior probability, 0.999024, exceeds the stopping rule's (arbitrarily) pre-selected value of 0.999, so the experiment stops and H_G is accepted with more than 99.9 % confidence, i.e. the chance of making an error when concluding that H_G is true, is less than 1 in 1024.

The table confirms our common sense reasoning that results become more convincing as an experiment progresses. This particular experiment required 13 draws to reach a conclusive result. But it is interesting to reflect on how long this experiment would run on the average if it had been repeated many times. A little reasoning leads to the observation that the posterior probabilities of H_G and H_R depend only on the difference between the number of green and red draws. Letting N be the difference between the number of green and red draws, then the posterior odds of H_G:H_R equal 4^N. This exceeds 999:1, favouring H_G, when N = 5, or is less than 1:999 favouring H_R when N = −5. Thus we can keep track of the progress of the experiment simply by starting at N = 0 and adding 1 for a green draw and -1 for a red draw, stopping when N = 5 or −5. In our case the scoring would have led to 1, −1, 1, 2, 1, 2, 3, 2, 3, 4, 3, 4, 5.

This simple experiment demonstrates the strength of a well-structured application of theory compared to relying on common sense. While common sense does not enable us to know how many draws should be performed before we can arrive at a reasonably well-founded conclusion, nor how probable the resultant conclusion is, the application of probability theory does.

8.2.2   Bayesian Decision

When making decisions there is a need not just for probabilities as to outcomes, but also for measures of "costs" and "benefits". This enables combining the information on probabilities with the costs and benefits so as to compute decision criteria, such as the gains and risks that can be expected to result from a decision. While *inference problems* pursue true belief in the form of good descriptions, *decision problems* pursue effective actions.

Although clearly of interest for students and practitioners of social sciences such as economics and management, the reflective reader may ask why students and practitioners of the natural sciences should be interested in theory regarding decisions. After all, their research is often considered by them to be purely descriptive or explanatory; decisions on *how* to use their research are up to others. But this is a rather narrow and naïve perspective. First of all, one continually faces choices when performing research; budgets are not unlimited, nor does a researcher have unlimited time to carry out the research (it is only seldom that I have met a Ph.D. student who completed her or his research and wrote the thesis on time without experiencing a time-crunch). For example, when designing an experiment the researcher must weigh the potential contribution that will result against the costs and other demands of the experiment; *designing an experiment is a very practical decision making problem.* Secondly, experience shows that it is not uncommon for a research project to be directly related to a specific problem that someone (e.g. a funding organisation that sponsors the research) is trying to solve—and this orients the research (and often the interests of one's advisor or superior) in the direction of more applied research, where the aim of the research is not simply to provide better descriptions and explanations, but also better (more useful, less costly) results. And finally, today's Ph.D. student may be tomorrow's expert/consultant, or a researcher in a private company, or the Head of Department at a university, or a corporate manager concerned with how to apply research results, or an administrator in a governmental regulatory body. All such positions presume that the person has the ability to combine skilfully elements of uncertainty as to the outcomes of possible actions and the potential gains and costs associated with these outcomes, so as to make good decisions.

Without going into detail, I note that Bayesian inference provides a powerful basis for decision theory in that it permits both prior probabilities, including subjective beliefs and learning, to be directly included in the analyses of the expected benefits of decisions. According to (Jeffreys 1983; 30–31) in his development of Bayesian inference theory, "… we have in practice often to make decisions that involve not only belief but the desirability of the possible effect of different courses of action. … The fundamental idea is that the values of expectations of benefit can be arranged in an order; it is legitimate to compare a small probability of a large gain with a large probability of a small gain."

The reflective scientist will note that this implicitly assumes that the value to the decision maker of an extra unit of "benefit" referred to, say money, is independent of how much of that "benefit" she possess—that say the value to her of an additional Euro is the same no matter whether she is poor or rich. This implicit assumption may be false, in which case using expected values as a guide to decision making (and as are used below), may lead to less than optimal decisions. This is often referred to as the "St. Petersburg paradox".

Decision theory requires the addition of only one more axiom to those considered earlier in connection with the development of inference theory; (Gauch 2003; 243), Jeffreys (1983; 80–81). Letting u(Z) denote the utility of an action Z, we introduce the following axiom:

Axiom of utility

$$\text{If} \quad P(X \wedge Y) = 0 \quad \text{and} \quad P(X \vee Y) \neq 0, \quad \text{then}$$
$$u(X \vee Y) = [u(X)P(X) + u(Y)P(Y)]/[P(X) + P(Y)]$$

In ordinary English, the axiom states that the utility of an action equals the average of the utilities for its various mutually exclusive outcomes, here two possible outcomes X and Y, weighted by their probabilities. It should be clear that the axiom can be extended to deal with any number of possible actions and their associated mutually exclusive outcomes.

To illustrate, consider the following very simple example of a decision problem whose structure is inspired by (Gauch 2003; 242). A tourist is considering which route to follow in taking the long trip by car from city A to city B. One of the routes, Route 1, is based on using a highway; although this route will be the longest, under ordinary conditions it will require the least driving time. However, as it is vacation time, many others may also use the highway which may lead to traffic congestion and increased driving time. The tourist therefore also considers a second route, Route 2, based on a combination of the highway and secondary roads; the travelling times on these secondary roads are unaffected by the time of the year. A third route being considered, Route 3, is based only on secondary roads, so the expected travelling time here is totally unaffected by the fact that it is vacation time. A fourth route, the scenic Route 4, tends to be congested when the highway is uncongested, and vice versa.

There are three possible states of nature (three possible degrees of congestion on the highway), all of which are outside the tourist's control; the highway can be highly congested, reasonably congested or uncongested.

The goal of the tourist is to find the route that will minimize his expected driving time (and thereby maximize his expected utility).

Given this information, the problem can be structured as shown in Table 8.3.

Let us begin with the information on the consequences, the expected driving times for each of the four possible routes dependent on the state of congestion of the highway. This data (the twelve driving times listed in the first table) is assumed here to be readily available from the Automobile Association's historical records. For example, if the outcome is that the highway is Reasonably Congested and if Route 2 is chosen, it is estimated that the total driving time will be 6 h.

Table 8.3 Bayesian analysis of the tourist's decision problem

Consequences (Driving times)

Actions/States	Highly Congested	Reasonably Congested	Uncongested
Route 1	14 hours	9 hours	4 hours
Route 2	8 hours	6 hours	5 hours
Route 3	7½ hours	7½ hours	7½ hours
Route 4	7 hours	8½ hours	9½ hours

Old and New Data on States

Prior	0.35	0.45	0.20
Likelihood	0.60	0.30	0.10

Probability of States

Posterior	0.575	0.370	0.055

Expected Driving Times (Nr. Hours)

11.60
7.10
7.50
7.69

In addition, we assume here that the tourist knows something about the probabilities of the three possible states of congestion and this is summarized in the second table showing the priors and likelihoods for each of the three states. The priors/old data could be based on extensive historical records provided by the Automobile Association. They could also simply be the tourist's best guesses based on his experiences of driving from City A to City B a number of times in the past. Or they could be some combination of these two. In other words, the priors could be either what one might refer to as more objective (the historical records), more subjective estimates (his best guesses)—or some combination of these two. In any case, his prior probabilities are: (0.35, 0.45, 0.20) for the three congestion states of the highway: (Highly Congested, Reasonably Congested, Uncongested). The new data, the likelihoods, could be based on recent forecasts by the Automobile Association giving likelihood probabilities of 0.60, 0.30, and 0.10 for the three states of congestion and indicating rather severe congestion on the highway.

We can now apply Bayes' theorem to combine the prior probabilities and the likelihoods in order to derive the posterior probabilities of the states of traffic congestion on the highway as shown in the table: "Probability of States". Multiplying each prior by its corresponding likelihood gives the values 0.210, 0.135, and 0.020 for a total of 0.365, and division of each of those three values by the total of 0.365 so as to normalize the probabilities (so they sum to 1.000) yields the posterior probabilities listed in the table for (Highly congested, Reasonably congested, Uncongested): (0.575, 0.370, 0.055).

Finally, the last table shows that the decision that maximizes the criterion of expected utility (here, minimizing the expected number of driving hours) is to choose Route 2. This is arrived at by applying the Axiom of Utility: Determine the expected utility of any action by multiplying the utility for each possible state by its corresponding probability and summing over the states. For example, the expected driving time associated with choosing Route 1 is $(14 \times 0.575) + (9 \times 0.370) + (4 \times 0.055) = 11.6$, and this is the least desirable route as it maximizes the expected driving time, The smallest of the three values of expected utility is 7.10, which indicates that Route 2 will be the best decision, i.e. the decision that minimizes the expected driving time/maximizes the expected utility.

Clearly this is an extremely simplified problem formulation. For example, it is assumed that it is meaningful to characterize the tourist's decision problem by only three states of nature—Highly congested, Reasonably congested, Uncongested— while in practice, there can be large differences in the way that the traffic unfolds on the highway, the secondary roads and the scenic route, dependent on the day of the week, the time of day, etc. Nevertheless, a number of interesting reflections can be made as to the differences between decision problems and inference problems. First of all, the example above clearly illustrates a significant distinction; the best decision can differ from the best inference. In the analysis above, the state "Highly congested" had the greatest posterior probability, and this favours the choice of Route 4 while the decision analysis points to Route 2, primarily because "Reasonable congestion" is also rather likely and thus the combination of highway and secondary roads becomes more attractive than a route based solely on the highway, the secondary roads or the scenic road.

A second such distinction between decision problems and inference problems is that although decision problems are more complex, in practice they are often easier to solve. The necessity of making a decision may permit even weak/"fuzzy" data and small probability differences to force a decision to be made while, since inference problems aim at truth, it may be difficult to evaluate just how strong a probability will be required to justify a conclusion. For example, comparatively strong odds, say of 30:1 in favour of a given alternative hypothesis compared to a null hypothesis, might satisfy one scientist, but not another, particularly if incorrectly rejecting the null hypothesis and accepting the alternative hypothesis can have serious repercussions for that scientist's research or for his reputation.

A third reflection deals with the fact that in many problems that are more complex than the extremely simple illustrative problem above, the quantities can be better modelled by distributions rather than by numbers. For example, the various consequences (Route-State combinations in the table above) are characterized by numbers, while in fact they might better be modelled by distributions due to the uncertainties/fluctuations in congestion that can arise from many factors, such as an accident on the highway or on one of the secondary roads. Assuming statistical data are available, the numbers in the table could then be replaced by probability density functions. Similarly, also the probabilities could be given in the form of distributions rather than by numbers. This of course would lead to a more complex analysis than was performed here—an analysis that could also provide those making the decision not only with information on expected utilities, but also with probabilistic measures of risk. In such situations simulation models and so-called Monte Carlo techniques might be required to provide estimates of the resulting posterior probability distributions as well as of the distributions of the outcomes of any Route-State combination.

Fourthly, and building upon the last comment above as to risk, because decision makers may have different attitudes towards risk, it may be better to supplement or replace the single criterion of expected utility with other criteria that also reflect the risk associated with a decision. For example, a decision maker who has to catch a plane at City B might not only be interested in minimizing his expected driving time, he might also be interested in minimizing the maximum possible expected driving time. Should he therefore focus on security rather than on expected time this might prompt him to choose Route 3. It provides an expected driving time of 7½ h no matter what the congestion is, while in the worst case Route 1 can result in expected driving time of 14, Route 2 can result in an expected time of 8, and Route 4 can result in expected driving time of 9½ h, all of which are greater than the relatively certain 7½ h when choosing Route 3.

This leads directly to a fifth consideration: Many decisions are better characterized by a multiple of criteria than just one, as in the driving example presented here. For example, if the tourist wants to somehow balance his desire for a minimum expected driving time with his desire for avoiding risk, he will have to consider these two criteria simultaneously. In addition, he might also be interested in following a scenic route in which case he might choose Route 4 even though the expected driving time is more than a half hour greater than Route 2. Furthermore, since the calculations indicate that there is relatively little difference in the expected driving times on the two routes with the least expected driving times, Routes 2 and 3, if the tourist feels that Route 3, with no highway driving is more relaxing, he might prefer this route, even though the expected driving time would be increased by 24 min. If the formulation of the decision problem is to permit inclusion of such preferences, it will be necessary to operationalize additional criteria. Dealing with multiple criteria implies a greater complexity but also permits greater realism with respect to the analysis of many decision problems (Bogetoft and Pruzan 1997).

A sixth, and final, consideration here deals with the fact that in the above problem it was assumed that the decision was to be made by only one person, the tourist. In fact, many decisions are characterized by a multiple of decision makers (and a multiple of criteria)—and even by the need to consider the preferences of a multiple of "decision receivers" or stakeholders, who are affected by the decision; (Bogetoft and Pruzan 1997; chap. 11). A more thorough treatment of such situations is far beyond the ambitions of the reflections provided here.

8.2.3   The Frequentist Paradigm

Although the Bayesian paradigm for inference from 1763 preceded the frequentist paradigm by about a century and a half, the frequentist paradigm is far better known today amongst most researchers in the natural sciences (as well as economists, engineers etc.) and is the approach to inductive logic that is most frequently taught by teachers of statistics. As its name indicates, the frequentist paradigm is based on the view that probabilities reflect frequencies—interpreted as the long term likelihood of the occurrence of an event.

To be more precise, according to the frequentist interpretation, the probability of the occurrence of an event is defined as the limit of the relative frequency of the occurrence of the event (such as the outcome of an experiment or the occurrence of heads when tossing a coin) in an infinite sequence of trials (experiments/tosses of a coin). Thus, according to this frequentist interpretation of probability, when we say that the probability of getting heads with a fair coin is ½, we mean that in the potentially infinite sequence of tosses with the coin, the relative frequency of heads converges to or has the limit ½.[6] In practice, since we cannot ordinarily observe an infinite sequence of trials, a measure of the probability of the event occurring is provided by the relative frequency of its occurrence in a number of repetitions of the trial. More precisely, if n_t is the number of trials and n_e is the number of trials where the event e occurred, the probability $P(e)$ of the event occurring is approximated by the relative frequency: $P(e) \approx n_e/n_t$, where it is implicitly understood that this relative frequency will converge to the true probability We briefly considered this topic in Sect. 8.1.1 where we discussed "probability rules".

We have already met some of the major concepts of this paradigm in Chap. 6 when we considered validity and reliability in experimentation. I note as well that the literature on experimentation that we referred to, as is the case with most such literature, employed a frequentist approach, with its emphasis on Type I and Type II errors. Presumably this is due to a strong tradition that developed as to the frequentist paradigm in the last century, particularly since the 1950s. This is often attributed to what the proponents of the frequentist approach refer to as its more "objective" nature as compared to the Bayesian approach.

As we have seen, a Bayesian approach permits prior probabilities to be other than observed frequencies; they can be such frequencies but can also be beliefs, inspired by one's own experiences or those of others, or even a uniform distribution between zero and one in the case where no prior information at all is available. Scientists have been trained to emphasize the objective nature of their investigations and therefore can find it difficult to accept a procedure that permits subjective estimates of the priors—and thus that these subjective estimates can influence the posterior inferences.

---

[6]To be more precise, we say that a sequence f_n ($n = 1,2,3, \ldots$) has the limit L as n goes to infinity if and only if, for every positive μ, no matter how small, there exists a number N such that, if $n > N$, $|f_n - L| < \mu$.

In addition, frequentists also tended to reject the Bayesian approach to inference for practical reasons; information/estimates of the prior can be difficult to obtain, inaccurate, questionable. Furthermore, if different scientists working on a project employ different (subjective) priors, the use of the Bayesian approach can lead to different posteriors, which could lead to all sorts of complications and require a procedure for how to obtain a single posterior for the group as a whole.

So the motivation was to remove subjectivity from inference procedures and to replace it with a paradigm characterized by greater objectivity and less complexity.

To introduce the frequentist paradigm, the following is a brief repetition of the information provided earlier when we considered external validity in connection with experimentation. In frequentist statistics, one of the hypotheses under consideration is designated the null hypothesis, typically designated H_0. Ordinarily the null hypothesis is to be compared to the alternative or research hypothesis, H_1, which is the hypothesis the researcher really wants to test. The null hypothesis is used to predict the results which would be obtained from an investigation if the alternative hypothesis, the one the researcher is really investigating, is *not* true. For example, the null hypothesis could be that a new treatment of an illness is not better than the existing treatment while the alternative hypothesis is that the new treatment does have a positive effect.

A null hypothesis is considered to be either true or false. For example, if the alternative hypothesis is that variable A affects variable B, the null hypothesis is that A does not affect B, and this can then be tested by manipulating the independent variable A and measuring the corresponding variations in the dependent variable B. If statistical tests do not reject the null hypothesis that A does not affect B, then the research hypothesis H_1 should be rejected. If on the other hand the null hypothesis that B is independent of A is rejected, then more confidence can be placed in the alternative hypothesis that A affects B and it is accepted as being true.

Thus there are four possibilities when testing a hypothesis, and these include two types of errors:

Type I error: This error occurs when one rejects a null hypothesis that is in fact true.
Type II error: This error occurs when one accepts—or perhaps more precisely, fails to reject—a null hypothesis that is in fact false. I note that this "fails to reject" wording is more in line with a falsification understanding of how to corroborate/justify scientific statements.

	H_0 is true	H_0 is false
Accept/fail to reject H_0	Success	Type II error
Reject H_0	Type I error	Success

Since low Type I and II error rates permit reliable learning from experiments, the basic consideration of a frequentist approach to hypothesis testing should be to minimize the likelihood of both types of errors. Typically, however, the emphasis or the weighting is placed on minimizing the likelihood of the error that causes the greatest "damage", usually of rejecting a null hypothesis (and thus of accepting the alternative hypothesis) when in fact the null hypothesis should be accepted. After all, the alternative hypothesis deals with something "new", that say A affects B or that a treatment A is better than a treatment B, and the researcher wants to feel very certain about this, i.e. have strong statistical evidence, before presenting the results of his research.

Note that Type I errors can be totally avoided by simply accepting the null hypothesis no matter what data obtains, and note too that Type II errors can be totally avoided by rejecting the null hypothesis no matter what data obtains. Thus there is an inherent trade-off between these two error types. Therefore, in principle, some compromise should be made between the weighting placed on their importance. Ideally, this is done by evaluating the "costs" (not only monetary costs or penalties, but all possible negative results, including such non-tangibles as e.g. danger to users of a new drug that is hypothesized to be more effective than an existing drug, loss of a scientist's prestige due to accepting or rejecting a hypothesis incorrectly, etc.), and then somehow balancing the weights so as to maximize the overall expected "utility" or minimize the total expected "costs" due to both kinds of errors. In other words, in principle, decision theory should guide the choice of weights to be placed on the two types of errors.

In practice however, scientists tend to disregard such weighting or balancing and simply set the Type I error rate—the conditional probability of rejecting the null hypothesis given that it is true—at some convenient level. In so doing, they do not take into account the accompanying Type II error rate and thus ignore the overall expected "costs" and "benefits".

A plausible interpretation of the apparent logic behind such an (arbitrary) choice is that most scientists consider their efforts as purely descriptive and/or explanatory; they are interested in inferences, in the pursuit of truth, rather than in applications and decisions. In addition, and this may be the major underlying, though subconscious, reason, is that choosing an acceptable level of risk as to a Type I error simplifies things immensely—no additional independent reflection is required as to the best way of statistically testing one's hypotheses and this is almost universally accepted; other scientists who use the frequentist approach justify their statements in the same way.

In determining such an acceptable level of risk as to making a Type I error when carrying out an experiment, the so-called p-value is used. It is the conditional probability of getting an outcome at least as extreme compared to its expected value as is the actual observed outcome, given the assumption that the null hypothesis is true; using the terminology from Bayes' theorem, the frequentist approach provides $P(D|H)$, the probability of obtaining the observed data assuming that the null hypothesis is true—and not $P(H|D)$—the probability that the hypothesis is true given the evidence, which is what reflective scientists really want to determine.

To compute the p-value, envisage that you repeat an experiment an infinite/very large number of times and then, based on your observations, estimate the probability of getting an outcome at least as far from the expected value as the actual experimental outcome, under the assumption that the null hypothesis is true. In other words, the p-value is the probability of making a Type I error (rejecting the null hypothesis when it is true) when an observation is at least as far from what is expected given the null hypothesis. The smaller this p-value is, the more strongly will a frequentist test reject the null hypothesis.

To illustrate the logic involved, consider an experiment to determine whether or not to accept a null hypothesis that a given coin is a fair coin; the researcher suspects that the coin is "rigged" so that the probability of a head is greater than 0.5. The coin is tossed 100 times and the researcher has (somehow) decided that the null hypothesis will be rejected if the number of heads is such that the p-value is less than or equal to 0.01. The number of heads has a binomial distribution and the researcher determines that should the 100 tosses lead to 63 or more heads, the null hypothesis will be rejected; if the coin is a fair coin, 62 or more heads will occur with probability 0.01049 while 63 or more heads will occur with probability 0.00602[7] so the cutoff is 63.

If the result turns out to be at least 63 heads and the researcher rejects the null hypothesis, this is *not* statistical evidence that the coin is an unfair coin. If ten thousand people flip identical fair coins 100 times each, it becomes highly likely that quite a few of them will observe at least 63 heads. So one of those events on its own should not be interpreted as statistical evidence that the coins were somehow rigged; we can only conclude that if the null hypothesis is true, in less than one percent of the time we can expect to get 63 or more heads—and commit a Type I error.

It is common in the literature to see such a result referred to as being "significant at the 0.01 level". Due to the many misunderstandings regarding the interpretation of such a statement, the following repeats what we have considered in the above reflections on p-values. Such a statement only means that if, based on the observed data, one rejects the null hypothesis, there is at most a 1 % chance of committing a Type I error. In other words, the researcher is willing to infer from the experimental outcome that if the null hypothesis is true—so that we are sampling exclusively from cases for which the null is true (here that the coin is fair)—the probability of obtaining an observation as extreme or more extreme than that actually observed in the experiment (here at least 63 heads) is at most 1 %.

It is common in the assessment of test results to refer to rejection at a p-value of 0.05 as a "significant" result, and rejection at the 0.001 level as a "highly significant" result. I note however that in high energy physics, where huge sets of data can be collected, in order to avoid statistical anomalies, a so-called "five sigma" criterion is often employed in connection with the discovery of a new particle. This criterion corresponds to a p-value of 3×10^{-7}, or about 1 in 3.5 million. This was the threshold value of p employed by researchers at CERN in Geneva, Switzerland when on July 4, 2012 they announced the discovery of the so-called "Higgs boson". Repeating the logic above, a criterion of the p-value being at most 3×10^{-7} cannot be understood as the probability that the Higgs boson does or does not exist. 3×10^{-7} is simply the probability that *if* the particle does *not* exist (which was the

---

[7]The binomial probability of 63 or more heads is computed as: $(100!/63!37!)(0.5^{63})(1 - 0.5)^{37} + (100!/64!36!)(0.5^{64})(1 - 0.5)^{36} + \cdots + (100!/100!0!)(0.5^{100})(1 - 0.5)^0 = 0.00602$.

null hypothesis), the data that the scientists collected at CERN would be at least as extreme as what they actually observed.

Such more or less standard cut-off points (e.g. rejection of a null based on a "significant" result or a "highly significant" result) are completely arbitrary—but make life simple for the researcher who thereby does not need to justify a weighting of Type I and II errors. It is my experience based on reading/guiding Ph.D. theses and reviewing journal articles that reliance on such arbitrary standards, together with the use of standard software packages for performing statistical tests, tends to significantly reduce the reflection and reasoning of researchers as to the role of uncertainty in their analyses. This holds both for those who perform the research and for those who read about it in a publication. A result is that dealing with uncertainty takes on a more technical or mechanical nature whereby insight is lost and poor(er), less reliable, conclusions are drawn.

Consider for example the fact that there may be moral consequences of rejecting a true null hypothesis and thereby of accepting an alternative hypothesis. Suppose a research group has developed a new "wonder drug" that appears far more effective in treating a minor heart disease than the standard method used at present. Suppose too that the new drug may also have some serious side effects such that, in spite of its apparent effectiveness with respect to the specific heart disease, it may lead to an increased overall risk of death. The research group develops a null hypothesis that the existing treatment, although far less effective in treating the specific minor heart ailment, does *not* lead to a higher overall risk of death than the new "wonder drug". Presumably, rejecting such a null hypothesis, and thereby accepting the alternative hypothesis that the new drug is, all in all, more effective in saving lives than the present treatment, should require far stronger evidence (a smaller p-value) than a null hypothesis that butterfly type A does not experience a decrease in fertility due to the presence of butterfly type B that recently has started breeding in the same region as butterfly type A. The moral implications of making a Type I error and rejecting the null hypothesis in the first case are presumably far greater than the implications of rejecting a hypothesis regarding the fertility of butterfly type A dependent on the presence of butterfly type B. Yet statistical testing in both cases might use the same levels of significance as a basis for determining whether or not to reject the null hypothesis.

Relying on computation of the p-value and of standard cut-off points, although far easier to deal with than the consideration of such moral implications, removes the analysis—and the analyser!—from the reality she or he is immersed in

To give a more detailed feeling for the concept of the p-value and inference based on the frequentist approach, the table below (Table 8.4), following the structure of the experiment described in (Gauch 2003; 247–249), reconsiders the marble experiment presented earlier, this time from a frequentist perspective. Let H_R once again be the null hypothesis that the bag is a Red bag, i.e. that it contains four red marbles and one green marble, and let the alternative hypothesis be H_G— i.e., that the bag is a Green bag, that it contains one red marble and four green marbles. Note that in this investigation, where neither H_G nor H_R refers to a

Table 8.4 Frequentist analysis assuming the null hypothesis H_R is true and the experiment stops after 13 draws

Green draws (out of 13)	Probability	p-value P($\geq$ g draws)
0	0.0549755813888	1.0000000000000
1	0.1786706395136	0.9450244186112
2	0.2680059592704	0.7663537790976
3	0.2456721293312	0.4983478198272
4	0.1535450808320	0.2526756904960
5	0.0690952863744	0.0991306096640
6	0.0230313621248	0.0300353232896
7	0.0057579405312	0.0070035611648
8	0.0010796138496	0.0012456206336
9	0.0001499463680	0.0001660006784
10	0.0000149946368	0.0000160604160
11	0.0000010223616	0.0000001657792
12	0.0000000425984	0.0000000434176
13	0.0000000008919	0.0000000008192

treatment, H_R has been arbitrarily chosen to be the null hypothesis, but this has no effect on the procedure and results.

We recall that the experiment depicted in Table 8.2 had 13 draws that resulted in 4 red marbles and 9 green marbles. Furthermore that the conclusion of the Bayesian analysis was that the probability that H_G was true was 0.999024 and that the probability that H_R was true was the complement, 0.000976. Unlike that analysis with its initial coin toss and the prior probabilities of H_G and H_R, both of which were 0.5, we assume here that we know nothing at all about the prior probabilities (priors are not permitted in a frequentist approach).

The first column in the table lists, for an experiment with 13 draws, the 14 possible outcomes, namely, 0–13 green draws (and correspondingly 13–0 red draws). Since the null hypothesis is H_R, the probability of a green draw equals 0.20 (under the assumption of H_R there are four red marbles and one green marble in the bag). Thus, if the experiment with 13 draws were to be repeated many times, and if the null hypothesis H_R is in fact true, we would expect relatively frequent outcomes with two and three green draws out of the 13 draws, while we would expect outcomes with say more than eight green draws to be extremely infrequent. To make such intuitive results more precise, let us calculate the probability of g green draws and r red draws from a total of $n = g + r$ draws given the null hypothesis that the bag is a Red bag (contains 4 red marbles and 1 green marble). Such an outcome can result with $n!/(g! \times r!)$ combinations,[8] and the probability of each such outcome is $0.20^g \times 0.80^r$. For example, the probability of four green and nine red draws is

---

[8] *Permutation*: Distinct ordered arrangement of items. E.g. 3 items, A, B, C can be ordered in 6 ways/permutations $(3!/1!1!1!) = 6$, and items A, A, B, C can be ordered in 12 ways/permutations,

$[13!/(4! \times 9!] \times 0.20^4 \times 0.80^9$ which is roughly 0.1535451. The probabilities for all the possible outcomes are listed in the second column; clearly they must sum to 1. The third column gives the p-value obtained for g green draws by summing the probabilities for all outcomes with g or more green draws. For example the p-value for 0 green draws is 1, since it is the sum of all 14 probabilities, and for 10 green draws it is sum of the last four probabilities—and can be computed as 1 minus the sum of the first nine probabilities that have already been computed.

We can use the results in the table to define test procedures that will control Type I errors at a given rate. For example, a procedure that rejects the null hypothesis H_R if there are five or more green draws will result in a Type I error rate of about 0.099130609 while if the procedure rejects the null hypothesis if there are six or more green draws the error rate will be about 0.030035323. Therefore, to obtain significant results at the 0.05 level, the appropriate procedure would be to reject H_R if there are six or more green draws; similarly at the 0.01 level, reject H_R if there are seven or more green draws, and at the 0.001 level, reject H_R if there are nine or more green draws.

For the particular marble experiment we considered earlier using a Bayesian approach, the actual outcome was nine green draws. According to the above table, nine green draws would have a corresponding p-value of roughly 0.000166001. Therefore, the conclusion from a frequentist approach is to reject the null hypothesis H_R; the probability of getting an outcome of nine or more green draws under the assumption that H_R is true is seen to be less than one in a thousand, which is said to be highly significant.

Sample size

The above analyses assumed without question that the sample size was 13. In fact, the choice of sample size can play an important role in one's analyses. Nevertheless, aside from more practical matters such as the cost or difficulty of obtaining a sample, many researchers pay minimal attention to this aspect of an investigation. So let us conclude the present discussion of the frequentist paradigm by reflecting briefly on the question of the size of a sample, a theme that we have touched on earlier, for example in connection with the design of experiments.

In many scientific investigations, researchers are looking for new, surprising, unexpected results. After all, it is far easier to publish an article in a scientific journal that provides evidence of an unexpected relationship between phenomena than the opposite—that nothing unexpected was found; editors and those awarding prizes or making promotions are far more interested in being able to present

---

(Footnote 8 continued)

$(4!/2!1!1!) = 12$. The general formula with a total of N items where N_1 are of type 1, N_2 are of type 2 ... N_R are of type R is: $(N!/N_1!N_2!...N_R!)$.

Combination: Order does not matter. N items taken R at a time lead to $(N!/(N - R)!R!)$ combinations. E.g. the number of ways to choose 5 cards out of a standard deck of cards is $(52!/(52 - 5)! \times 5!)$. Note for R = 2 (only 2 types of items), the number of permutations equals the number of combinations; this is the case here with only green and red marbles.

exciting news rather than a demonstration of the fact that what could have been exciting was not supported by the evidence. In other words, there exists a pervasive, but strong, bias favoring the publication of claims to have found something new/exciting/unexpected. Therefore, scientists are likely to test hypotheses that are unlikely in the sense that the results are new and potentially unexpected and thereby worthy of publication. And therefore they often are interested in carrying out investigations that enable them to measure the effect of a potential "cause", even when its effect on the observed data is small. This can lead to pressure on carrying out investigations requiring relatively large data sets—which may result, for example, from running an experiment many times. However, the larger the investigation, the greater the resources required. So there is a need to somehow determine the "optimal" sample size that is to be used in an experiment—and we once again see that a reflective scientist must also be a practical decision maker.

For example in Chap. 4 we considered the concept of verification and the argument that if an inductive inference from observable facts to a scientific statement is to be justified, the number of observations forming the basis of a generalization must be "large". To provide some food for thought as to the influence of the size of a sample in relation to the concept of statistical significance, which is central in a frequentist approach to statistics, consider the following reflection from (Radin 1997; 47). Based on a hypothetical genetic-engineering experiment he shows why we cannot have high confidence in an experiment if we have too few trials: "… (suppose) we find that 70 out of 100 births that were genetically designed to produce boys actually resulted in boys. This is a success rate of 70 % instead of the 51 % average rate expected by normal population births. This experiment with a 70 % success rate would result in odds of 10,000 to 1 that the genetic-engineering was better than the chance expected rate. This would convince most scientists that the genetic engineering method was effective. Now, suppose that a sceptic came along and tried to replicate this experiment, but only checked 10 births. To his surprise he found 7 boys, also for a 70 % success rate. The problem is that because the smaller experiment had less statistical power (fewer trials provide less confidence in the point estimate), this would result in odds against chance of only 5 to 1. Since by convention we need odds against chance of at least 20 to 1 to claim a statistically significant result, the sceptic could loudly proclaim that the replication was a failure. … even though it found exactly the same male birth ratio of 70 % as the first study!"

This example illustrates the need for reflection on the relationship between sample size and the trustworthiness of the inferences drawn. In an experimental setup with a test group and a control group we would expect that the degree of confidence we can have as to the results—e.g. the degree of confidence with respect to an acceptance or rejection of the null hypothesis—would depend on the amount of data collected. While this will hold if the data collection is done correctly, it will not hold if there are errors in the collection. For example, if the sample is unbiased then the greater the sample size, the less the sampling error (the difference between the characteristic measured in the sample and the measure that would have been obtained had the total population been examined). However, if the sample is biased, no matter how large it is and how small the sampling error, there will still be errors due to the sample not being representative of the population it is presumed to reflect.

A typical example of how such error is independent of the sample size is if a measuring instrument has not been properly calibrated. Assume that a special thermometer is to be used to provide precise measures of the temperature in a Petri dish used to culture cells in a biology experiment and that the thermometer measures 0.1 °C too high as compared to a standard thermometer. Then, although the sampling error will be reduced if the sample size is increased, no matter how many measurements are made, both the individual measurements and statistical measures based on the whole sample, say the average of all the measurements, will have an error of 0.1°.

To demonstrate how important such an error can be, consider the following story, not from the natural sciences, but from a presidential election in the USA. In 1936 there were two major candidates for the presidency, Franklin D. Roosevelt and Alfred Landon. According to a huge survey carried out by the influential weekly magazine, *The Literary Digest*, Landon was highly likely to win with an overwhelming majority; the prediction was Landon 57 %, Roosevelt 43 % while the final results were Landon 38 % and Roosevelt 62 %. The magazine was so totally discredited that it eventually went out of business. It was the sampling technique that led to this amazingly incorrect result. The magazine had polled an astronomical 10 million people (!) of which roughly 2.4 million (!) responded by returning a post card with a tick beside the name of which of the two candidates they intended to vote for. In spite of the huge number of replies, the sample was strongly biased and did not reflect the preferences of the total population of voters; the survey was only carried out amongst its own readers, as well as registered owners of automobiles and telephones. Such a sample population had, on the average, a significantly higher income than the average American voter, and therefore the result of the poll was strongly in favour of Landon, whose political party, the Republicans, received greater support from more wealthy citizens than did the Democrats.[9] The link http://historymatters.gmu.edu/d/5168/ provides a copy of the magazine's article based on its survey and where it predicted (incorrectly) a landslide victory to Landon.

Another question to be considered when designing one's data collection is that of the absolute size of a sample compared to the size of the total relevant population. For example, suppose one is to investigate the potential benefit of a new drug for a sickness that is both serious and rare. Since there are relatively few people who suffer from the disease, one could consider attempting to test the drug by involving all those who are recorded as having the illness and who are physically capable (and willing) to participate in the test. One half could be in the test group, the other half

---

[9]But there is yet another lesson about bias to be learned from this story, although it may not be directly applicable to data collection in the natural sciences. The size of the actual sample (2.4 million potential voters), huge though it was, was only about one-fourth of what was originally intended (more than 10 million potential voters were contacted). People who respond to surveys may be different from people who do not respond and thus may not be representative of the population they are assumed to represent.

could be in the control group. However, a smaller but well-designed representative sample might be able to provide accurate data with significantly lower costs than a complete survey. Involving all the patients (a total sample) corresponds to what many countries do when they periodically carry out a census and attempt to survey all the citizens rather than relying on a statistical sample. I note that in 2000 the US Supreme Court ruled that a census be performed rather than relying on a very large and well-designed sample. According to http://www.factmonster.com/ipka/A0905361.html this led to a budget estimate increase from $1.723 billion to $4.512 billion; the final costs of the census, that employed roughly 860,000 temporary workers, were about $6 billion.

8.3 Bayesian Versus Frequentist Approaches to Inference

Statistics (or "applied inductive logic") contributes to scientific research in two principal ways: (1) it helps us to design efficient experiments for gathering data, and (2) it helps us to formulate inference procedures for analysing the data. But preceding the experiments and analyses is a more basic task: formulating the questions to be answered/the hypotheses to be tested. As we saw in our discussions of scientific method in Chap. 7, this, in practice, is not (should not be) just a linear process whereby scientists start a research project by formulating their research questions/hypotheses, then design their data collection processes/experiments, and then, after the data are collected, choose an appropriate statistical paradigm to analyse the data. There often is an interaction between how one chooses to formulate one's questions/hypotheses and how one chooses to answer them! Unfortunately, there appears to be little conscious awareness as to this among younger researchers. Often it is only after Ph.D. students have collected their data that they consider how to analyse them. Thus, there is often very little reflection as to the choice of one's approach to inductive logic. And as a result, for many researchers the statistical paradigm is not questioned—it is unconsciously taken for granted to be the frequentist approach. And since the frequentist paradigm pervades contemporary scientific research as well as the teaching of statistics, most scientists are either not aware of an alternative approach, the Bayesian paradigm, or if they are, do not consider replacing the frequentist approach with the Bayesian approach since it is not "standard" in mainstream scientific discourse.

The following presents food for thought on how a well-grounded and rational choice between these two dominating approaches can/should be made by a reflective scientist. At the risk of some repetition, the following are the major elements in the debate as to the relevance and utility of the two paradigms.

Hypotheses and data

Although Bayesian and frequentist paradigms have the same objective of drawing inference and learning from observations and experiments, and although they both structure their approaches based on three key elements: hypotheses, data, and inference procedures, they in fact ask very different questions.

Bayesians ask: What is $P(H|D)$—the probability that a hypothesis H is true given the data collected D as well as any knowledge prior to the investigation? Frequentists ask: What is $P(D|H)$—the probability that the data would have been observed given the hypothesis and use this to determine the resultant error rates (of rejecting a true null hypothesis or accepting a false null hypothesis).

Assume data is available from an experiment involving two different treatments. The Bayesian approach allows researchers to ask and answer questions such as: Given the experimental data and prior knowledge or belief as to pre-experimental data, what is the probability that treatment A is more effective than treatment B? The frequentist approach enables researchers to ask and answer questions such as: Given the assumption that the null hypothesis of "no treatment differences" is true, what is the probability of an outcome as extreme as or more extreme than the actual observations? In this case, there is no specification of an alternative hypothesis—so the analysis *cannot* provide evidence for or against the null hypothesis since its competitor is not specified.

The subjectivity debate

As the discussion so far has indicated, the best known element of the debate is the concept of prior probability. In the Bayesian approach, the researcher requires prior probabilities for the hypotheses; they are evaluated prior to observing the actual experiment's data. Frequentists argue that such priors may be difficult to specify accurately, burden the scientist(s) who are to supply them, and may introduce undesirable subjectivity in the form of beliefs into what would otherwise be an experiment characterized by objectivity, often referred to as a hallmark of good science. They argue that their approach does not presume the existence of a prior and only focuses on data from the current experiment, data that are specific and objective, there for everyone to see.

Frequentists can also argue that in a research team, different individuals may have different priors based on their different experiences and beliefs, whereby the resulting posterior will differ depending on whose prior was used. In addition, since the frequentist approach pervades contemporary scientific research, as well as the teaching of statistics, they also argue that it would be unwise to replace it with an approach which is not a "standard" in scientific discourse.

Bayesians reply that frequentist methods are also based on subjective input since no matter what methods are used, the values and aspirations of the researcher affect the investigation's design and data collection; it is impossible to investigate without a prior somehow affecting the results. According to (Berger and Berry 1988; 159, 165), frequentist methods "… depend on the intentions of the investigator, including intentions about data that might have been obtained but were not. … few researchers realize how subjective standard methods really are … common usage of statistics seems to have become fossilized, mainly because of the view that standard statistics is *the* objective way to analyse data. Discarding this notion, and indeed embracing the need for subjectivity through Bayesian analysis, can lead to more flexible, powerful, and understandable analysis of data."

I note though that (Chalmers 1999; 189) draws the exact opposite conclusion in cases where the prior is a subjective estimate: "It would seem that, provided a scientist believes strongly enough in his or her theory to begin with (and there is nothing in subjective Bayesianism to prevent degrees of belief as strong as one might wish), then the belief cannot be shaken by any evidence to the contrary, however strong or extensive it might be."

Furthermore, Bayesians argue that if there is substantial prior knowledge efficiency is gained by using a well-founded prior.[10] In addition, if there is negligible previous knowledge, a non-informed prior can be used (as was done when starting the marble experiment, where we started with ignorance as to which bag was being used, and therefore assigned odds of 1:1 as the prior odds) and that the learning process will eventually lead to a well-documented decision as to the truth or falsity of the null hypothesis being tested.[11] Furthermore they can argue that the robustness of the results can be checked by varying the prior and seeing how the results change. According to Davies (2010; 127): "The Baconian idea (he refers to Francis Bacon, 1561–1621, whom he considers to be the progenitor of the scientific method; my comment) that objective evidence should precede the drawing of conclusions has been found not to work as well as a combination of data and expert judgement, modified by later experience. Indeed it is now being suggested that this might be how the human brain works, and might be the reason that we are able to cope in ill-defined situations that current computers cannot handle."

With respect to the criticism of Bayesian analysis in situations where a team is involved, making it potentially difficult to arrive at a reasonable prior, Bayesians reply that even if the beliefs of individual scientists may differ in the beginning, as the evidence starts to appear the beliefs will tend to converge.

The concept of convergence is also important when considering the question of whether researchers will get the same answers to their questions no matter whether they follow a Bayesian or a frequentist approach. In the case of experiments that generate few data, different conclusions may very well result, so in such cases, the

---

[10]Although this is often the case in the natural sciences it is even more so in the social sciences, where it tends to be the rule rather than the exception. For example, this is very often the case in problems faced in economics or management, where decisions are to be made and where a decision maker typically has such prior knowledge based on experience (and intuition), often supported by expert evaluations and the like.

[11]A counter argument provided by some of those who deny the reasonableness of using subjective priors is that in the absence of a firmly established prior, it is necessary to list all the possible hypotheses and distribute probabilities among them, for example by ascribing the same probabilities to all of these using the principle of indifference (as we did with the marbles example). They then ask where such a list can come from—and argue that in principle the number of hypotheses can be infinite leading to a probability of zero for each, whereby the analysis cannot take place; (Chalmers 1999; 178).

choice of statistical framework can be very important. However, as the volume of data increases, the influence of statistical differences diminishes and the same conclusions will be arrived at independent of which framework is used.

Learning

Bayesians argue that their paradigm has an advantage compared to the frequentist because it permits learning via the sequential progression of the analyses. For example, in experiments with structures similar to the trivial marble example presented earlier, once sufficient confidence is obtained regarding the null hypothesis, the experiments terminate.

Such flexibility is e.g. of significance in clinical trials where a Bayesian approach is well-suited to adapting to information that accrues during a trial; results can be continually assessed which enables modification of the design of the trial. Prior to the start of a clinical trial, a trial protocol is prepared. Often this will include stopping rules with respect to interim analyses of trial data as it accumulates. For example, a stopping rule might be established in order to prevent participants from continuing to receive a drug that the statistical tests indicate is unsafe, or to stop the testing if statistical tests show it to be highly successful, in which case permission will be sought to make the drug generally available (Whitehead 2004). An example of the latter rule is the trial of the AIDS drug AZT which appeared to have a dramatic effect on the participants in a double-blind test, leading to a suspension of the test and provision of the drug to the members of the control (placebo) group.

According to (Berry 2006; 27) who comments on this potential of Bayesian clinical trials to improve the effectiveness of drug development: "A Bayesian approach is ideally suited to adapting to information that accrues during a trial, potentially allowing for smaller more informative trials and for patients to receive better treatment. Accumulating results can be assessed at any time, including continually, with the possibility of modifying the design of the trial …."

So Bayesians argue that their paradigm permits learning via the sequential progression of the analyses and that their approach remains valid even in situations characterized by new and surprising information, thereby allowing much greater flexibility than frequentist methods. In contrast, they contend that when frequentist measures, such as *p*-values are used, "every detail of the design, including responses to all possible surprises in the incoming data, must be planned in advance." (Berger and Berry 1988; 165).

Interpretations

Bayesians and frequentists differ in their interpretations of probability. For frequentists, probabilities cannot include or be based on subjective estimates or beliefs but are exclusively long-run frequencies for an event or outcome of an experiment that is repeated numerous times, with the repetitions being either actual or imaginary. I note in this connection that there are contexts where the notion of probability is divorced from frequency and thereby prohibit the use of a frequentist approach. For example, in the introduction to this chapter we considered a number of common sense usages of the concept of probability such as: "the probability of rain tomorrow", "…the probability of being invited to a classmate's birthday party" and

"… the odds are in favour of team X winning over team Y in a highly publicized soccer match". While these are all meaningful from both a common sense and a Bayesian perspective, they are meaningless from a frequentist perspective as they all refer to one-off, non-repeatable events.

For Bayesians, probabilities can be both personal degrees of belief in hypotheses, given the researcher's relevant knowledge, and relatively objective estimates based say on a series of earlier, documented experiments/observations.

Another area where Bayesians and frequentists differ in their interpretations deals with the analyses performed. Frequentists tend to interpret statistical analyses, which are of the form $P(D|H)$ with Bayesian meanings—as though they were of the form $P(H|D)$. In other words, they tend to be unaware of the fact that the frequentist inference is *not* founded on the influence of data on hypotheses, while the Bayesian paradigm is!

Continuing this line of thought, scientists also tend to misinterpret frequentist reports based on the *p*-values As we demonstrated, the *p*-value is *not* a measure of an experiment's evidence for or against a hypothesis—it only answers the question: Assuming the null hypothesis is true, i.e., that we are sampling exclusively from a population where the null hypothesis is true, what is the probability of observing an outcome at least as far from what is expected as that actually observed in the experiment? For example, if we consider observed differences between two treatments of an illness, a *p*-value of 0.05 does *not* mean that there is only a 5 % probability that the observed differences between the two treatments is due to chance and that there is a 95 % probability that there are real differences. Instead, it simply tells us that if the null hypothesis of no treatment differences is true, we can only expect to observe the differences 5 % of the time—nothing more, nothing less. If the *p*-value were to say anything about the probability that a hypothesis is true, as we have just seen from our treatment of the Bayesian paradigm, one must specify the hypothesis set and specify the prior probabilities of the hypotheses—and this is *not* done in the frequentist approach.

We can see this by comparing the two tables from the marble experiment. The Bayesian analysis (Table 8.2) has H_R (four red marbles and one green marble) compete with the alternative H_G (one red marble and four green marbles) while the frequentist analysis (Table 8.4) attempts to assess H_R as the null hypothesis on its own, i.e., without *specification of its competitor*. It should be clear that an experiment *cannot* provide evidence for or against a hypothesis unless its competitor is specified!

To illustrate this strong statement, suppose we know nothing about the contents of the bag other than that it contains marbles that are either green or red—but nothing about how many green and red marbles there are. Then the probability of a red draw, $P(red)$, can be anywhere in the range [0, 1], including the earlier null hypothesis H_R that the bag contains four red marbles (and one green marble), corresponding to $P(red) = 0.80$. Suppose now that 13 draws result in four red draws (as was the case in the analysis in Table 8.2), providing us with an estimate of P (red) as 4/13 or roughly 0.31. Does this support or go against a null hypothesis H_R that the bag contains four red marbles and one green marble? It depends on the alternative hypothesis. If it is $P(red) = 0.15$, then the result of the experiment provides evidence against the null and for the alternative (since the result 0.31 of the experiment is closer to the alternative hypothesis of 0.15 than to the null hypothesis of 0.80). On the other hand, if the alternative hypothesis is that P (red) = 0.95, then the same experiment gives evidence that supports the null (0.15 is closer to 0.80 than to 0.95)! So if an alternative hypothesis is not specified, as in a frequentist approach, then the outcome of an investigation *cannot* provide evidence either for or against the null hypothesis.

The stopping rule

As indicated earlier, the Bayesian and frequentist paradigms have strikingly different views about the influence that the stopping rule should have on statistical inferences. Many scientists would be surprised to learn that changing a stopping rule such as the rule used in Table 8.4, "the experiment stops at 13 draws", could change *p*-values, even if the data from one's experiment remained exactly the same. For example, consider an additional stopping rule for the marble experiment: stop after nine green draws. Both of the rules would have led to the same data that were a result of the experiment presented in Table 8.2. However, the rules would result in different *imaginary* outcomes that would occur since a frequentist analysis presumes an infinite number/a very large number of replications of an experiment. The first stopping rule (used in the analysis shown in Table 8.4) has exactly 14 possible outcomes, while the second rule has an infinite number of possible outcomes, ranging from zero to an infinite number of red draws. What is to be noted here is that the probabilities that are a result of the additional stopping rule are different from those listed in Table 8.4. In fact numerous other stopping rules could be chosen that would all stop this experiment after 13 draws (for example, stop after four red draws followed by two green draws), but lead to different *p*-values for the exact same experimental data! (Gauch 2003; 277)

So different stopping rules can generate different imaginary experiments, thereby generating different *p*-values and different conclusions as well. In contrast, the Bayesian paradigm, via its reliance on the likelihood, P(D|H), is based on the principle that statistical procedures and inferences should depend only on the actual outcome of an experiment, and therefore not on possible but imaginary other outcomes, as does the frequentist paradigm. This is in stark contrast to the argument by frequentists that their paradigm is objective while they criticize a Bayesian paradigm for permitting subjective evaluations via the prior. Both sides of the Bayesian-frequentist debate can chastise the other for introducing subjectivity into what are supposed to be objective statistical methods for inference.

Finally here, Bayesians argue that their paradigm has an advantage compared to the frequentist because while the frequentist approach presumes that stopping rules are determined before an experiment commences, their approach permits learning via the sequential progression of the analyses. As we have commented on earlier, such flexibility is e.g. of significance in clinical trials where a Bayesian approach is well-suited to adapting to information that accrues during a trial.

The influence of computer software

Having just considered the potential influence of stopping rules, it is depressing to note that standard software packages for frequentist tests provide automatic, supposedly objective answers without ever asking the user to specify a stopping rule. In addition, the output of such software packages is based on implicit assumptions about the experiment being analysed without ever making these assumptions explicit or asking the user whether they are acceptable given the specific characteristics of the investigation! My observations of the statistical analyses provided by Ph.D. students in their theses strongly indicate as well that many of them know little, if anything at all, about the statistical tools that they are using! I note in this connection that there has been a huge increase in the development of more refined statistical tools in recent years.

Nevertheless, it appears that the greatest influence on the statistical practice of many younger as well as experienced scientists is not what they have learned about uncertainty, probability and statistics, but rather the statistical software they use more or less routinely, without really understanding the outputs or the underlying theory and assumptions.

I note in this connection that many young researchers ask why, given the availability of such software for carrying out statistical analyses, they should bother to learn even the fundamental aspects of probability and statistics. After all, isn't statistics just a tool, so why is it important to think about the foundations for such a tool, as we are doing here?

This is a good question. After all, a person doing medical research may use a scanner to measure the effect of a new drug on tumors, an astronomer or a chemist may use a spectrometer to measure properties of light to identify materials, a physicist may use a Geiger counter to detect the emission of nuclear radiation, a biologist may use an electron microscope for 3D tissue imaging. In all these cases it may not be a requirement that the researcher has a deep understanding of how the tool/technique works (and therefore also of the theories that underlie its use) for her results to be accepted by the scientific community, although it will certainly help her to have background knowledge of how and why the instrument she is using works. So there is a similarity here to the use of statistics in research. But there is also a significant difference: The tools just referred to—scanners, electron microscopes etc.—are used to *produce* data while the researcher's statistical tools will be used to *analyze* the data and draw inferences.

To conclude these reflections on the need for scientists to be more active participants when they use software packages to analyse their data, consider the following thought provoking reflections from an article on "Unreliable Research" in the *Economist*, Oct 19, 2013, also available as (http://www.economist.com/news/briefing/21588057-scientists-think-science-self-correcting-alarming-degree-it-not-

trouble). The article presents some rather shocking statistics that serve as the basis for the following conclusions: "Statistical mistakes are widespread. The peer reviewers who evaluate papers before journals commit to publishing them are much worse at spotting mistakes than they or others appreciate. ... There are errors in a lot more of the scientific papers being published, written about and acted on than anyone would normally suppose, or like to think. Various factors contribute to the problem. Statisticians have ways to deal with such problems. But most scientists are not statisticians. ... scientists' grasp of statistics has not kept pace with the development of complex mathematical techniques for crunching data. Some scientists use inappropriate techniques because those are the ones they feel comfortable with; others latch on to new ones without understanding their subtleties. Some just rely on the methods built into their software, even if they don't understand them."

Decision making

Bayesian inferences lead readily to decision procedures whereas frequentist inferences do not. Bayesian analysis of prior and new data yields posterior probabilities of the occurrence of various states of nature. These can be multiplied by the entries in a consequence matrix, where each entry specifies the benefit or loss (utility or disutility) of each action-state combination, to calculate each potential action's expected utility. This provides the input to a decision criterion (for example maximizing the expected utility or maximizing the minimum utility) that can be used to determine an optimal decision. This natural connection between inference and decision works because Bayesian inferences provide probabilities of the states of nature, whereas frequentist inferences provide no satisfying connection to decision problems.

It is also interesting to note here that practical decision making, particularly in matters dealing with management and economics, often takes place in contexts characterized by considerable uncertainty. In such situations, it is natural that decision makers draw upon their subjective opinions/evaluations, based on their experience as well as more objective evidence. In other words, decision making is more tuned into the learning aspects of Bayesian statistics than is a purely inferential investigation.

Is the Bayesian paradigm used in practice by scientists?

In spite of the frequently cited criticism of the use of the Bayesian prior, such an approach is gaining renewed recognition, not only in the social sciences, but also in the 'harder' sciences, in particular physics. For example, the Bayesian approach has been used for years in connection with the re-estimation of parameters at CERN (The European Council for Nuclear Research; in French: Conseil Européen pour la Recherche Nucléaire), whose purpose is to operate the world's largest particle physics laboratory). D'Agostini (1998; 1) reflects on the use of statistical reasoning in high energy physics as follows: "The intuitive reasoning of physicists in conditions of uncertainty is closer to the Bayesian approach than to the frequentist ideas taught at universities and which are considered the reference framework for handling statistical problems." Similarly, Lyons (2013) provides analyses and illustrative examples of the relative strengths and weaknesses of Bayesian and frequentist approaches in dealing with parameter determination and the testing of hypotheses in particle physics. He concludes by referring to analyses at CERN where both approaches are used in the search for new phenomena: "...similar answers would strengthen confidence in the results, while differences suggest the need to understand them in terms of the somewhat different questions that the two approaches are asking."

Davies (2010; 127) presents the results of a recent study which showed that the proportion of Bayesian articles in statistics journals has increased from 10 to 35 % since 1970. He concludes: "The Bayesian approach has now been so well integrated into statistics … the once fatally subjective element of Bayesian theory has become a merit—its ability to make use of expert judgement *as well* as objective evidence when judging the likelihood of various outcomes." A similar observation is provided by the former president of the American Statistical Association and former editor of the *Annals of Applied Statistics Journal*: "Maybe a quarter of the papers (sent to the journal for publication) used Bayes' theorem." (Efron 2013; 1178).

8.4   Brief Overview of "Uncertainty, Probability and Statistics"

1. There exists a complex and intimate relationship between observing/testing, probabilistic reasoning, statistical analyses, and the interpretation of the results of one's observations/experiments in the sciences.
2. *Probability*: Deductive approach to uncertainty; simple rules together with mathematically coherent and physically sensible axioms lead to the deductive derivation of theorems. Probability serves science in the same way that a single unified theory of arithmetic does.
3. *Statistics*: Inductive approach to uncertainty.
4. It is easy and commonplace to reason poorly as to probabilities; e.g. the "blood test" example where we ignored prior information, and the "two girl" example where we had difficulty dealing with conditional probabilities.
5. *Bayes' Theorem*: $P(H|D) = P(D|H) \times (P(H)/P(D))$. For ease of presentation we interpreted H as a hypothesis and D as data that has been collected to test the hypothesis. Since $P(D)$ is unconditional and therefore does not affect the relative probabilities of different hypotheses we obtain the simpler form: $P(H|D) \propto P(D|H) \times P(H)$, i.e. Posterior $\propto$ Likelihood $\times$ Prior.

   The Posterior, $P(H|D)$, is the probability of hypothesis H being true given the new evidence/data D; the Prior, $P(H)$, is the previous/initial/old belief/knowledge as to the probability of H being true; the Likelihood, $P(D|H)$, summarizes the impact that the new data has on the probability of the hypothesis H being true.

   Confusion about conditional probabilities can lead scientists to confuse $P(H|D)$ and $P(D|H)$.
6. Alternative formulations that follow directly from the formula for Bayes' Theorem:

   Enquiry involving n mutually exclusive and jointly exhaustive hypotheses H_i, $i = 1, …, n$. For all j, $j = 1, …, n$,

$$P(H_j|D) = (P(D|H_j)/\sum_{i=1}^{n} P(D|H_i) \times P(H_i)) \times P(H_j)$$

For two mutually exclusive and jointly exhaustive hypotheses this reduces to:

$$P(H_1|D) = [P(D|H_1) \times P(H_1)]/[\{P(D|H_1) \times P(H_1)\} + \{P(D|H_2) \times P(H_2)\}]$$

For two mutually exclusive and jointly exhaustive hypotheses we obtain the Ratio form:

$$P(H_1|D)/P(H_2|D) = (P(D|H_1)/P(D|H_2)) \times P(H_1)/P(H_2)$$

7. *Bayesian approach to inference*: Determine the probability a hypothesis is true given prior knowledge and new evidence/data. Major criticism: Use of subjective priors. Example of use: Bayesian Decision that combines measures of utility (costs and benefits) with probabilities as to outcomes of alternative decisions, where these are updated using Bayesian approach, so as to obtain expected utility of alternative actions/decisions.

8. *Frequentist approach to inference*:

   (a) Null hypothesis and an Alternative/Research hypothesis (what the researcher really wants to test—the opposite of the null hypothesis).

   (b) Type I error: Reject null hypothesis when it is true/correct. Type II error: Accept/fail to reject null hypothesis when it is false. Logically there exists a trade-off between these—but this is almost always ignored in testing (software packages).

   (c) Problems here: (i) Scientists often interpret frequentist analyses as though they are Bayesian: the latter provides, and the former does not, data-based evidence on hypotheses; (ii) influence of stopping rules in frequentist approach; these can introduce subjectivity into what is considered an objective approach. (iii) users' lack of knowledge as to influence of computer software on one's analyses.

9. *Summing up*:

   (a) Statistics (applied inductive logic) can contribute in two major ways: (i) in designing efficient experiments for gathering data, (ii) in formulating inference procedures for analysing data.

   (b) Bayesians ask: What is P(H|D)? Permits not only inferences but also decision making given data/evidence and any prior knowledge in form of prior probabilities.

   (c) Frequentists ask: What is the probability of an outcome at least as extreme as is obtained from an observation or experiment under assumption that the null hypothesis is true? Basic idea of frequentist inference is that procedures with low error rates (in principle both Types I and II but in practice only Type I) are suitable for sorting true/acceptable from false/not acceptable hypotheses.

Chapter 9
Research

At this point it is appropriate to specifically focus on the concept of research, which is at the very heart of the natural sciences and of scientific method. In the preceding chapters 'research' as such has not been defined and delimited, although it has continually been referred to indirectly via consideration of the formulation of hypotheses and theories, measuring, testing, experimenting, scientific method, etc. So while the earlier chapters provided a broad conceptual framework for empirical research, the present chapter goes into further depth and deals with the following topics: (1) types of research, (2) multidisciplinary and interdisciplinary research, (3) the value of research skills, (4) formulating a research problem, and (5) research in relation to teaching and publishing. Just as was the case with the discussion of scientific method in Chap. 7, these topics are relevant for all domains of the natural sciences.

9.1 Basic, Applied and Evaluation Research

A simple and straightforward definition that has characterized our reflections so far is:

> Research in the natural sciences is a systematic process for developing new knowledge of the physical world that can be shared and contested.

The definition underscores that research is not performed just for the sake of the individual researcher and that its quality is determined by members of the scientific community via the sharing and testing that is so fundamental to science. In other words, even though some of the world's most famous theoretical developments have been the result of intuition and imaginative suppositions (we referred earlier to Einstein's so-called "gedankenexperiments"), these activities are not included here; when we refer to research in this section, we refer to *empirical* research.

By the word "systematic", I mean following principles and employing procedures that are in harmony with what is considered "normal science" (in the Kuhnian sense, as described in Chap. 2, Sect. 2.2) within one's field. "Systematic" also implies here that a researcher in the natural sciences attempts to the best of his or

© Springer International Publishing Switzerland 2016
P. Pruzan, *Research Methodology*,
DOI 10.1007/978-3-319-27167-5_9

her ability to be neutral and objective, while being aware that this can only be an ideal[1] due to the influence of one's values, culture etc.

In addition to considering research as a process, the word research also refers to the product of that process—the new knowledge generated from a research investigation. In the context of the natural sciences such new knowledge is of physical reality—and focuses on description, explanation and prediction of physical phenomena and processes.

Research in the natural sciences has typically been classified under two broad headings: *basic research* and *applied research*. In addition to these two categories, we also discuss a newer type of research that has in a fairly short period of time become widespread, influential and contested: *evaluative research*.

However, before proceeding to consider these various categorizations of research, it is appropriate to introduce a subtle distinction not yet considered: between science and research. Up until now, it has been more-or-less implicitly understood that "science" is the more general term and that "research" is an activity within science. However, it can also be argued that "research" can be considered to be a broader term since not all research lives up to the norms of science while all science is research-based. For example, although living up to the norms of scientific method considered earlier, considerable research that is performed in private companies and in the military may be kept secret and thus not live up to other, more tacit, sociological norms for science, including that the results of one's research are to be published, be available to testing by one's peers and contribute to the scientific community's development of a shared knowledge reservoir. Perhaps one can say then that research is a necessary aspect of science—that science is research-based scholarship such that research is a necessary condition for the performance of science, while not all research lives up to the norms of science.

Basic research

Basic research involves the generation of new knowledge via the study of some phenomenon of nature without regard to the utility of the research. Ideally, it is based entirely on curiosity, upon being able to describe and/or understand the phenomenon. It could, for example, involve investigations of the many possible implications of a hypothesis regarding the phenomenon. In such cases, the investigations may be entirely theoretical, using deductive logic and mathematics together with library research. This has certainly characterized much of research in physics, where creative intuitive thinking has led to some of the greatest advances in science, e.g. Einstein's special theory of relativity or Planck's quantum hypothesis. However, even a hypothesis that represents an intuitive leap also eventually leads to the design, construction, and execution of an experiment to test

---

[1]This is not an ideal for many scientists in the social sciences, particularly those who employ a constructivist approach; see the discussion in Chap. 2, Sect. 2.7 on 'Realism and anti-realism. In addition, so-called 'action researchers' employ a participative or collaborative approach to research, considering themselves as change agents, often in pursuit of social justice (Brydon-Miller et al. 2003).

the implications that follow from the hypothesis. This has been the case with curiosity inspired investigations that eventually led to technological breakthroughs such as lasers, the World Wide Web and LED (light-emitting-diodes). Of course, in cases where the research is purely explorative, the investigations will not explicitly be based on hypotheses, but will be oriented towards providing a description of the phenomenon.

As examples of basic research, we may consider research regarding the expansion of the universe and investigations into the fundamental constituents of matter. Modern investigations in either of these fields can be very expensive without an immediate return on investment of money, time and effort—and without any indication or consideration of the possible benefits to mankind. But scientists nevertheless engage in them (and receive financial support for their research) because such basic research contributes to the storehouse of knowledge. It is of course less clear what motivates institutions and societies to support such extremely costly basic research projects. Prime examples are the high-energy particle accelerators employed to probe the quantum world; we have several times referred to the huge international CERN laboratory, the site of the Large Hadron Collider, the world's largest and most powerful particle accelerator.

This raises the issue that what a scientist may consider to be basic research may very well be financed by sources that expect some kind of long-run benefits/returns from their investments in the research. An example could be basic research into the habits of a particular insect, where the research is sponsored by a pharmaceutical company, or by a chemical company producing pesticides, or by a governmental health agency that is interested in knowledge regarding the insect's possible role in the spread of a disease. Another example could be basic research regarding the changes to the structural characteristics of a new form of plastic when it is subjected to extremely low temperatures. Such research may be designed by scientists so as to build on existing knowledge and, via the use of sophisticated experiments and computer simulations, lead to the development of new knowledge, insights and perhaps theories. On the other hand, such basic research may be considered by say NASA as being highly relevant for the development of spacecraft for the purpose of collecting rare minerals in outer space; from NASA's perspective, it can "pay" to support such basic research as a long term investment.

Applied research
In contrast to basic research, applied research has as its primary focus, not the generation of new knowledge, but the exploitation of existing knowledge to develop new technologies and artefacts—ideally for the benefit of mankind and nature.

There are innumerable examples that can illustrate the nature of applied research. Here we mention only two. One example is photovoltaic research. Basic research has shown that when light shines on a silicon crystal prepared in a special way, positive and negative charges are separated in the material, leading to a build-up of electrical voltage. This basic knowledge is used to develop solar cells that are made of semiconducting materials. Given the huge potential of solar energy, applied research here can lead to the development of inexpensive solar panels that are so efficient as to enable the wide-scale generation of "clean" electrical power when methods based on burning fossil fuels become economically inefficient and environmentally destructive. Clearly, such research is applied research and is not motivated by the creation of new fundamental knowledge. A specific example is the applied research at the National Renewable Energy Laboratory in the USA that focuses on boosting solar cell conversion efficiencies, lowering the cost of solar cells, modules, and systems, and improving the reliability of components and systems. An aim is to contribute to the goal of making large-scale solar energy systems cost-competitive with other energy sources by 2020; http://www.nrel.gov/pv/.

A second example of applied research deals with the development and use of laser technology. Basic research has shown that when atoms are in excited energy states they tend spontaneously to emit electromagnetic radiation. This results when an atom makes a transition from an excited energy level to a level of lower energy. When this occurs, the energy given to the light emitted is exactly equal to the difference in the energy states of the atom before and after the decrease in excitation level. It has also been shown that an atom not only can emit light spontaneously in going from a higher to a lower energy level but it can also be induced to emit light by light incident on it. There are atoms which possess metastable states, that is, de-excitation by spontaneous emission is extremely slow on atomic time scales. Materials containing such atoms are 'pumped up' to metastable levels by a powerful lamp. Then they can all be induced to go down to the lower energy state simultaneously resulting in an avalanche of radiation, all of the same energy and same direction (coherent). That is referred to as *l*ight *a*mplification by *s*timulated *e*mission of *r*adiation, the emphasized first letters of the words spelling the word laser.

Thus far in this story about lasers, we have considered the results of basic research. To develop a real device which exploits the principle in a manner that is technologically applicable and economically efficient requires applied research. Such research commenced in the early 1960s and technologies were developed that could produce intense and coherent (highly unidirectional) beams of light. Building on that early applied research, many types of working lasers have been perfected and are in use in diverse areas of human activity.

The distinctions as to basic and applied research are in practice often rather unclear. Very often scientists who perform more basic research are directly or indirectly affected by visions of the possible applications of the results of their research. Furthermore, since as indicated earlier, much basic research today is funded by institutions that are not simply philanthropic, there is often a bias in the selection of projects in favour of those that may lead to long term benefits via applied research. Finally, there is a personal aspect of this unclear distinction between "basic" and "applied". Many "basic research" projects do not just appear out of the blue. Often they are part of a larger "project"—say at a university department, where the individual researcher has responsibility for only one well-defined part of the project. To the younger researcher, this may appear to be basic research, while to those who support the overall project it is just a part of a larger applied research project.

Evaluation research

This is a rather new perspective on research. Although it can be considered to be a special type of applied research, evaluation research has become so widely used that it can be considered an independent category of research.

Evaluative research deals primarily with the systematic evaluation of institutions, projects and interventions (e.g. the performance of research institutions, vaccinations against influenza, use of safety belts by back seat passengers in cars, the capacity of global climate models for projecting future climate change). Such research is often classified into three types of studies: *Descriptive* (describes e.g. the goals, implementation processes, anticipated outcomes of a programme), *Normative* (evaluates e.g. a programme's goals based on a number of criteria), and *Impact* (evaluates e.g. a programme in terms of its effects on society and nature, its costs and benefits).

Given that the focus of this book is research in the natural sciences, evaluative research might be considered irrelevant; its subject matter is not "nature", but rather institutions, projects and interventions—that may or may not be oriented towards the natural sciences. However, since so much research is performed to evaluate research institutions and research projects, and since, in the context of the natural sciences, the evaluative research is typically performed by scientists, and since evaluative research can directly or indirectly affect *your* research, it deserves consideration by a reflective scientist.

At the level of the individual institution and society, a typical example of evaluation research is the evaluations being made in many countries throughout the world today of the research performed at their universities. Typically these evaluations are sponsored by governmental bodies and can serve as a basis for research grants by e.g. funding institutions such as the University Grants Commission in India, the United Kingdom Research Councils, or the National Science Foundation in the USA. Such evaluations of research are even being made at the international level. For example, as early as 1997 the OECD (the Organisation for Economic Co-operation and Development with 34 member countries) published the report *The Evaluation of Scientific Research: Selected Experiences*, focusing on the evaluation of government-supported research and emphasizing that "Governments need such evaluations for different purposes: optimizing their research allocations at a time of budget stringencies; re-orienting their research support; rationalizing or downsizing research organizations; augmenting research productivity, etc. To this end, governments have developed or stimulated research evaluation activities in an attempt to get 'more value for the money' they spend on research support."

At the department level, evaluation research is typically performed so as to enable the leadership of a university or research institution as well as of the individual departments to improve the research environment, the effectiveness of the research performed and the teaching/publications generated by the department.

No matter at what level of organization the evaluations are made, they are typically performed by a group of experts, most of whom are not associated with the institution, department or project whose research is being studied. Nevertheless, such evaluative investigations ordinarily assume a high degree of participation and cooperation (e.g. as to the provision of factual data as well as interviews) with the unit/activity being evaluated. Often the evaluation also includes a self-evaluation by the organisation or project being reviewed. Such self-evaluation can have a dual purpose: it provides documentation and serves to inspire the organisation/project to improve its performance.

Evaluative research dealing with research is heavily based on quantitative measures. For example, in a university context the evaluations typically include the number of various types of publications produced (e.g. the number of articles in respected journals, the number of monographs, conference presentations, and so on), the number of citations in respected journals as to publications by members of the institution,[2] the number of media reports, the number of patents received/applied for, the number of awards and prizes received, the number, type and value of research contracts, and so on and so on. In addition, questionnaires are often employed to collect quantitative data. But also qualitative measures may be employed, for example based on interviews (of peers, representatives of donor organisations, and students), the reading of media reports and other relevant documents, etc.

Larger individual research projects are also being evaluated, for example in connection with a midway report. The evaluation is becoming increasingly complex today as many larger research projects are being carried out by teams from a number of universities and other research institutions, often located in different countries/cultures. An example deals with the evaluation of models for climate change. The international interest in climate change/global warming led to the establishment in 1988 of the Intergovernmental Panel on Climate Change under the auspices of the United Nations and to the development of a number of global climate change models. Considerable evaluative research has been performed to assess the capacity of these models for projecting future climate change; see e.g. Randall et al. (2007).

As to more specific tools that can be employed when carrying out such evaluative research I will only here refer to the models developed by economists and management scientists for so-called "efficiency analyses" and "benchmarking". These can provide relative performance evaluations via the systematic comparison of the performance of one "production entity" against other comparable entities. Such entities can be commercial firms, hospitals, projects—and with our focus here, research organisations, universities and university departments, as well as individual projects and researchers (Bogetoft and Otto 2011). Such evaluations can be employed to provide intra-organisational comparisons (e.g. when a headquarters wants to promote efficiency in its subunits such as departments in a university), inter-organisational comparisons (e.g. between universities) and dynamic comparisons (e.g. when the performance of one or more research organisations is compared

---

[2]See for example 'Science Citation Index Expanded'. As of September 2013 it covered more than 8500 of the world's leading journals of science and technology across 150 disciplines and is available online through the "Web of Science" database, a part of the "Web of Knowledge" collection of databases (an online service covering 10,000 leading journals of science, technology, social sciences, arts and humanities and over 100,000 book-based and journal conference proceedings).

over different time periods) and, as just considered above in connection with climate change, comparisons of the efficacy of models.

A special example of evaluation research that deals with universities, but is not focused on their research, is the development of a ranking that can help prospective students to choose which university to apply to within their anticipated field of study. For example, the website: http://www.shanghairanking.com/FieldSCI2012.html provides a yearly "Academic Ranking of World Universities in Natural Sciences and Mathematics". It also provides an overview of the methods used to develop the ranking. I note in this connection that evaluations and rankings of service institutions in general, for example hospitals, can provide perspective users with a basis for making a choice as to service-provider in a far more systematic manner than was possible earlier.

Due to the rapidly growing importance of evaluative research, also in comparing and evaluating organisations/projects in an international context, independent governmental bodies increasingly are being established to provide quality evaluative research. For example, in my own country, the Danish national knowledge centre in the field of evaluation, EVA (Danish Evaluation Institute), assesses and develops the quality of educational institutions and programmes; (http://english.eva.dk/).

Since evaluation research is rather new, at least as compared to the more traditional classifications of research, and yet can have a great influence on many individuals, projects and institutions, it is itself the subject of considerable research; evaluation research is increasingly being evaluated! For example, since the mid-1990s, the international and interdisciplinary refereed journal *Research Evaluation* (published by Oxford Journals) has covered the emerging methods, experiences and lessons for evaluating research, ranging from the evaluation of an individual research project, through portfolios of research and research centers, up to inter-country comparisons of research performance. There exist several other such international journals that focus on more specific disciplinary fields, for example *Evaluation: The International Journal of Theory, Research and Practice* (published by SAGE) that focuses on evaluation of research in the social sciences and related disciplines, and *Educational Research and Evaluation* (published by Routledge) that addresses research and evaluation in education.

9.2 Multidisciplinary and Interdisciplinary Research

Originally I considered treating this subject in the preceding chapter's Sect. 8.1 on "Types of research". However, the categories to be treated here are not "types" of research; they deal instead with different ways of *organising* research that are playing an increasingly important role in many fields of science. Let us start by considering the most well-known and least demanding of these two forms of cross-disciplinary research: multidisciplinary research.

Multidisciplinary research

Multidisciplinary research involves teams of scientists who come from various disciplines and who work together on a research project. The various teams continue to work *within* the frameworks and traditions of their disciplines. In this way they continue to draw upon a broad pool of shared knowledge, skills, technologies and paradigms that they are familiar with. At the same time, their research activities are pooled together with those of other teams. Such research has characterized both basic and applied research within larger and more complex research projects where two or more teams (e.g. departments within a university,[3] units within a large company, or teams at different universities and companies) each work on one aspect of a research project. A typical example of a very large applied multidisciplinary research could be the development of a new type of spacecraft or high-energy particle accelerator, where a number of different teams each work with one of the many design and engineering problems that have been identified based on earlier more fundamental scientific, technical and political/business goals.

Such research is *not* characterized by an attempt to integrate the teams or to develop a new hybrid discipline. Typically, the individual teams are located in the departments they are employed in and when the multidisciplinary research project has been completed or otherwise terminated, the scientists simply continue their work within the environments they come from; no new permanent organisational structures have been created, no new paradigm or more inclusive scientific domain has resulted. The research in ecological and evolutionary sciences carried out in the USA by the National Institutes of Health, the National Science Foundation, and the United States Department of Agriculture provides a large scale example of such multidisciplinary research. According to the website, (http://grants.nih.gov/grants/guide/notice-files/NOT-TW-14-009.html), released August, 2014 and inviting the submission of proposals for grants, "…the purpose of the Ecology and Evolution of Infectious Diseases Program is to support multidisciplinary teams in the development of predictive models that integrate ecology and evolution with the goal of discovering principles governing the transmission dynamics of infectious disease agents to humans and other hosts. Proposed projects should include research and associated expertise in diverse disciplines including, for example, modelers, bioinformaticians, genomics researchers, social scientists, economists, epidemiologists, entomologists, parasitologists, microbiologists, bacteriologists, virologists, pathologists or veterinarians, as relevant to understanding the disease transmission system proposed."

Common to such multidisciplinary research projects is the need for a coordinating body that can design, follow and evaluate the progress of the overall project as well as the progress of the individual teams, including the many decisions as to budgets. An example of such an overarching organisation in the USA is the

---

[3]For example, in order to even provide an awareness among its seven faculties of its many and diverse research activities, Stanford University produces a quarterly journal called *Interaction*; http://multi.stanford.edu/interaction/.

National Human Genome Research Institute (http://www.genome.gov/) which led the uncovering of the full human genome sequence in 2003 and now continues to coordinate genomic research aimed at improving human health and fighting disease.

Interdisciplinary research

In contrast to multidisciplinary research, interdisciplinary research is a more recent development/ambition and is characterized by an attempt to *integrate* the perspectives/methodologies/technologies of two or more disciplines to create a new hybrid discipline. This shift in organizational design is driven by the need to address complex problems that cut across traditional disciplines, and by the capacity of new technologies to transform existing disciplines and generate new ones. Clearly, the demands of such attempts are huge and complex, dealing as they do with such basic matters as the vocabulary, traditions and world-views of the domains involved. But the challenges are not just to the practice of science. Perhaps one of the greatest challenges is to the mind-sets of those who organise and fund research; the success of interdisciplinary research can be impeded by policies on hiring, promotion, tenure, proposal review, and resource allocation that favour traditional disciplines and more traditional organizational forms and environments for research.

Since we have just considered multidisciplinary research within medical and biomedical domains, it will be instructive to focus here as well on developments within these domains to illustrate the potentials and problems associated with interdisciplinary research. The following is an excerpt from a newsletter from the National Institutes of Health in the USA dated September 6, 2007 announcing the launching of interdisciplinary research consortia to be financed by the NIH: "The National Institutes of Health (NIH) Roadmap for Medical Research will fund nine interdisciplinary research consortia as a means of integrating aspects of different disciplines to address health challenges that have been resistant to traditional research approaches. The funding of these consortia represents a fundamental change in both the culture within which biomedical and behavioural research is conducted and the culture within the NIH where research projects are normally managed by an individual Institute or Center. … As opposed to multidisciplinary research, which involves teams of scientists approaching a problem from their own discipline, interdisciplinary research integrates elements of a wide range of disciplines, often including basic research, clinical research, behavioural biology, and social sciences so that all of the scientists approach the problem in a new way. The members of interdisciplinary teams learn from each other to produce new approaches to a problem that would not be possible through any of the single disciplines. Typically, this process begins with team members first learning the language of each other's discipline, as well as the assumptions, limits, and valid uses of those disciplines' theoretical and experimental approaches. Experiments are then designed in ways that cut across disciplines, with, for example, an experiment based in one discipline producing data that can be correlated—or otherwise connected to—data generated in experiments based in another discipline. The common understanding by the team of the disciplines involved assures that this tight linkage across the disciplines is valid. These consortia will not only develop new ways to think about challenging biomedical problems, but will provide a stimulus for academic research culture change such that interdisciplinary research becomes the norm. The consortia address directly several current barriers to interdisciplinary research. The strategies for accomplishing this include: (1) dissolving departmental boundaries within institutions; (2) providing recognition of team leadership within the projects; (3) cross-training students in multiple disciplines; and, importantly, (4) changing the NIH approach to interdisciplinary research administration." http:// www.nih.gov/news/pr/sep2007/od-06.htm

There appears to be an increasing need for scientists from diverse fields to collaborate and to work together in new, more flexible organisational structures suited to evolving hybrid domains that cut across existing disciplines. At the same time, new technologies are continually being developed that can potentially stimulate and enable the necessary organisational and individual transformations. However, the task of developing environments for the interdisciplinary research—and teaching as well—will present huge challenges to the structures and mind-sets that characterize existing research institutions, as well as to the mind-sets of the individual scientists. Perhaps those commercial and nationally funded laboratories that have experience with organising complex research projects where goals are established and pursued in terms of projects rather than scientific discipline, will be well-suited to contribute to the development of environments enabling and promoting interdisciplinary research.

9.3 The Value of Having Research Skills

Younger researchers tend to be so focused on the shorter term aspects of their research that they tend not to consider the value of their research skills in a broader perspective and in relation to their longer term career paths. This is particularly true of Ph.D. students whose focus, quite naturally, is on fulfilling the requirements for obtaining a Ph.D.

Therefore, and to inspire reflection regarding such broader and more long term considerations, the following is a brief presentation of five roles/positions that are highly dependent on having well-developed research skills:

1. A researcher in an organisation.
2. A research-based decision maker, for example as a leader of a research project or a project that is dependent on research (e.g. in a company developing new pharmaceuticals or in a space research organisation such as NASA, the National Aeronautics and Space Administration in the United States, or ISRO, the Indian Space Research Organisation).
3. An evaluator of research services provided by others. An example could be the on-going evaluation and control of the quality and effectiveness of research that has been outsourced by a company seeking more cost-effective research.

4. A leader in an organisation that is highly dependent on the quality of its research; here the leader is not directly responsible for the performance of the research, but has a general leadership position. An example could be where the organisation's production of goods and services is based on research and development, no matter whether the R&D is produced internally (as in 2) or externally (as in 3).
5. A university professor with responsibility for carrying out research, for mentoring younger researchers, and for developing and implementing research-based teaching.

Each of these positions is characterized by the need for a high level of research skills—even though only numbers 1 and 5 directly deal with the performance of research.

9.4   Formulating a Research Problem

In Chap. 3 I offered considerable space to a discussion of the role of hypotheses in science and to reasonable criteria for ascertaining the relevance of a hypothesis for a research project. I mentioned there that while the literature on research offers attention to the general question of *how to test hypotheses*, it pays only limited attention to the more fundamental question of *how to develop hypotheses*. By this I do not refer to the content of the specific formulation but to the far more general question of how to formulate hypotheses that are of scientific relevance/interest—which is closely related to the overall question: how to go about selecting and formulating a research problem.

Although the creative aspects of this task can be extremely important, since in many cases the actual formulation of hypotheses involves far more than logic, knowledge and a systematic approach, we will not consider here the role of creativity or methods for stimulating creativity in research. Instead, we will simply stick to a more "logical" presentation.

As underlined in Chap. 7 on Scientific Method and the Design of Research, the first requirement before choosing to design one's research is to be familiar with the earlier work done in the area of investigation. This is easier said than done. To appreciate how difficult this task is, it is sufficient to note that there are more than ten thousand research journals (both in print and e-journals) where researchers can publish their findings; in Footnote 2 of the present chapter we referred to the "Web of Science" database that contains information on more than 8500 journals of science and technology across 150 disciplines.

Many of these are highly regarded journals while others may not be as well regarded—and publish articles whose quality can be seriously challenged, as was shown in Sect. 4.4 of Chap. 4 on Peer review. Thus, assuming the researcher has access to publications and data bases via a good library and the internet, he or she can very quickly be overwhelmed by the potential sources of data and inspiration, many of which can be worthless or even counter-productive. This is where a good research guide can be very helpful. Ideally, such a guide has both a broad and deep knowledge of the particular field you are considering and is well acquainted with many of the recognized researchers in the field as well as with their scientific production. Such a guide will have a finger on the proverbial pulse, and will not only be able to help the younger researcher in the demanding job of seeking, evaluating and sorting the available literature—s/he can hopefully also help in establishing contact with other researchers and networks, something that is becoming increasingly important in an increasingly globalized research world. The quality of the research work done in close consultation and collaboration with such a knowledgeable and helpful guide will thus, ceteris paribus, tend to be of superior quality. And if the advisor is also endowed with good communicative skills, is creative, patient, persevering, empathetic, and takes the job of being an advisor seriously, the guide can be a most powerful and inspiring mentor.

A word of caution is called for though. Many younger researchers/Ph.D. students experience that their research guide is *not* characterized by the qualities just mentioned. In fact the young researcher should not be surprised if the research guide turns out to be quite different from the kind of mentor referred to above. The guide may be extremely busy with his or her own research and teaching as well as administrative responsibilities, or even uninterested in the student's research and scientific development. He or she may also lack the humane qualities of kindness, generosity, patience and empathy that can be of great importance in establishing and maintaining a well-functioning advisor-student relationship. If this is the case, the advisor may seriously restrict the curiosity, creativity and independence of the younger, aspiring scientist; rather than being a most powerful asset, the advisor may be the greatest challenge the budding scientist faces. So it appears that a task that is important—at times crucial—and that precedes the formulation of the research problem and hypotheses is the choice of a knowledgeable, talented and dedicated research guide.

Alas, in many cases the student does not have very much say regarding the choice of an advisor, for example in situations where the advisor has been appointed by the department head even before any serious work has been done as to problem formulation, or where your project is funded from external sources and is part of a larger research project that your advisor coordinates. So in practice, often the problem is not one of choosing but of being *chosen* by a mentor!

In any case, no matter how the advisor is appointed, an extremely important challenge facing the young researcher is to develop a fruitful teamwork with the advisor and to draw upon the very best that the advisor can offer. How this can be done goes well beyond the limits of these reflections. I limit myself here to underlining that the researcher should not consider himself to simply be a servile lackey; the researcher has a great responsibility for actively contributing to the establishment of fruitful research teamwork while demonstrating a positive attitude, scientific maturity and personal integrity. As emphasized several times earlier (see in particular Sect. 2.2.4), science is also a social activity, something that seems to be ignored in the literature that describes the organization of research and teaching programs.

Once the researcher has become reasonably familiar with the work previously done on the specific problem and/or with work done on related problems, he or she can "proceed to the next stage" (I put these last few words in citation marks to underscore, once again, that the process is not linear, whereby one stage begins when an earlier stage is completed; the process is in fact dynamic, characterized by feedback, retracing one's steps etc.). This "next stage" in the evolution of one's problem formulation is to focus on questions that previous investigators have *not* investigated or that they have *not* successfully answered, or on apparent *disagreements* between the results obtained by different investigators. With respect to the last of these, it is legitimate and can be challenging and productive for researchers to formulate a problem based on experiments performed by others. Such replication need not only be an identical repetition of what was previously performed; it may also involve extending the results already obtained to a more inclusive context, thereby potentially extending the generalizability of existing knowledge.

Although it may appear to be an activity that is a result of and follows the problem formulation, the design of the experiment is another major activity that, in principle at least, should be considered prior to developing one's problem formulation. The reason is that there is always a relationship between the design of experiments to be performed and the data to be collected; it can be disastrous to formulate a problem that cannot reasonably be solved within the constraints (time, technology, budgets and expertise) that determine one's ability to collect evidence in the form of data.

Integral to this problematic is the choice of the apparatus to use in one's experiments/observations. We briefly considered this specific matter in Chap. 6 in the section on "Epistemological Strategies". The choice of apparatus can be a delicate matter with important limitations imposed by the expense involved in acquiring and using the equipment, as well as by the skill level needed in working with the equipment. A good research guide can be an invaluable help in such matters. If the equipment one would like to use is not available, there can be many possibilities of sharing equipment at other universities or research institutions. Once again, the advisor's contacts and networks can be of great importance here.

9.5 Research in Relation to Teaching and Publishing

Having just emphasized the role of "a good advisor"—something of importance not only for the Ph.D. student performing research, but also for the same person when he or she becomes a more senior faculty member and has the responsibility of guiding others—a few thoughts are now called for regarding "a good teacher" of science. Such a teacher, and here I refer mainly to teachers who give courses to post-graduate students, in particular to Ph.D. students, is characterized by keeping abreast of the research developments in his/her field and by informing the students of the latest advances in the field. Although it is in general a big advantage, it is not necessary for the teacher to be personally engaged in research activity as long as s/he keeps up to date by studying and assimilating the newest results in the field. It will of course be far better if the teacher is personally involved in research activity because then his or her personal hands-on experience can inspire the students, even if the teacher's research is not directly related to the particular subject matter being taught. After all, not all teachers are so privileged as to be assigned responsibility for teaching courses (and for guiding Ph.D. projects) that are directly related to their own research. Generally speaking, it is important that teachers of post-graduate students are active researchers themselves, or, at a minimum, keep abreast of the research in their own and related fields—their students will not be served well otherwise.

But perhaps of even more fundamental significance than keeping abreast of one's own field is the teacher's continued interest in and love for science and for teaching. Many senior teachers may not have had the privilege of having studied research methodology. And it is not unlikely that they primarily, or perhaps exclusively, specialized in a particular "narrow" field. But younger university teachers today, no matter where they are located, are becoming better versed in research methodology and the philosophy of science as this is a requirement in many Ph.D. programmes. Such knowledge and the reflective mind-set that can be a result are great assets for reflective scientists when they are playing the role of teachers/mentors. As we have seen in our considerations of multidisciplinary and inter-disciplinary research, so much science is transcending traditional disciplinary and geographical boundaries. Scientists are increasingly becoming both "edge-walkers" who work on the boundaries between different fields, and "glob-alizers", who participate in international conferences, write in international journals and 'network' via the World Wide Web.

A teacher of science must have access to some measure of the quality and effectiveness of his work. As mentioned earlier in this chapter in the section on evaluative research, such measures may be provided if his work and/or his department are subject to an external evaluation. The peer review of articles submitted for publication and/or of proposals for support of one's research is another important source of such measures—but this assumes that the teacher is in fact active as a researcher and author. However, one source of such measure is directly available via a rather special kind of peer review—one's students! Increasingly teachers are being evaluated internally by their students. In many institutions of higher learning, such feedback is not only available to the teacher, but also to the department head who can use it in connection with periodic evaluations of the scientist's progress. Such evaluation typically not only includes matters dealing with teaching, but also with research, administration and, increasingly, the ability to generate funds via grants.

It is also important to realize that research does not only deal with producing new knowledge and that the teaching of science does not only deal with producing new scientists. There are other vital goals one can attribute to research and teaching in science, amongst these being the enhancement of the "quality of life" of human beings and other sentient creatures, as well as of the world we inhabit. Most often this enhancement depends not so much on creating new knowledge, but on the use of existing knowledge through applied research and development. It can be noted that far more scientists work with applied research than with basic research. This does not only apply outside of universities, as you might expect, but also at many universities as well, since research at universities increasingly is being funded from external sources, where the provider of funds often has specific applications in mind. This is particularly true as regards the huge research expenditures throughout the world by the military, as well as by the space, energy and pharmaceutical industries.

These more humanitarian goals of science raise questions as to ethics and the responsibility of scientists—which are the major themes to which we now turn.

Chapter 10
Ethics and Responsibility in Scientific Research

With the advent of technologies that have the potential for mass destruction, for modifying genetic structures/codes, for intervening in one's private life, for polluting the environment, for colonizing other heavenly bodies, for creating artificial life, for … we are all witnesses to an increasing, and often disturbing, connection between science and technologies that can seriously affect fundamental existential conditions for life on this planet and for the well-being of all its inhabitants. This relatively recent result of the amazing growth and power of science to model, predict and control natural phenomena and to develop technologies that increase its ability to model, predict and control, has not been accompanied by a concomitant increase in our attention to the resultant ethical and values issues that have arisen. We have been so enticed and excited, bewitched and bewildered, by the ability of science to develop technologies that can increase productivity, standards of living, life expectancy, abilities to communicate, compute and to analyse the micro and macro levels of physical reality, that we have embraced these apparent boons of scientific research without seriously questioning whether they can be accompanied by negative, destructive, unacceptable results. When I write "we" in the above, I refer not only to the proverbial man-on-the-street, but also to members of the scientific community. According to (Radin 1997; 253): "History has shown that decisions affecting millions were made on the basis of industrial expediency, technological imperatives, and economic pressures." To Radin's list should also be added "political and military rationale".

Fortunately, questions are increasingly being posed as to the value-neutrality of science and as to the ethical responsibility of the individual scientist—questions that present new and vital challenges of significance for scientists and for science as an institution. In this chapter we will reflect on a number of these challenges with respect to ethics and responsibility in scientific research in general as well as with regard to the behaviour of the individual scientist in particular. We will consider fundamental questions such as:

1. What do we really mean by ethics? Why is the focus today not on ethics as a set of socially constructive principles, norms and guidelines for the behaviour of scientists, but on the avoidance of (being caught for) *un*ethical behaviour?

© Springer International Publishing Switzerland 2016
P. Pruzan, *Research Methodology*,
DOI 10.1007/978-3-319-27167-5_10

2. Do human beings have an inherent competency to behave virtuously and wisely
   —to make choices that are not simply the result of calculations as to rational
   self-interest or of norms based on tradition or of evolutionary/biological processes that somehow have led us to develop mutually supportive behaviour?
3. Do scientists have special ethical responsibilities due to their profession?
4. What are the barriers that prevent or delay the acceptance by scientists of ethical
   responsibility for their professional work?
5. Do scientists working on basic research have less responsibility than those
   working on applied research?
6. What are the special ethical issues that must be considered when dealing with
   experimentation involving sentient beings?
7. What is the relationship between (a) ethics in science, (b) the search by science
   for truth, and (c) the search by spirituality for Truth?

The discourse that follows does not attempt to give clear and precise answers to
these questions—attempting to do so would be an expression of arrogance—nor
does it treat them in a sequence, providing answers to the first question before
continuing to the second, and so on. Rather, these questions form the leitmotif that
permeates the exposition.

The chapter commences by considering the concept of ethics in research, and
then continues on to the concept of responsibility in research. The presentation of
ethics will, by and large, be abstract, referring primarily to principles, while the
discussion of responsibility will, by and large, be more practical and personal. Of
course these two concepts are closely intertwined, so some degree of repetition is
unavoidable.

10.1 Ethics

Clearly it is important in science that the specific scientific and technical terms one
uses are well-defined, particularly as regards concepts that are to be operationalized
via variables and measurement. However, sometimes, and in particular when
dealing with broader concepts that do not require measurement in order to be
meaningful, it is better not to strictly define concepts than to do so. Much of our
ability as humans to communicate ideas depends on the use of our daily language
with its lack of precision and therefore with room for interpretation and inspiration,
which also means with room for misunderstandings, disagreements, and conflict.
Words are only formally defined via other words. In other words, words are elements in a closed, self-referential system, whose physical representation is a dictionary. But that which gives meaning to words is far more than other words; it is
also experience in the form of correspondence between words and what they refer
to; although we cannot be burned by the word "fire", we learn from experience to
avoid contact with things that burn—and this can be communicated using words.

Continuing in this frame of thought, although the noun "table" may be well defined in a dictionary (a typical definition is: "a piece of furniture with a flat top on one or more legs providing a surface for eating, writing, or working at"; Concise Oxford English Dictionary, 11th edn., revised 2009), our understanding of the word may go far beyond such a lexical definition. I myself once had the following experience. I had been teaching a small group of Danish scientists and leaders from a research-based organization. This took place at a lovely old manor house that had been converted into a conference centre in beautiful surroundings. After several hours of concentrated discussions, we decided to take a break and have a light picnic in a nearby wood. We came across the stub of what must have been a huge oak tree. Without hesitation one of the scientists said: "wouldn't this be a fine table for the picnic?" We all agreed. Clearly, the stub of a tree in no way corresponded to the dictionary definition of a table. It was not a manmade artefact, as is furniture. It was not flat but was quite uneven. It certainly did not have legs. Yet we all could immediately and quite easily agree that it was a "table", at least given the specific context of a picnic. This is the beauty of language; it provides a firm structure and content (grammar, vocabulary), yet permits context-dependent usage and inter-pretation that go beyond the apparent limits of semantic and syntactic rules[1]—this is what enables a poet to touch our hearts.

Thus I will not attempt to provide a tight definition of "ethics" or "responsi-bility" in science; this might unnecessarily delimit our ability to reflect upon their essential meanings and to integrate such meanings in our world views, beliefs and behaviour. Nor will I develop this chapter by presenting a traditional philosophical distinction between various types of ethics, such as virtue ethics, deontological (duty) ethics and consequentialist ethics. Instead, and in line with the major motivation underlying this book, various aspects of ethics and responsibility are presented with the object of stimulating the reader's reflections on why and how to integrate an ethical dimension into one's research—and how to build upon this reflection so as to be an ethically responsible practitioner of science.

Therefore, let us start with some broad reflection on how we tend to understand these concepts in order to develop a shared frame of reference. While philosophers provide extensive definitions and interpretations of ethics, for our purposes it will suffice to use the term as it is loosely defined in a day-to-day sense—as the norms and standards that guide choices about behaviour and relationships with others.[2]

---

[1] *Syntax*: Deals with sentence structure—with the relationship of words to form structure according to rules in a language. *Semantics*: Deals with meaning in language and with understanding human expression through language—with the relations between symbols/words and what they refer to.

[2] Philosophers tend to consider *morals* as norms/standards for conduct based on reflection as to what is right and wrong, good and bad, virtuous and non-virtuous, while *ethics* tends to be considered as reflection on how such morals can be justified. Such justification is provided from over-arching concepts such as duty, justice and utility (the consequences of one's actions). A modern approach to morals and ethics is based on the fact that society today is increasingly heterogeneous, and therefore lacks shared conceptions of what is moral. This leads to an understanding of morals as social norms within a group while ethics is seen as a second-order morality, dealing with how different groups, each with their own morals, can communicate and coordinate their actions (Pruzan and Thyssen 1990).

Most of the time the word is used when referring to the behaviour of individuals—but we also hear reference to e.g. "business ethics", indicating that this term also is used at a collective level (Pruzan 2001). We even hear the word used in connection with more universal frames of reference. An example from the life sciences is "bioethics", often referred to as including shared, reflective examination of questions as to the rights of living creatures and of ethical issues in health science/care/policy. A concrete and more general example than provided by science is the *Universal Declaration of Human* , approved by the General Assembly of the United Nations on December 10, 1948. The document's preamble states: "Whereas recognition of the inherent dignity and of the equal and inalienable rights of all members of the human family is the foundation of freedom, justice and peace in the world ..."; http://www.un.org/Overview/rights.html.

Often the word ethics is used to refer more to *how* actions are chosen than to the actual *content* of a specific action, that is, to the processes leading up to a decision rather than to the content or results of the action. For example, in the late 1980s and early 1990s, together with a colleague I developed the concept of "ethical accounting" based on the premise that an act is ethical if all parties affected by it give their approval; (Pruzan 1995) and (Zadek, Pruzan and Evans 1997).

Probably most of us have observed that in typical day-to-day activities we experience that we make "ethically acceptable" decisions with confidence and without rational reflection—and that this confidence is based on an inner guidance. This observation deals primarily with routine matters (e.g. not telling a lie, spontaneously assisting an elderly person who has stumbled or avoiding stepping on a frog).

In other words, in most daily situations we tend to deal with whatever confronts us in a more or less spontaneous manner, drawing upon what one can refer to as "embodied knowledge" in contrast to more deliberate and willed action that is the result of logical generalizations and prescriptive principles. According to the Chilean biologist and neuroscientist, Francisco Varela (1946–2001), such spontaneous actions "... do not spring from judgment and reasoning, but from an *immediate coping* with what is confronting us. And yet these are true ethical actions, in fact, in our daily, normal life they represent the most common kind of ethical behavior." (1999: 3–5; Varela's emphasis).

With these introductory remarks, let us now reflect on the rather fundamental question: what is the source of our tendency to behave ethically? What is the source of our competency as human beings to behave virtuously and wisely? Attention to such a fundamental question can, hopefully, contribute to our ability as human beings to better appreciate, emancipate and empower our propensity as humans to be empathetic, altruistic and to promote the common good.

10.1.1 Western and Eastern Perspectives on the Source of Ethics

As we have seen, particularly in Chap. 2 where we considered the question "What is science?", a presupposition of dualism underlies the natural sciences and the philosophy of science as they have developed in the "West". Dualism, with its dichotomy between a "subject" that observes and an "object" that is observed, is the basis for the metaphysical assumption in science that scientists can objectively investigate an external physical reality that exists independently of the observer. But also "Western" approaches to ethics are, for the main, implicitly grounded in such a dualism. Here the presumption is that individuals can and should evaluate their own behaviour and that of others with respect to the possible positive or negative effects on other sentient beings and on nature. As we will see, "Eastern" approaches tend to provide an alternative perspective with greater focus on what it means to be human than on either the motivations for our actions or the consequences. There is an emphasis on our inherent capacity for self-reflective choice and our propensity as human beings to behave ethically—on what one might call our moral sense—and not just on rational reflection or on tradition-based norms.

From a "Western" neuroscientific perspective, ethical norms are a product of human consciousness. And since consciousness is implicitly assumed, both by scientists and ethicists, to be a result of activity in the brain, it follows from such a neuroscientific perspective that ethical (as well as unethical) behaviour can be ultimately analysed and explained from a materialist, reductionist perspective whereby matter is the fundamental substance in nature and all physical reality is built up of individual units of matter that obey the laws of nature.

From such a neuroscientific perspective it is relevant to ask whether there are biological processes that somehow have led humans to develop brains that promote mutually supportive behaviour. Are our apparently intuitive/spontaneous tendencies to respect the needs of others and of the common good a result of evolution—of cultural and biological processes that have led humans to evolve into a cooperative species whose members are genuinely concerned about the well-being of others? Not according to the physicist Heinz Pagels, at least regarding our genetic conditioning: "No doubt animal and human behavior is materially conditioned by the microscopic organization of genes … But it is not possible to reduce all human behavior to just microscopic genes. Genes do what they are told to do by the laws of chemistry. The point is that genes can't be selfish but human beings can." (Pagels 1982; 133).

Other perspectives on this matter are provided in the relatively new fields of evolutionary psychology and sociobiology. Investigations in these fields indicate that humans do have a propensity to participate in mutually beneficial activity and altruistic cooperation and this is said to be the result of evolutionary processes where groups of individuals who were (somehow) predisposed to cooperate tended to survive and expand relative to other groups (Bowles and Gintis 2011). Similarly,

Grassie (2010; 81–85) presents potential explanations for the existence of other-regarding behavior within an evolutionary paradigm.

In contrast to such "Western" dualistic perspectives on ethics where individuals can and should evaluate their own behaviour with respect to others and to nature, and where, from a neuroscientific perspective, the source of human ethical competency is our materially-founded human consciousness, a more "Eastern" (Buddhist and Hindu) approach to ethics is based on a monist spiritual perspective. Reality is considered as an organic whole and the primary source of human ethical competency and our promptings to act in accord with this competency, our moral sense, is our inherent and latent divine nature and not external stimuli in the form of tradition and moral prescriptions. From this perspective, our "knowing" what to do in a specific situation when facing a moral dilemma is not just some combination of intuition, personal experience and rational reflection as to what is the "right" thing to do; it is not just a balance between: (a) promptings that are a result of cultural tradition, (b) rational evaluation, supported perhaps by principles from moral philosophy, and (c) utility calculations as to 'what's in it for me?' Instead, such 'knowing' at a very foundational level is considered to be latent and independent of cultural tradition, logical reasoning and reflection on individual (as well as group) benefit, such that *all* human beings have an inherent capacity to know what is the "right" thing to do in a given situation, no matter how complex and characterized by uncertainties and moral dilemmas that situation may be. Which is *not* the same thing as saying that this inherent capacity is, in general, readily available or that we can rely on it alone. The ability to draw upon this intrinsic capability is related to the degree to which an individual has developed what one may refer to as a higher level of consciousness, a higher level of self-awareness and self-knowledge, what in a more spiritually oriented vocabulary is often referred to as self-realization or enlightenment (Pruzan 2015).

In particular, the ancient Vedantic concept of *advaita* provides a fundamentally different perspective on "ethics" than provided by traditional Western perspectives. Vedanta is often referred to as the "end" of the Vedic texts called the Upanishads, where "end" can be understood both as a goal and as its being the culmination of the Vedas. According to the renowned Hindu sage Swami Vivekananda (1863–1902), the perspectives on ethics provided by Vedanta[3] are based on a fundamental metaphysical presupposition of *advaita*, of unity in existence where every being and every thing has divinity as its core: "Vedanta claims that man is divine, that all this which we see around us is the outcome of that consciousness of the divine ... all that we call ethics and morality and doing good to others is also but the manifestation of this oneness. ... This is summed up in the Vedanta philosophy by the celebrated aphorism *Tat tvam asi*, 'Thou art That', Thou art one with this Universal

---

[3]According to Swami Vivekananda, best known in the West for his presentations of the philosophy of Vedanta and for his work on interfaith awareness, Vedanta is "...the culmination of knowledge, the sacred wisdom of the Hindu sages, the transcendental experience of the seers of Truth ... the essence or conclusion of the Vedas, so it is called *Vedanta*. Literally, *Veda* means knowledge and *anta* means end." (Swami Vivekananda 1987; editor's note, 16).

Being, and as such, every soul that exists is your soul, and every body that exists is your body. And in hurting anyone you hurt yourself. In loving anyone you love yourself." (Swami Vivekananda 1987; 58–9).

This non-dual 'Eastern' (Indian/Vedantic) perspective does not refer to an external source, to a philosophical first principle, to autonomous normative, practical principles (such as provided e.g. by a Kantian perspective) or to a social constructivist perspective whereby historical, social and economic forces determine what we find to be "good" or "bad", "right" or "wrong". Instead, it has its roots primarily in existential ontological perspectives on the very nature of reality whereby ethical competence is and always has been embodied in all sentient beings; we are all physical manifestations and agents of a divine source.

However, even though it follows from this Vedantic concept of *advaita* that ethical competency is latent in all that exists, the extent to which this competency can be manifested in sentient beings depends on the level of conscious awareness possessed. Such reflection leads to a conclusion that with increasing levels of consciousness follow increasing ethical responsibilities to respect and to help emancipate and empower the divinity that is inherent in all. Such a perspective on ethics is foreign to traditional Western thinking!

It also follows that since there can be different moral intuitions that have the same source, this perspective does not challenge the position that rational discourse and a search for consensus are necessary in our post-modern times, characterized by a pluralism of traditions that can support different moral intuitions. Nor does it challenge the position that although particular morals, the specific content, cannot be justified from a universal perspective, the phenomenon of moral norms can be universally justified; such norms are required at all levels of human organization—an individual, a family, an organization, a community, a society cannot *be*, cannot have an identity that enables coherence, without shared values and norms.

10.1.2 Unethics

It is clear from the media, no matter whether in the "West" or the "East", that most references to the word "ethics" in our daily language deal not with socially constructive guidelines for behaviour but with its apparent absence. Just as "cold" can be considered to be the absence of "hot", and "light" can be considered to be the absence of "dark", what I refer to as *un*ethics is behaviour that is characterized by the absence of an ethical foundation or motivation.

For example, the following are three areas where concerns often arise as to *un*ethics:

1. Harm to sentient beings and to the environment
2. Lack of informed consent and the invasion of privacy
3. Deception and coercion

In the sequel we will often refer to these three broad areas when characterizing and exemplifying ethical and *un*ethical behaviour in research.

Laws, governmental regulations and codes of ethics (developed by corporations, professional and scientific associations and universities) are often developed with the aim of preventing such unethical behaviour. But ethics goes beyond living up to the law and rules—and in certain cases ethical reflection actually can lead to one's breaking the law. A widely recognized example was the behaviour of Mahatma Gandhi, both in South Africa (refusing to leave his seat in a train compartment reserved for whites[4]) and in India (the "salt march"). However, while we might applaud Mahatma Gandhi's principles and actions, considerable reflection and discrimination are required before taking the law in one's own hands when faced with something one feels is unethical. To illustrate, consider the actions of groups of fundamentalist Christian in the USA who, in their moral indignation over the fact that it is legal in certain circumstances for women to have an abortion, have attacked and even murdered doctors offering such services and have bombed clinics where abortions take place. Most of us would probably strongly condemn such violent behaviour as unethical—while the attackers and bombers feel that their violent behaviour is ethically justified since they "have God on their side". One of the Ten Commandments (principles regarding worship and ethics) in the Bible is: "You shall not murder/kill", which they apparently interpret as giving them the right to attack such doctors—and even, ironically and tragically, to kill them so as to stop them.

While obeying the law cannot be taken as an indication of ethical behaviour, not obeying the law, with exceptions like that of Gandhi, is generally considered to be unacceptable and perhaps unethical as well. I write "perhaps unethical as well" since there are many illegal acts that we would not tend to classify as unethical. An example could be the driver of a car who, while driving on a highway with no other vehicles in sight exceeds the speed limit; if caught, he could be fined by the police for speeding, but most of us would probably not consider his behaviour as unethical. On the other hand, we might do so if he was driving on a crowded road so that his speeding could seriously harm others. In fact, even if he was driving below the speed limit but in a way that could harm others, most people would call that behaviour unethical even though it was legal.

Of course it is necessary to have societal standards, including laws, dealing with such matters as noise levels, safety, pollution, working conditions, experiments involving sentient beings, genetic manipulation, etc. But these can never simply replace the guidance provided by one's conscience. On the other hand, when making a decision in situations characterized by complex ethical dilemmas, one's conscience and ability to discriminate between ethical and unethical very likely will require qualified input from regulations, traditions and standards as well as,

---

[4]This can be considered a forerunner for the behavior on December 1, 1955 of the, now-famous, African-American civil rights activist, Rose Parks. Her refusal to give up her bus seat in the 'coloured section' of a bus for a white passenger, after the white section was filled, became an important symbol for the Civil Rights Movement in the US.

perhaps, from experts. Nobel laureate Bertrand Russell (1961; 224–5) provides the reflective scientist with food for thought on the relationship between rules of conduct and conscience: "One of the ways in which the need of appealing to external rules of conduct has been avoided has been the belief in 'conscience', which has been especially important in Protestant ethics. It has been supposed that God reveals to each human heart what is right and what is wrong, so that, in order to avoid sin, we have only to listen to the inner voice. There are, however, two difficulties in this theory: first, that conscience says different things to different people; secondly, that the study of the unconscious has given us an understanding of the mundane causes of conscientious feelings."

Such ethical dilemmas as referred to above are typically met in decision situations characterized by the following: individuals or groups that may be affected by a decision, a number of possible outcomes that may result from the decision, and considerable uncertainty as to the relationships between the decision, its possible outcomes, and the utility/disutility that may result for all parties affected by the outcomes. An example of such a dilemma, which increasingly frequently confronts all hospital administrators, as well as politicians involved in health matters, is whether to purchase highly specialized and extremely expensive high-tech equipment that can save the lives of a relatively small number of people, if this means that such resources will not be available to treat a far larger number of people who, for example, could benefit from operations to prevent blindness or from the replacement of damaged and painful knees.

In such situations, conscience, which in more day-to-day situations is our most reliable guide, must be supplied with "input" in the form of information and calculations (e.g. as to risks, uncertainties, and potential benefits and harms) and guidance from others. In such situations we cannot rely solely on unaided intuition, as people normally do when they speak of their conscience.[5]

The fields of health care and health policy are arenas where many such ethical dilemmas are played out, sometimes with very powerful forces at work. In (Campbell and Campbell 2007; Chap. 13) T. Colin Campbell, a world renowned expert in the relationship between health and nutrition, demonstrates that to heed the voice of one's conscience, even when it is supplemented by reasoning, calculations etc., may also require conviction, steadfastness and courage when facing powerful vested interests (in his case, the dairy and meat industries). No matter whether one agrees or not with his conclusions, Campbell provides an extremely challenging perspective on (the lack of) ethics when scientists serve such interests

---

[5]Some dilemmas have no "answers". An example is provided by the best-selling book by W. Stryon, *Sophie's Choice*, from 1979, where a Polish Jewish woman on arriving at the Nazi concentration camp Auschwitz during World War II is forced to choose which of her two young children is to stay with her, and which is to be sent immediately to death. I contend that no amount of reasoning and no listening to her conscience could provide her with an "ethical choice" and that only the virtually utopian state in such a situation of non-attachment and surrender to a divine source in the form of the biblical "thy will not mine be done" can provide a "solution" in the form of acceptance amidst such indescribable suffering.

more than the search for truth: "There are some people in very influential government and university positions who operate under the guise of being scientific
'experts', whose real jobs are to stifle open and honest scientific debate. ... I was on
the inside of the system for many years, working at the very top levels, and saw
enough to be able to say that science is not always the honest search for truth that so
many believe it to be. It far too often involves money, power, ego and protection of
personal interests above the common good. Very few, if any, illegal acts need to
occur. It doesn't involve large payoffs being delivered to secret bank accounts or
private investigators in smoky hotel lobbies. It's not a Hollywood story; it's just
day-to-day government, science and industry in the United States." (Ibid.; 267–8).

Campbell's description is strongly supported in a thought provoking and rather
depressing article dealing with the enormous influence of the pharmaceutical
industry on the behaviour of medical researchers and thus on the information that
flows to our doctors: "Why you can't trust medical journals anymore". See the link:
http://www.washingtonmonthly.com/features/2004/0404.brownlee.html.

Scientists are accustomed to working with highly reliable instruments in their
research. Ethics can be considered to be a very special, personal, and non-tangible
"instrument" in science. Normally an instrument can be used by a researcher
without his or her becoming personally involved. The instrument and the researcher
are two different things; this is in line with the dualistic distinction in science
between the subject and object, between the objective and the subjective and
between the researcher and his tools. But with ethics things are different. As clearly
illustrated by T. Collin Campbell's experiences, the user of the "instrument" ethics
must personally stand up for its use—for his or her ethics, for his or her values.

As considered in Chap. 4, where we considered the justification and acceptance
of scientific statements if a scientist is not able to obtain the trust of others, in
particular her peers, they will tend not to believe in her research findings. Over and
above the quality and clarity of the findings and of the description of the scientist's
methodology, an important key to such trust is ethical behaviour—or, as just noted
and more in tune with the apparent attitudes in science (and business) today, the
avoidance of *un*ethical behaviour. In the report *Responsible Science, Volume I:
Ensuring the Integrity of the Research Process* (National Academy of Sciences,
National Academy of Engineering, and Institute of Medicine 1992) a panel composed of leading scientists identified a number of ethical problems (which in fact
are examples of *un*ethical behaviour) in science. These included:

- Failing to retain significant research data for a reasonable period of time
- Maintaining inadequate records, especially for results that are published or
  relied on by others
- Not allowing peers to have reasonable access to unique research materials or
  data

- Inadequately supervising research subordinates or exploiting them[6]
- Conferring or requesting authorship on the basis of services/contributions not significantly related to the project reported on—and omitting someone from the list of authors who did make a significant contribution
- Using inappropriate methods to enhance the significance of findings[7]
- Misrepresenting speculations as fact or releasing preliminary results, particularly in the public media.[8]

It is ironic that while trust based on ethical behaviour can be important for the acceptance of one's research findings, trust itself can also contribute to, or at least enable unethical behaviour! Sheldrake (2012; 311–12), after presenting several amazing cases of fraud in physics and biology, refers to studies—see for example (Broad and Wade 1985)—that demonstrate that "Fraudulent results are likely to be accepted in science if they are plausibly presented, if they conform with prevailing prejudices and expectations, and if they come from a suitably qualified scientist affiliated with an elite institution.". Sheldrake concludes: "Scientists usually assume that fraud is rare and unimportant because science is self-correcting. Ironically, this complacent belief produces an environment in which deception can flourish."

The role of trust amongst members of the scientific community is closely related to another aspect of an ethical perspective on science and research; holism. Ethics is oriented towards a holistic/macro evaluation (who and what will be affected? how? is this acceptable?), while science in an age of ever increasing specialization tends to focus on the micro aspects, the particular. Ethics has a demand: the scientist must assume responsibility for the way that his or her research can affect others, including nature itself. Therefore he or she must reflect not only on how results will be accepted by the scientific community, but also on how they will be accepted by all those who may be affected, directly or indirectly, by both the investigations and the results.

---

[6]When considering "epistemological strategies" for designing experiments in Chap. 6 we considered the oil-drop experiments credited to Robert Millikan, for which, together with his work on the photoelectric effect, he won the Nobel Prize in physics in 1923. After his death and the death of one of his Ph.D. students, Harvey Fletcher, it was revealed that the oil-drop experiment was performed together with Fletcher but that Millikan used his position as Fletcher's advisor to forge an agreement with him that Millikan should be the sole-author to the publication that resulted from their teamwork. Millikan was also accused of scientific fraud in connection with that publication; he had written that the article contained all the observations that had resulted from the experiments, while in fact a number of observations that did not strongly support his arguments as to the magnitude of an electron's charge were omitted. Both of these apparently unethical practices have been subject to much debate as the circumstances characterizing both these matters were quite complex (e.g., the student never complained and remained a friend of Millikan all his life; and omitting the less favourable observations did not seriously affect the conclusions).

[7]For example, if several statistical methods are available and if the most reliable of these does not support a hypothesis, it is unethical to ignore this and then refer to another method that supports the hypothesis.

[8]Earlier we referred to the case from 1989 of 'cold fusion' that was announced to the media before the results had been made available in (or even submitted to) a peer-reviewed journal.

To make things clearer as to the complexity of many ethical dilemmas in science let us consider another example from medical research; I have chosen (once again) to use medical research so readers can understand the example no matter what their scientific discipline. Consider a so-called double-blind experiment where neither doctors nor patients know who receives a standard treatment and who receives an alternative treatment, while this information is of course confidentially recorded. For example, consider an experiment involving 200 seriously ill patients where 100 receive an experimental medicine and a control group of 100 receives medicine that reflects the current standard of care for that particular illness.[9] Assume too that both groups have been informed as to the design of the experiment and have agreed to participate. Such research using double-blind experiments is typically justified as being necessary in order to develop new medicines that benefit patients. However, in searching for a new drug, if it works, the doctors involved may have withheld the promise of life from the 100 patients who did not receive the new medicine, but who received the standard treatment (and vice versa: the doctors may have withheld the promise of life from the 100 patients who *did* receive the new medicine, if the new medicine was in fact ineffective compared to the standard treatment). This could clearly result in an ethical dilemma for the doctors involved, since they have all taken the so-called Hippocratic Oath whereby they commit themselves to do their very best to save *every* life, not just life on the average. So the individual doctor may, or may not, consider such a test to be "ethical". Most doctors in fact do not protest against such testing. They consider it to be necessary to carry out such research when developing new drugs that may help a great number of people in the future—and can point to the fact that no deception takes place since both groups knowingly and willingly agree to participate in the investigation. This is a typical example of ethical reasoning based on the general consequences of one's actions (not on the specific consequences).

A dilemma met at times in connection with experimentation involves to what extent (if any) it is acceptable to use *deception* to obtain knowledge that would otherwise be difficult to obtain. A classic example from the behavioural sciences is the series of controversial experiments in the 1980s by Stanley Milgram (Milgram 1983). These experiments involved humans and were performed to investigate the concept of obedience, in particular to what extent ordinary humans would disregard their feelings and common sense when being strongly persuaded by an authority, here a scientist, to do something. In these experiments an actor pretended to receive electric shocks that were supposedly administered to him by the person whose obedience was being studied and who believed that he was participating in a scientific experiment designed to evaluate how people learn. He was given instructions to administer shocks of increasing intensity to a person sitting across

---

[9]Note that it is an ethical requirement in some countries that the control group does not receive placebo; in such situations the research question cannot be "What is the effect of the new medicine relative to no medicine?", but must be "What is the effect of the new medicine relative to the medical options that are considered to be standard for the illness?" (Stern and Lomax 1997; 291–92).

from him behind a glass panel. In fact, when he pushed on a button to give the person (the actor) a shock, nothing really happened, but the actor simulated receiving a shock. A researcher (also an actor) talked with the person being studied, encouraging him to continue giving 'shocks' of increasing intensity even though the person receiving the 'shocks' appeared to be in great pain and stress, and even though the person whose behaviour was being observed wanted to quit. These scenes have been filmed. It turned out that a number of the persons who were studied suffered psychological harm when they realized, to their amazement, that they would willingly hurt another human being when influenced by a scientific authority. These experiments are notorious for their use of deception, for their lack of informed consent, and for raising questions of how far the acquisition of scientific knowledge can justify risk to a human being whose behaviour is being studied. We will return to the special case of experiments involving humans later in this chapter.

Closely related to the question of the legitimacy of using deception to obtain information that otherwise might be impossible to obtain is the use of *coercion*. In recent years, the media have presented us with numerous stories of cases where torture is accepted by political and military leaders as part of the "war on terror". The question faced here is whether it is acceptable to use a means (torture) that is known to cause significant physical and psychological harm to an individual in order to obtain information that might save a greater number of people from suffering and death. Although this question is not directly related to scientific matters, the issues are very similar. In both the case of Milgram's experiments and the case of torture that is motivated by saving lives, the main issue is the question of when, if ever, humans may legitimately be deceived (Milgram) or coerced (torture) to act against their will. The matter of coercion can be extended to more daily situations such as the disciplining of children and the use of force/weapons in defence of persons or property. In each case, arguments can be made that coercion, here in the form of physical force, is required to bring about legitimate ends. But such arguments must be evaluated against a backdrop of fundamental questions regarding whether humans have inalienable rights. Here we can recall the fundamental metaphysical perspective on existence provided by Swami Vivekananda: "Thou art one with this Universal Being, and as such, every soul that exists is your soul, and every body that exists is your body. And in hurting anyone you hurt yourself. In loving anyone you love yourself."

Or consider the ethical dilemmas of a completely different nature that faced the scientists in the USA who worked to develop the first atom bomb. The risk of an uncontrolled chain reaction could not be neglected. And the perspective of unleashing huge amounts of energy for military purposes was awesome. Yet the potential benefits, in terms of developing such weaponry before both Germany and Japan and of then using it to help end a war that otherwise could continue for a long time and result in untold suffering and destruction, were enormous. Many scientists participated directly in the research, based on the so-called consequentialist or utilitaristic arguments in its favour. Yet a number of prominent scientists, including Niels Bohr, protested against the whole idea of developing an atomic bomb due to both humanitarian and pacifistic arguments as well as to the long term risks to civilization that could result from a world that accepted such weapons of mass destruction. Although Einstein prior to the entrance of the USA into World War II promoted the development of such weapons by the US, he later on decried their use. In the midst of the Cold War and just shortly before his death in 1955 Einstein together with the philosopher Bertrand Russell and other prominent scientists and intellectuals signed the so-called *Russell-Einstein Manifesto* that emphasized the dangers posed by nuclear weapons and the need for peaceful resolution to international conflicts.

A final example here could be the conflicting views amongst both scientists, laymen and political leaders as to genetic research involving GMOs, cloning, and stem cells, where those promoting such research and its applications can point to great potential benefits for mankind—and where those who protest against such technologies point to the great potential harm that might result due to our lack of knowledge as to the long-run risks involved.

It should be clear that it is of great importance that reflective scientists develop the ability to deal with ethical dilemmas, both at the individual and the collective level. There is a need for training and reflection that can contribute to the development of an awareness of, and skill in dealing with, the ethical implications of one's research. However, the question can be raised whether mature people, for example graduate students of science or practising scientists, can refine or alter their values and ethical standards. There is widespread belief that this is not possible; a nationwide survey amongst science and engineering faculty in the USA revealed that "over 40 % of the respondents strongly agreed or agreed with the statement that 'by the time students enter graduate school, their values and ethical standards are so firmly established that they are difficult to change.'" (Swazey and Bird 1997; 2) Nevertheless, this belief is not justified; considerable research evidence refutes this belief and demonstrates that the way that people resolve ethical dilemmas continues to change throughout their formal education (Ibid.; note 3, p. 18). The evidence indicates that graduate students can learn to be sensitive to ethical issues in research, that they can learn skills that will enable them to reason about ethical issues and formulate ethical judgments, and that they can learn to apply their skills to their work as scientists.

To contribute to such reasoning and skills, and to establish guide lines that can help individuals in their work, many professional organisations, commercial enterprises, research institutions and universities have developed codes of values/ ethics, some of which are specifically oriented towards scientific research. The following section discusses such a code with respect to research that has been developed at a major university in the USA. The guidelines present a series of down-to-earth definitions and reflections that most certainly can stimulate the thinking of a reflective scientist. In addition, underway I provide a series of comments on the individual topics treated that hopefully will contribute to such reflection.

10.2 Guidelines for Ethical Practices in Research

We consider here the extensive *Guidelines for Responsible Conduct of Research* (2011), which is an updated edition of *Guidelines for Ethical Practices in Research* (2007), developed by the Office of Research Integrity at the University of Pittsburgh in the USA; http://www.provost.pitt.edu/documents/GUIDELINES FOR ETHICAL PRACTICES IN RESEARCH-FINALrevised2-March 2011.pdf.

The Guideline's Table of Contents provides an overview of its areas of concern:

"MATTERS OF ETHICAL CONCERN IN RESEARCH

1. Plagiarism
2. Misuse of Privileged Information
3. Data

   (a) Integrity of Data
   (b) Use and Misuse of Data
   (c) Ownership of and Access to Data
   (d) Storage and Retention of Data

4. Authorship and Other Publication Issues

   (a) Criteria for Authorship
   (b) Order of Authors
   (c) Self-citations
   (d) Duplicate Publication
   (e) Accessibility of Publications
   (f) Early Release of Information About to be Published

5. Interference
6. Obligation to Report

   (a) Reporting Suspected Misconduct
   (b) Correction of Errors

7. Curriculum Vitae
8. Conflict of Interest
9. Responsibilities of a Research Investigator
10. Responsibilities to Funding Agencies
11. Special Obligations in Human Subject Research
12. Laboratory Animals in Research
13. Research Involving Recombinant DNA (rDNA)"

The guidelines, though not an official policy of the university, supplement existing policies and procedures governing research. No abstract ethical-philosophical arguments are provided for motivating the guidelines. Instead, they appear to be motivated solely by a pragmatic rationale that following the guidelines will: (1) help researchers maintain integrity in research (what we can refer to as a personal, moral argument), (2) lead to better scientific results in the form of more attention to details, more thoughtful collaboration amongst scientists who work together, and improved credibility with the public (consequentialist/utility arguments), and (3) lead to less research misconduct (a duty argument: avoid behaviour that can damage the university's reputation). I note that the guidelines define research misconduct as: "fabrication, falsification or plagiarism, including misrepresentation of credentials, in proposing, performing, or reviewing research, or in reporting research results. It does not include honest error or differences of opinion. Misconduct as defined above is viewed as a serious professional deviation that is subject to sanctions imposed both by the University and by a sponsoring federal agency."

In addition, since this is an American university, it is quite likely that two other motives also play an important role: (4) the avoidance and mitigation of potential law suits arising from unethical research behaviour,[10] and (5) living up to increasingly severe demands from the American government as to research ethics.[11] In this connection I note that the guidelines referred to below are based on the structure and recommendations of the report (National Academy of Sciences, National Academy of Engineering and Institute of Medicine 1992). The third edition of that report from 2009 that provides analyses and cases is available free of charge on line: (Committee on Science Engineering and Public Policy 2009) and www.nap.edu/catalog.php?record_id=12192.

Let us now reflect on several of the topics considered in the guidelines.

Plagiarism

The guidelines define "Plagiarism" as: (a) the theft of intellectual property, i.e. presenting the "words, data, or ideas of others with the implication that they are their own, without attribution in a form appropriate for the medium of presentation" and (b) the misuse of privileged information (information taken from grant proposals made by others or from manuscripts received for peer review). In other words, according to the guidelines, plagiarism involves not giving explicit credit (in the form of quotes, references) to sources one uses, no matter whether they are published or not and whether they are written or oral or material on a website. Note however that according to the guidelines it is not necessary to provide citations "in the case of well-established concepts that may be found in common textbooks or in

---

[10]Institutions in the USA tend to pay far more attention to regulations, codes, and laws than is the case in most other parts of the world, presumably due to the tradition in the USA for litigation in connection with disagreements. For example, if a university or company has a code of ethics, and if an employee behaves in a way that can lead to the organisation being sued, there is a tendency for the judge or jury to reduce the fine that otherwise would have been levied if the organisation can: (1) refer to its code of ethics that clearly forbids such behaviour and then (2) show that the employee was familiar with the code.

[11]The following is the very first sentence in (Elliott and Stern 1997) that deals with research ethics: "In 1991, Dartmouth College, like other research institutions around the country, was struggling with the need to deal with increasingly strict federal requirements to include ethics in the research training of its students of science and engineering.".

the case of phrases which describe a commonly-used methodology." Furthermore, the need to cite the work of others and to provide reference to the original source only applies if there is "word-for-word copying beyond a short phrase or six or seven words of someone else's text…"

The guidelines emphasize that such explicit recognition is required in all publications, grant proposals and student papers. I note in this connection that it is my experience that plagiarism need not be a result of conscious behaviour; it is not uncommon that students simply, more-or-less, copy parts of what others have written without giving explicit credit, and without thinking over the unethical nature of their behaviour. Most of the time this goes unnoticed, but on rare occasions a reader (for example an evaluator of a thesis) may recall that what he is reading is in fact a copy of the writings of someone else. In such cases the author risks very serious consequences.

Plagiarism is apparently becoming so common amongst students when writing papers—and is so easy due to the ability to download publications on the internet, that software is now available for checking papers for plagiarism. For example, a widely used software is "Turnitin" that helps teachers prevent/catch plagiarism. The software allows educators at more than 15,000 institutions to check students' work for improper citation or potential plagiarism; over 50 % of unoriginal work comes from other student papers. It does this by comparing papers submitted against three continuously updated databases (45+ billion web pages, 337+ million archived student papers, and 130+ million articles from 110,000+ journals, periodicals and books). According to its website: http://www.turnitin.com/en_us/about-us/our-company, Turnitin processed nearly 100 million papers in 2013 (300,000 daily on average) and supports 19 languages. The average turnaround time for checking a paper was 23.9 s.

Referring to our earlier considerations of a wide-spread emphasis on avoiding unethical behaviour rather than on behaving ethically, I note that the arguments in the University of Pittsburgh's guidelines regarding such research misconduct appear to be based more on the potential negative consequences than on an appeal to duty, to values to be promoted, or to doing what is "right". Although such negative consequences are not spelled out in the document, the following can be said to be potential negative results of plagiarism. The first is for science as an institution; plagiarism can lead to a gradual loss of faith by the public, as well as by those organisations and politicians that support scientific research, in the trustworthiness of the members of the scientific community. Plagiarism can also lead to a lack of trust among scientists themselves—and this can lead to more difficult conditions for teamwork, which is so vital for modern science, particularly with the growth in multidisciplinary and interdisciplinary research (Chap. 9, Sect. 9.2).

At the individual level, being caught plagiarising can be disastrous for a student/scientist, including loss of regard amongst colleagues and the scientific community, loss of one's job, and fines or even imprisonment in some countries. According to http://www.plagiarism.org/ask-the-experts/faq/, in the USA "Most cases of plagiarism are considered misdemeanors, punishable by fines of anywhere between $100 and $50,000—and up to one year in jail. Plagiarism can also be considered a felony under certain state and federal laws. For example, if a plagiarist copies and earns more than $2500 from copyrighted material, he or she may face up to $250,000 in fines and up to ten years in jail."

Of course, over and above these pragmatic reasons, there is most certainly a more or less universally accepted norm among scientists (and the rest of us) that plagiarising is simply wrong and goes against one's sense of justice. Unfortunately of course, here as in many areas of human endeavour, apparently it is not sufficient to rely solely on such tacit norms, and this has led to the development of guidelines (such as those we are considering now at the University of Pittsburgh), codes, rules and laws—and also tools such as Turnitin.

Integrity of data

Closely related to the concept of plagiarism, the guidelines underscore the need for the "Integrity of data", including the avoidance of charges of the fabrication of data and of falsifying data so as to support one's hypotheses, both of which are considered to be serious forms of misconduct. This leads to a focus on meticulous record-keeping, a matter emphasized several times earlier in this book since it is my observation that this does not appear to be common practice among many researchers. The guidelines draw upon another document here, the university's *Guidelines on Data Retention and Access*: "Records should include sufficient detail to permit examination for the purpose of replicating the research, responding to questions that may result from unintentional error or misinterpretation, establishing authenticity of the records, and confirming the validity of conclusions."

To obtain such credible detailed record-keeping, especially in many fields of laboratory research, the guidelines note that it is standard practice to record data in ink in an indexed permanently bound laboratory notebook with consecutively numbered pages and where each entry/page is dated and initialled. Furthermore, computer printouts should be labelled and pasted in the notebook or stored securely and referenced in the notebook as to their location.

If they are unique, the guidelines emphasise that critical materials (such as cell lines, synthetic chemical intermediates, artefacts) should be preserved and labelled. Furthermore, primary data (clinical/laboratory records, questionnaires, tapes of interviews, field notes) should be stored so as to be available for review by other scientists.

To protect human subjects, data regarding them should be stored in such a manner as to insure privacy/confidentiality. In particular, in order to maintain the anonymity of human subjects, the guidelines suggest that questionnaires be stored using code numbers to link them to computerized files and that transcripts of taped interviews are to be stored without names and other key identifiers.

Furthermore, the Guidelines emphasize the need to manage data that have been generated and stored electronically. It is suggested that electronic lab notebooks be used that allow the direct entry of laboratory observations etc. and that permit the user to "lock down the data and prevent subsequent data manipulations … (and) the ability to add electronic signatures for further validation."

Use and misuse of data

The guidelines also focus on the "Use and misuse of data", emphasizing that *all* relevant observations must be reported, including data that contradict or fail to support one's conclusions. I refer here to the notorious example of the misuse of data treated earlier in Footnote 6—the oil-drop experiments credited to Robert Millikan, as well as to the discussion in Sect. 5.3.3 in Chap. 5 on whether to discard an apparent anomalous observation.

Ownership of and access to data

The guidelines emphasize that research data belong to the university, not to the individual researcher. It is argued that it is the university that can be held accountable for the integrity of the data and that it is the university that is the recipient of sponsored research awards, not the individual researcher.

If there is possibility of a research project leading to a copyright or patent application, the guidelines note that a written agreement must be made within the group as to rights to the intellectual property. The university patent policy allows the sharing of revenues (licensing/sale/royalties) between inventor(s) and the university. Written agreements must also be made between Ph.D. students, postdoctoral fellows or other researchers on the one hand, and the research director or the principal investigator on a project on the other hand, as to what one may continue to work with after a researcher leaves the research group—particularly as to unique materials. These agreements should specifically consider which parts of the project the researcher is permitted to continue to explore after leaving the research group, including a specification of the extent to which a copy of research data may be taken. I note in this connection that it is considered to be in accord with generally accepted practice that other researchers have access upon request to all the data after the results are published.

Storage and retention of data

The guidelines state that data must be stored securely for at least seven years after the completion of a project, submission of the final report to a sponsor, or publication of the research, whichever comes last. In particular, some types of data are to be deposited in national/international databanks; this is particularly relevant when the data is so extensive that it precludes publication (e.g. X-ray crystallographic data on protein structures, human genomic data, and DNA micro array data).[12]

Authorship and publication practices

The guidelines emphasize that sponsors may not be granted a veto right as to what may be published, but they may be allowed to delay publication for at most six months to permit filing a patent application. Terrorism is also considered. In recognition of the potential harm of information that could contribute to bioterrorism, it has become a norm that journal editors may stop publication of an article

---

[12]Guidelines are also provided as to archiving of data from research in the social sciences; reference is made to the list of websites for social science archives available thru the University of California at San Diego: http://odwin.ucsd.edu/idata.

if they consider that the potential societal benefits are outweighed by the potential risks of publication due to terrorism. Therefore, researchers are advised to consult editors about the procedures to follow in such cases, for example modifying or withholding publication.

The guidelines as to authorship and publication practices also provide criteria for the authorship of publications; I note that many professional associations and research journals also have established such criteria. A common standard is that each person listed as an author should have participated in formulating the research problem, in interpreting the results, and in writing the paper, and should be prepared to defend the publication against criticism; authors who do not meet the criteria should not be listed. Following such a standard might exclude many senior scientists from being listed as a co-author in those cases where a publication is primarily authored by a junior scientist (say a Ph.D. student) and where the senior scientist is the student's guide, but has not actively contributed to the research or to writing the paper.

As regards the order in which authors are to be listed, customs vary with the particular discipline, so no specific guidelines are provided. It is underlined however that it is important that all co-authors understand the basis for the ordering and agree to it in advance. My own experience is that a rather broadly accepted procedure is that authors are listed in alphabetical order unless one (or more) of the authors are to be considered as having made significant contributions compared to the others, in which case they will be listed first, and the rest in alphabetical order. There also appears to be a tendency in some research environments that a senior scientist is named first, even though he or she has not been a major contributor. Such a practice would not be in agreement with the guidelines being considered here and would not give proper credit to a more junior author who was in fact the major contributor, something of increasing importance in a world where so much emphasis is placed on "who published what, where, when?".

Also as regards publications, the guidelines state that it is "poor practice to allow the same manuscript to be under review by more than one journal at the same time." Not all scientists agree as it often takes many months before a refereed journal replies to the authors of an articles that has been submitted. If a journal turns down a submission, additional months will be required for the next journal to evaluate it— and then it may never be accepted due to its being "dated". On the other hand, other writers on science have strong feelings that it is unacceptable to submit to more than one journal: "No, don't ever do it. ... duplicate submissions are absolutely prohibited." (Locke et al. 2007; 36–7)

The Guidelines also note that it is unethical to contact the media regarding information that is contained in an accepted publication prior to the publication; we can refer to the earlier discussion in Chap. 3, Sect. 3.2 of "cold fusion" that was announced to the media before the results had been made accepted by a peer-reviewed journal. An exception may be where a public health issue is involved and where the editor of the journal or book agrees to such advance release.

Finally, I note that not only do research institutions such as universities have guidelines dealing with the ethical aspects of authorship and publication practices.

A number of associations that publish scientific journals also have clear guidelines. An example is provided in the note: (Editors of the Publications Division of the American Chemical Society 2010). The preface to this note emphasizes: "… the editors of journals published by the American Chemical Society now present a set of ethical guidelines for persons engaged in the publication of chemical research, specifically, for editors, authors, and manuscript reviewers. … the observance of high ethical standards is so vital to the whole scientific enterprise that a definition of those standards should be brought to the attention of all concerned."

Obligation to report

The University of Pittsburgh's guidelines emphasize that it is a serious responsibility of all members of the scientific community to report suspected research misconduct. In the specific case of the University of Pittsburgh this means that such suspected misconduct must be reported to the relevant dean or to the university's Research Integrity Officer, where allegations are handled using procedures described in the university's Research Integrity Policy.

I note that such a guideline could certainly lead to a difficult personal ethical dilemma for a scientist who suspects a colleague of such research misconduct. On the one hand, she or he would feel obligated to report research misconduct (just as one would in the case of criminal behaviour), while on the other hand the scientist might very well feel uncomfortable about reporting a colleague, particularly if the colleague is a friend. I note in this connection that at a number of leading universities throughout the world, including Princeton, Stanford and Oxford, examinations of all kinds are conducted under an honor code. For example, at Princeton University, the students are to sign the following pledge after each such examination: "I pledge my honor that I have not violated the honor code during this examination." In addition, every student is obligated to report to the Honor Committee any suspected violation of the code that he or she has observed: www.princeton.edu/honor/.

Conflict of Interest

Conflict of interest is considered by the university to be of such major importance that it is treated in depth in a separate *Conflict of Interest Policy*. An example of how such a potential conflict of interest can arise is the case where a researcher's personal financial interests may lead him to compromise the integrity of the research or to research misconduct, for example by the distortion of research outcomes. Preventative measures mentioned in the Policy include the mandatory annual disclosure of outside interests, the possible divesture of such financial interests and the public disclosure of such outside interests; many journals and funding agencies also require such disclosures. Another example of a situation where a conflict of interest can arise is when a researcher is asked to enter into peer review of a manuscript or a proposal (for example, in cases where the reviewer is writing on the same topic or is seeking financial support for research in the same area). In such cases the researcher should disclose any conflict of interest with respect to the matter under review. In addition, the guidelines emphasize that in a situation where researchers from the university are considering/are involved in the commercialization of an invention (e.g. via a start-up company or by licensing of technology), they should consult the university's *Conflict of Interest Policy* and the *Commercialization of Inventions through Independent Companies Policy* (dealing with limitations on a faculty member's equity holdings, role in such a company, etc.).

Although a number of other aspects of research ethics are considered in the guidelines, including Curriculum Vitae, Responsibilities of a (Principal) Research Investigator, Responsibilities to Funding Agencies, and Laboratory Animals in Research, these topics will not be treated here; some have been briefly considered elsewhere in the book. Instead, I will conclude the presentation here by reflecting on the broad implications of the guideline's section on *Special Obligations in Human Subject Research* that deals with the special situations one meets when carrying out experiments with human beings. This is of interest in fields such as clinical psychology/psychiatry, medicine (where new drugs are developed and tested),[13] space research and research as to new weaponry, as well as in the social sciences in general. I note that in many such situations one can meet ethical issues dealing with all the three categories we introduced earlier when considering *un*ethics: harm to sentient beings and the environment, lack of informed consent and invasion of privacy, and deception and coercion.

The guidelines underline that "Research protocols involving human subjects must be approved in advance by the University Institutional Review Board (IRB) which determines whether information describing risks posed to subjects are acceptable and whether information describing risks and benefits of subject participation is conveyed to subjects in an accurate and intelligible manner." The guidelines also underscore that protocols "submitted to the IRB must include a plan for data and safety monitoring" and that "every research protocol involving human subjects should receive a scientific review and written approval as specified in the investigator's academic unit prior to the submission to the IRB for review."

Although in many countries there is a strong focus on the ethical aspects of research that involves experiments with human beings, this is particularly the case in the USA, where there are well-established rules and regulations regarding the use of IRBs. This may be due to the earlier referred to focus on law suits that can lead to huge costs and fines to those whose behaviour is the cause of injury or death to others.[14] IRBs in the USA have the responsibility to ensure and the authority to require that research programmes live up to a series of criteria dealing with some or

---

[13]Realizing the increasing focus on the ethical aspects of human subject research by customers, regulators and the public at large, some far-sighted pharmaceutical companies have developed a pro-active stance as to how they deal with the potential ethical dilemmas that can result. An example is provided by the major Danish pharmaceutical company, Novo Nordisk, which makes public its policies dealing with clinical trials: www.novonordisk-trial.com (as well as its policies dealing with bio-ethics in general: www.novonoridsk.com/science/bioethics).

[14]An example of another country where such boards are increasingly being used is India. See for example "Human Research Participant Protection in the Indian Health Service; http://www.ihs.gov/research/index.cfm?module=hrpp_irb.

all of the following: minimizing risk; how subjects are chosen; special attention to groups such as children, prisoners, pregnant women, the mentally disabled and the educationally disadvantaged; documentation of consent by subjects or their legally authorized representatives based on adequate information; privacy and confidentiality, etc.

For example, if we focus on the issue of obtaining informed consent, potential subjects must be informed in such a way that they understand the various issues involved and that might affect their decision to participate in an experiment. Typically, such information must include the purpose of the research, its duration, potential risks and potential benefits to participants as well as the probability of such risks and benefits. A simple example that one might not ordinarily think about could be where people who participate in a drug test must be informed as to the potential financial risks they could face if physical harm results from the test and their medical insurance refuses to cover their treatment due to their participation in the experiment.

In addition to the requirement of information, it is also a requirement that the informed consent must not be obtained using any form of coercion. In particular safeguards are required as to vulnerable populations (e.g. the terminally ill, prisoners, military personnel, and individuals who are in a subordinate position to the researcher such as graduate students).

Another aspect of human subject research is the issue of when to stop a project. Protocols are usually required to contain criteria for determining when research should be stopped if and when the risks appear to be too great compared to the potential benefits. An example could be the explicit statement of when to stop an experiment with a new drug if the subjects get sicker or evidence considerable stress or other physiological and psychological symptoms. (Stern and Lomax 1997; 289–291)

To conclude this presentation of several aspects of the University of Pittsburgh's *Guidelines for Responsible Conduct of Research*, I note that the underlying motivation behind this was not just to use the guidelines to introduce and reflect on a number of issues that are relevant as regards ethics in research. It was also to give an impression of the increasingly serious attention paid by research institutions to the ethical aspects of research.

10.3 From *Un*ethics to Ethics in Research

Before proceeding to a more direct consideration of the responsibility of the individual scientist, let us conclude the discussions of ethics in scientific research by considering how the prevalent focus on the negative, on *un*ethics, can be complemented with a focus on the positive, on ethics. As was mentioned earlier, when the word ethics is used today, most of the time it is preceded by the letters "un"; it is the word "unethical" that dominates the agenda today, not the word "ethical". Another way of saying this is that ethical guidelines (such as those presented above) tend to focus on what to avoid and how to protect one's self/one's institution, rather than on focusing on what one should do from a moral perspective. It is seldom that ethics is referred to in guidelines for research in a constructive or positive-normative sense. A likely reason for this is that while it is easy to appeal to things to avoid (deceiving, harming, coercing, polluting, etc.), scientists tend to avoid taking a standpoint regarding science in situations where they must invoke arguments that are considered to be *un*scientific. In particular, they are afraid of being met with questions as to values that may compel them to invoke philosophical, spiritual, religious or deeply personal explanations which are considered to be metaphysical. But there is no way of avoiding such reference to values; science cannot define what acceptable behaviour is and what unacceptable behaviour is solely by referring to its own institutionalized methods and frames of reference. Consider for example the provocative perspective on the relationship between values and science (1961; 243) by Bertrand Russell: "I conclude while it is true that science cannot decide questions of value, that is because they cannot be intellectually decided at all, and lie outside the realm of truth and falsehood. Whatever knowledge is attainable, must be attained by scientific methods; and what science cannot discover mankind cannot know." So a movement from *un*ethics to ethics might be considered as a movement from objectivity (considered to be impersonal—one of the most often used terms when referring to scientific research) —to spirituality (a most personal matter).

Earlier, we presented three areas where concerns as to *un*ethical research behaviour often arise: (1) Lack of informed consent and the invasion of privacy; (2) Harm to sentient beings and to the environment; and (3) Deception and coercion. We now reflect on the positive counterparts to these—on ethics rather than on *un*ethics in scientific research.

Complementing a focus on lack of informed consent and invasion of privacy with cooperation and participation

A shift in orientation from *un*ethics to ethics implies complementing a focus on *lack of informed consent and invasion of privacy* (the unethical) with a focus on *cooperation and participation* (the ethical counterpart). While ethical guidelines focus on avoiding surreptitious or clandestine behaviour—which in fact characterizes much research today due to the considerable commercial and military impacts of research as well as the prestige and rewards that accompany successful research— such a shift in orientation will imply a focus on the role of the researcher's awareness of and respect for the those who may be affected by her research— including nature and those not yet born.

The traditional ideal of the natural sciences is that of the neutral, objective observer. All personal values are to be removed or ignored; all biases are to be avoided. Opposed to this is an empathetic orientation towards service and towards loving all and serving all, which leads to the researcher seriously re-orienting her or his role. She or he is no longer just the "distant", neutral, objective investigator of and witness to phenomena. Rather she puts herself in the shoes, so to speak, of those who may be affected, directly or indirectly, by her research. This is a huge change in perspective. It emancipates the scientist from a strictly dualistic perception of her role (as objective observer) and provides her with an opportunity to consider herself a member of, and thereby to be "close to", those "stakeholders" (including nature) that are affected by the research. In this way, scientific research can become a powerful transformative practice. It permits the researcher to transcend the subject-object duality of science and, via an orientation towards service, to approach the unity ideal of spirituality.

Complementing a focus on not harming with a focus on serving
A transitional movement from *un*ethics to ethics implies complementing a focus on *not harming others* with a focus on *serving others*. This orientation affects all aspects of science. In particular, it can affect the identification and choice of research problems by focusing on problems whose solution potentially can benefit individuals, society, and the environment. Many scientists will react negatively to this suggestion, arguing that, for example, basic research cannot and should not consider the potential uses of one's research; such research, it is maintained, only aims at and only should aim at, discovering new knowledge and thereby at contributing to the progress of science. A counter-argument is that we should never forget that the purpose of life is not just progress in knowledge—it must also be to serve ourselves and others. Scientific knowledge is a means to an end, and what that end is, is a matter of human choice; scientists can never avoid the question of what purposes their research serves. Consider once again the perspective of the Nobel laureate Herbert Simon regarding the limits of reason as regards the choice of ends (Simon 1983; 7): "Reason, then, goes to work only after it has been supplied with a suitable set of inputs, or premises. If reason is to be applied to discovering and choosing courses of action, then these inputs include, at the least, a set of *should's*, or values to be achieved, and a set of *is's*, or facts about the world in which the action is to be taken. Any attempt to justify these *should's* and *is's* by logic will simply lead to a regress to new *should's* and *is's* that are similarly postulated. ... We see that reason is wholly instrumental. It cannot tell us where to go; at best it can tell us how to get there."

In other words, the very choice of one's hypotheses/research questions automatically implies a choice, consciously or unconsciously, based on underlying values; research, just as all of purposeful human activity, can never be value-free.

An example of research dealing with a fundamental value of non-violence and that pre-dated so much of the present research in environmental matters is the pioneering, ground-breaking work that was carried out by the American marine biologist, Rachel Carson on the widespread abuse in the use of pesticides. Her book, *Silent Spring*, from 1962 exposed the hazards of the chemical DDT and led eventually to making its production and use illegal in the USA in 1972. Later, its use in agriculture was banned world-wide.

In so doing, Rachel Carson eloquently questioned humanity's faith in the chemical industry and in technological progress in general and thereby set the stage for the environmental movement and in particular for the field of environmental ethics. Although her work was rather basic research, it dealt with a matter of great practical significance and was driven by a deep-rooted value of non-violence and service. She died fairly unnoticed in 1964, but today is considered to be one of the most influential women of the 20th century.

Note that in spite of the evidence as to its deleterious effects, DDT continues to be used in a number of countries in connection with disease control, e.g. in connection with malaria, which is highly controversial since mosquitos develop resistance to it. The choices to prohibit the use of DDT as well as to continue to use it were both responses, although highly different, to ethical dilemmas.

To illustrate the apparent difficulties that humans can face in discriminating between violence and service, consider briefly the following involving a scientist during World War II. Premo Levi, an Italian chemist of Jewish origin, has in a number of books provided deeply moving descriptions of his years in the huge Nazi concentration camp Auschwitz during World War II. In (Levi 1961) he described how Nazi leaders could be unimaginably brutal in serving the Nazi cause and yet could be gentle, affectionate and loving when serving persons in their families. Such "schizophrenic" behaviour, characterized by a dissonance between values when one is at "work" (although I truly hesitate to use this word to describe the activities of the leaders of the concentration camps) and when one is with family and friends, is *not* characteristic of an ethically inclined person. Many spiritual and religious texts emphasize that one should love one's neighbour; the challenge is to understand this not in a strictly limited and narrow physical/geographical sense (where "my neighbour" is generally understood to mean my family, those living close to me, members of my community ...), but ideally in a nonlocal, universal sense, whereby we are to love all and serve all. Fundamental human values cannot be context-dependent!

Another example, also from World War II, of the relationship in research between non-violence, violence and serving others is the behaviour of the renowned physicist Enrico Fermi during his work on the so-called Chicago Pile-1. On December 2, 1942, while World War II was raging, he led the operation of the first self-sustaining nuclear chain reaction, in what would later be referred to as the world's first nuclear reactor. This was a part of the so-called Manhattan Project whose activities eventually led to the development of the atom bomb.[15] As in many complex ethical dilemmas, there were good ethical reasons for actively contributing to this research and development, and good reasons for not contributing. The results of the work carried out by Fermi and his team led to the termination of the war with Japan far earlier than otherwise would have been the case, and thereby to saving the lives of a huge number of American and Japanese soldiers and ordinary Japanese

---

[15]Fermi's work was of great importance for further research on fission processes and paved the way for the far more massive reactors that would breed the plutonium employed in the so-called "Trinity" test atomic bomb and the bomb dropped over Nagasaki (the first atomic bomb was dropped over Hiroshima). Fermi is also known for his contributions to the development of quantum theory, nuclear and particle physics, and statistical mechanics. He received the Nobel Prize in physics in 1938.

citizen—and to the horrific destruction of life and property in the cities Hiroshima and Nagasaki. It led as well to the modern era characterized by increasing threats to humanity and all life by weapons of mass destruction.

A most thought provoking example of the movement to complement a focus on *not harming others* with a focus on *serving others* is provided by the web magazine *Edge* as reported on in the newspaper *The Guardian* in 2006 (day not available). *Edge* had interviewed more than 100 scientists and philosophers regarding "dangerous" issues in ethics. Professor Richard Dawkins from Oxford University referred to how increased understanding of the way the human brain works can lead to difficult questions when defining ethics: "As scientists, we believe that human brains, though they may not work in the same way as man-made computers, are as surely governed by the laws of physics. When a computer malfunctions, we do not punish it. We track down the problem and fix it, usually by replacing a component, either in hardware or software. Isn't the murderer or the rapist just a machine with a defective component? Or a defective upbringing? Defective education? Defective genes?" The fundamental question he really raises is: Do we have obligations to serve and help all human beings, even "the murderer or the rapist", rather than to punish them?

A similar challenging reflection is provided by (Kelly et al. 2007; 54): "If human beings are products of deterministic processes, how can they be held accountable for their actions under any social or ethical codes?" I note that, in contrast to Dawkins, Kelly et al. do *not* accept that we are solely "products of deterministic processes" and argue that we are accountable for our actions.

So whether or not human consciousness is a product of deterministic processes can be a pivotal issues in determining questions of guilt, ethics, responsibility and appropriate treatment/punishment.

Finally here, we can consider the field of leadership where this service-orientation has exerted considerable influence in recent years. Although rather removed from traditional scientific disciplines, leadership has become a field that has been increasingly researched due to its enormous influence on the effectiveness of organisations, including those dealing directly or indirectly with scientific research. Interestingly, although this concept is rather new in the West, as far back as the 4th century BCE in India, Kautilya dealt with the concept of the king (leader) as a servant in his treatise *Arthashastra* (*Arthaśāstra*) that deals with the qualities and disciplines required of a wise and virtuous king, a *rajarshi* (from Sanskrit: *rajan* "king" and *rishi* "sage"). In its modern expression, servant leadership emphasizes the responsibility of a leader to serve others while at the same time to achieve results that promote the organisation's values including "success". This modern formulation started with the writing of the highly respected former director of management development and research at the major American company, AT&T, Robert Greenleaf: In 1970 he published the essay "The Servant as Leader". This was followed up by (Greenleaf 1977) which presented a collection of his writings. In India, Dr. S.K. Chakraborty, former Convenor, Management Centre for Human Values at the Indian Institute of Management, Calcutta, has written extensively about the concept of *rajarshi* and its relevance for modern leadership;

see e.g. *Management by Values: Towards Cultural Congruence* (1993) New York: Oxford University Press.

Complementing a focus on not deceiving and coercing with honesty and transparency

Finally, a shift in orientation from *un*ethical behaviour to a socially constructive ethics is characterized by complementing a focus on avoiding *deception and coercion* with a focus on *honesty and transparency*. Here the focus is not just one of avoiding premeditated deception—for example, by suppressing data that does not support your hypotheses or by actively trying to suppress research by others that could hurt your own chances of recognition and of getting financial support. Rather it is on searching for and presenting the truth, the results of one's research, no matter whether this appears to benefit one's own situation or not.

Of course there are limits to the demands arising from a value such as transparency. Earlier we referred to the policy of certain scientific journals dealing with the publication of articles that could enable terrorists to develop biological weapons. In addition, the commercial and military funding of research also places limits on the transparency of one's research. An example is from the previously cited *Guidelines for Responsible Conduct of Research* at the University of Pittsburgh, where sponsors may delay publication of research results for up to six months. And of course a researcher who has sent an article to a journal or who has sought a patent on a new technology resulting from his research cannot be expected to publicize his results until the article has been published or the patent has been authorized/his property rights have been protected.

Similarly, in a scientific context, "honesty" and "presenting the truth" can be subject to interpretation. While there can be widespread agreement that it would be unethical to provide false information (e.g. to manipulate one's data so they lead to the acceptance of an alternative hypothesis that would otherwise be rejected), many scientists might disagree that an ethical demand as to honesty also implies that a researcher should attempt to publicize information that is true but that could harm oneself. For example, suppose a researcher earlier had committed fraud or otherwise behaved in a manner that could lead others to raise serious questions as to his or her trustworthiness—even though those acts were committed many years ago and well before he became a respected member of the scientific community. Does this imply that the researcher has a moral commitment to actively attempt to make others aware of such earlier unethical behaviour? Probably not—but that might be the case if the unacceptable behaviour was of recent vintage.

So a focus on transparency and honesty may be tempered by consideration of one's own interests, the interests of others and, at times, by the special interests of sponsors of research. It is just such a need for balance that can lead to ethical dilemmas in science.

Reflecting on this section on "From *un*ethics to ethics in research", one might say that a re-orientation from a focus on avoiding *un*ethical behaviour in science to one of promoting ethical behaviour can be considered as complementing scientific research and its search for truth about external reality with a spiritual search for truth about our internal, existential reality. It is the combination of these inner and outer searches that enables the reflective scientist to carry out his research with a passion that arises from a deeply ingrained desire for learning about the *external* world, while at the same time—no matter whether the scientist is conscious of this or not—searching for answers to fundamental existential questions as to who we are and why we are here, i.e. for learning about the *internal* world.

In this connection, consider the following statements by one of the scientists I have interacted with and whose wisdom has inspired me, the distinguished Indian aerospace scientist, Dr. A.P.J. Abdul Kalam (1931–2015), the 11th President of India (2002–2007). "The path of science can always wind through the heart. For me, science has always been the path to spiritual enrichment and self-realization." (Pruzan and Pruzan Mikkelsen 2007; 308) "The accepted view was that a belief in scientific methods was the only valid approach to knowledge. If so, I wondered whether matter alone was the ultimate reality, and were spiritual phenomena but a manifestation of matter? Were all ethical values relative, and was sensory perception the only source of knowledge and truth?... I had been taught that true reality lay beyond the material world in the spiritual realm and that knowledge could be obtained only through inner experience." (Kalam and Tiwari 1999; 19).

10.4 The Responsibility of Scientists and of Science as an Institution

The above reflections by the rocket scientist and former president of India pave the way for reflections on the responsibility of the individual scientist, i.e. on the special responsibilities that befall those who actively carry out scientific research. But what is meant by "responsibility" here? And how can a scientist determine how best to live up to such a responsibility? This is the focus of this chapter's concluding section; it should not surprise the reader that no simple answers are provided.

Just as with ethics, most of us have an intuitive feeling or understanding of what it means to be responsible—or to be irresponsible. Perhaps we can recall having the satisfaction of knowing we listened to our inner promptings and did the "right thing" in the "right way", when faced with a moral dilemma. Of course, "right" is generally open to interpretation and debate, both before and after we act, depending as it does on the motivations for our actions before undertaking them, and the results of our actions, including how they affected others.[16] Or perhaps we have experienced remorse when we behaved in a way we felt was "*ir*responsible". And

---

[16](Das 2009) presents a tour de force analysis of the moral dilemmas that characterize the ancient Indian epic, the *Mahabharata* and that nevertheless are strikingly relevant for a modern reflective scientist. According to its author, "The epic is obsessed with questions of right and wrong—it analyses human failures constantly. Unlike the Greek epics, where the hero does something wrong and gets on with it, the action stops in the *Mahabharata* until every character has weighed in on the moral dilemma from every possible angle." (p. xxxviii).

perhaps we have even felt anger or frustration when witnessing what we felt was irresponsible behaviour of others, including scientists who speak only of truth (from the mind) and not of its consequences (from the heart); in Chap. 6 on experimentation, I referred to the Nazi doctor, Josef Mengele, known as the Angel of Death, who during World War II carried out horrible, indescribably painful experiments on prisoners at the huge concentration camp Auschwitz in order to find the truth as to the cause of and cures for various diseases.

Having just referred to the former president of India, it is also relevant here to refer to words of warning and wisdom by the former president of the United States, Dwight D. Eisenhower who, prior to his presidency, was a highly respected general and commander of all the allied forces in Europe during the World War II. In his Farewell speech to the nation on January 17, 1961, he coined the phrase "the military-industrial complex". He warned the American people of the ambitions of an immensely powerful combination of industrial and military organisations to develop ever newer, ever more powerful, and ever more costly and refined weapons systems: "In the councils of government we must guard against the acquisition of unwarranted influence, whether sought or unsought, by the military-industrial complex. The potential for the disastrous rise of misplaced power exists and will persist." His warning was not only to the citizenry at large; it contained a message of huge importance for that relatively large percentage of scientists who work, directly or indirectly, for this "military-industrial complex" that they must be vigilant as to their responsibilities. There is *no* doubt that his warning was, and still is, called for, yet, for the main, unheeded.

According to *Webster's Deluxe Unabridged Dictionary*, 2nd ed. 1979, "responsibility" is derived from the Late Latin *responsabilis*—requiring an answer. The word can be seen as having two parts: response + able, i.e. the ability to respond, to be able to answer for one's conduct and obligations. Ultimately it means: "expected or obligated to account (*for* something, *to* someone)… involving duties …able to distinguish between right and wrong … trustworthy, dependable, reliable".

It is my fundamental position here that an inquiry into "Why be responsible?" is a precondition for a scientific organisation to successfully integrate "responsibility" into its self-awareness—and therefore into policies, processes and practices that promote responsible behaviour by the scientists it employs. This enquiry is a precondition as well for an individual scientist to transcend his or her self-interest (in terms of recognition, income, etc.) such that a search for scientific truth will be performed responsibly and with integrity—with harmony between one's thoughts, words and deeds.

Having said this, it will be clear that there is no simple, straight-forward, fool-proof way that the individual scientist can, in a concrete situation, rationally distinguish between responsible and irresponsible behaviour. We have already met a number of concrete ethical dilemmas where there is no simple, objective way of distinguishing responsible from irresponsible. Earlier, we considered briefly the ethical dilemma faced by Enrico Fermi in connection with his research that played an important role in the development of the atom bomb. The point to be made here, however, is not just the complexity of such ethical dilemmas in science. Rather, it is that ethics in research is also a very personal matter.

While leading the efforts to obtain a controlled sustaining nuclear chain reaction, to reduce the risks of a reaction that could lead to the spreading of radioactivity, Fermi carried out his tests under the stands of a stadium at the University of Chicago. Although he did not personally manipulate the control rods that absorbed neutrons, he personally monitored the neutron activity in developing the critical mass of the fissile material. Fermi's behaviour illustrates a particular aspect of human behaviour that is relevant to our reflections on the responsibility of scientists; our deep sensitivity to the here and now (in his case, reducing the risk of radioactivity in the area surrounding the pile by performing the tests in an isolated place), coupled with a more abstract relationship to potential results more distant in time and location (the consequences of developing the atom bomb). Fermi's activity perhaps should also be seen against the background of his having to flee Italy in the late 1930s due to his having a Jewish wife. Like Einstein (and many other brilliant scientists who had to flee Europe due to being a Jew or being married to one) who in 1939 signed a letter to the president of the United States, Franklin D. Roosevelt, urging that the bomb be built, Fermi was an active supporter of the development of the atom bomb. It is reasonable to assume that his own personal experiences, together with his more abstract philosophical reflections, played a major role in leading to his research decisions and behaviour.

Expanding on this story about Fermi, personal experience, reflection, listening to the voice of one's conscience, and the ability to discriminate—to distinguish between the inner voice of truth and the promptings of the ego—are indispensable for the individual scientist who is faced with an ethical dilemma; there is no simple checklist. And what appears to be intuition may be a poor guide in complex situations—unless it is the result of what one might refer to as pure consciousness. As we noted earlier, the history of science clearly demonstrates that not only the intuition of the individual scientist but also of the scientific community can change and progress; "Newton's notions of a field of force and of action at a distance and Maxwell's concept of electromagnetic waves were at first decried as 'unthinkable' and 'contrary to intuition'." (Feller 1968; 2).

Many scientists, particularly those who work with basic research, tend to disregard such reflection on a scientist's responsibility. They may say that basic research only deals with investigating the natural world, and such research is value-free. They tend to argue that the only relevant consideration of values and ethics deals with: (a) the way the research is carried out, and (b) the use (or misuse) of its findings. They also argue that for those who work with basic research, it can be very difficult to determine whether the research is for the detriment or the benefit of others, of society, of nature.

Although the sub-section that follows deals directly with this issue, I feel that it is appropriate here to repeat the following statement from earlier in this chapter:

Ethics has a demand: the scientist must assume responsibility for the way that his or her research can affect others, including nature itself. Therefore he or she must reflect not only on how results will be accepted by the scientific community, but also on how they will be accepted by all those who may be affected, directly or indirectly, by both the investigations and the results.

Is a distinction between basic research and applied research relevant for one's responsibility as a scientist?

Starting with above contention that basic research is and should be value-free, we can refer to Sect. 9.1 in the preceding chapter where we observed that the distinction between basic and applied research is not nearly as clear as text books imply. Consider for example biomedical research where there is a tight connection between a desire for particular applications (e.g. medical treatment of what appear to be incurable diseases via the use of stem-cells) and the definition of concrete basic research projects (e.g. within molecular genetics, protein chemistry, immunology, and neurology).

Similarly, we can refer once again to the earlier reflections on the development of the atom bomb under the Manhattan-project in the 1940s. The desire for a specific application, the development of the bomb, predetermined the theoretical research challenges to the group of physicists and chemists that carried out the basic research that enabled the development of the technology that eventually led to the A-bomb. The classical distinction between basic and applied research whereby there is a difference between the responsibility of a physicist working on basic theories as to the atom and a physicist working on the theoretical problems in connection with the design of an atom bomb became unclear; in 1945 basic research lost its innocence and could no longer be seen in isolation from political, military, social and commercial processes. It became clear that the individual scientist was a part of the world—and had responsibilities for the world, no matter whether his research was basic or applied.

Today this connection is even clearer. Research in fields such as high energy physics or the human genome is no longer just a matter of the acquisition of basic knowledge, but is part and parcel of huge systems financed by enormous military, political and commercial organisations with their own interests, often quite hidden from the individual scientists who carry out the research. The distinction between basic and applied research is becoming increasingly amorphous. And with it as well the normative attitude by those scientists who argue that they should leave their values outside the door before they enter their research laboratories, and that their only responsibility is to the advancement of knowledge.

Is there a clear distinction between beneficial and detrimental research?

History has shown us time and time again that what we think may be beneficial results of research, may also be destructive; a good example is DDT that we referred to earlier. On the other hand, research that knowingly leads to technology that is potentially destructive cannot simply be frowned on by a scientist, the world is not that simple. Knives can be used both by murders and cooks and surgeons. Animal testing that leads to harm to some living creatures, the animals, can in some cases also improve the conditions of other living creatures, humans. For example, such testing can be a pre-condition for the production of pharmaceutical goods, and e.g. the United States Food and Drug Administration has strict rules and regulations in this regard. It should be noted here that while such tests may be required, strict official rules and regulations to ensure the humane treatment of test animals do not exist.

Finally here, to indicate the complexity of such considerations, it should also be noted that animal testing is not only used in connection with research aimed at developing pharmaceutical products; very often animal testing is carried out in connection with the production of cosmetics; countless animals undergo immense suffering so that better lipstick, better nail polish, better perfume … can be produced and huge fortunes made in the process. The website of the Humane Society of the United Sates, http://www.humanesociety.org/about/hsus-transformational-change. html, provides information on the complex ethical issues associated with animal testing.

In an age of increasing specialization, it may be virtually impossible for the individual scientist to have an overview of how one's specific research contributes to a large and complex project. This is acerbated by the fact that with increasing specialization, the researcher learns more and more about less and less, developing tunnel vision in the process as though he was looking through a microscope. This results in a barrier to obtaining a holistic perspective as to the project, and thus to a lack of awareness as to both the potential effects of his research and his responsibility as a scientist.

Therefore there appear to be good reasons if a scientist concludes (if in fact he or she reflects on such issues at all) that it is not and cannot be the responsibility of the individual scientist to even attempt to consider the potential destructive and constructive aspects of one's research. One should, they might argue, simply do one's research in an ethical manner, leaving the use of one's research to others; that is the limits of one's responsibility. The explicit or implicit conclusion of such an argument is that the individual researcher is free from personal responsibility as to the nature of his or her research since one can never know how the results will be used and by whom.

It is indeed tempting to think this way. It reduces the responsibility of the researcher to simply deal with *how* the research is performed, and not with *what* the research can be used for or *why* it is financed. Furthermore, the individual researcher may have great difficulty in exerting influence on more strategic matters; decisions as to these matters are typically left to political/business/military leaders who are far removed from the influence of the individual scientists. After all, it is seldom that the individual scientist works in isolation on a project of his or her own design where one can assume ethical responsibility for the project and its results. Rather, it is the university or other research institution that organizes and leads the larger research project one contributes to that has the responsibility. So if one is not an idealist with a dominating sense of justice and responsibility, one simply carries out one's exciting work that may offer good opportunities for professional growth, recognition and earnings. In such situations there is a large risk of considering one's responsibility as dealing solely with the particular project one is working on, and not with the overall problem complex.

The paradox is that this is occurring at a time when the individual researcher more than ever before can be said to have a special responsibility to work not just so as to promote his own narrow interests, but those of his local society as well as of the world society, including the interests of nature in its broadest interpretation. In a fragmented world with global problems, science and the individual scientist are both part of the problem and part of the solution to the problems.

How the individual scientist can deal with these complex issues is, however, not at all clear. Science as an "institution" has not provided an institutionalized approach to these issues. Nor is there any generally agreed to statement of the rights and duties of scientists as can be found in e.g. the Universal Declaration of Human Rights regarding the rights of humans in general and the duties of the member states of the UN to protect these rights. It is also totally unrealistic to assume that the individual scientist has access to all the knowledge that would be required for him or her to rationally—or otherwise—determine whether and how s/he will work on a research project so as to behave in an ethically responsible manner.

So there are no simple answers, no checklists to follow, no rule books to refer to, no universal statements of principles. Based on "research into his research", seeking guidance from his inner compass, and utilizing his powers of discrimination, each scientist must determine as best he can whether or not to work for a particular research institution, whether or not to contribute to a particular research project, and if so how best to do this in a responsible manner. But having made such a choice does not relieve the scientist of the responsibility of continually reflecting on and reassessing his responsibilities. Life is complex. Science is complex. That is their beauty and challenge to us all. In spite of the difficulties and the lack of clear cut answers, scientists have a responsibility of using their heads and their hearts so as to determine, delimit, and live up to their responsibilities as scientists, human beings and citizens of the world.

Let us conclude by re-considering the concept of truth, a concept so fundamental to the institution and practice of science. Most scientists consider science to be concerned with the pursuit of true knowledge. In Chap. 2 we emphasized that science, and its accompanying rationality and methodology, is regarded (in particular by scientists!) as having a privileged access to truth—to true knowledge of the physical world and its inhabitants. We even provided a definition of science based on truth: "Natural science is a special way of looking at the universe—a method of discovering, generating, testing, and sharing *true* and reliable knowledge about physical reality".

The challenges and risks that have been outlined in this chapter indicate that this pursuit of truth, often within an increasingly specialized frame of reference, may lead one unconsciously to neglect the pursuit of the truth which transcends truth as to the material world and which is the underlying basis for our values and ethics. The huge yet subtle and implicit ethical challenge to the reflective scientist is to combine the search for truth in the external world of objects and phenomena with the search for one's inner truth and to be constantly aware of one's humaneness, even while probing the secrets of the smallest of the small and the largest of the large, the near and the far, the inner and the outer.

References

AAAS. (1990). *The liberal art of science: Agenda for action.* Washington, D.C., USA: American Association for the Advancement of Science.

Alexander, E. (2012). *Proof of heaven: A neurosurgeon's journey into the afterlife.* New York, NY, USA: Simon & Schuster.

Bayes, T. (1763). Essay towards solving a problem in the doctrine of chances. *Philosophical Transactions* of the *Royal Society* (of London), *53*, 370–418. Available online at www.stat.ucla.edu/history/essay.pdf.

Berger, J., & Berry, D. (1988). Statistical analysis and the illusion of objectivity. *American Scientist, 76*, 159–165.

Berger, P. L., & Luckmann, T. (1966). *The social construction of reality.* New York, NY, USA: Doubleday.

Berry, D. (2006). Bayesian clinical trials. *Nature Reviews Drug Discovery, 5*, 27–36. Available at http://www.nature.com/nrd/journal/v5/n1/full/nrd1927.html

Bogetoft, P., & Otto, L. (2011). *Benchmarking with DEA, SFA and R* (Vol. 157). New York, NY, USA: Springer International Series on Operations Research and Management Science.

Bogetoft, P., & Pruzan, P. (1997). *Planning with multiple criteria: Investigation, communication, choice* (2nd ed.). Copenhagen, Denmark: Copenhagen Business School Press.

Bohannon, J. (2013). Who's afraid of peer review? *Science, 342*(6154), 60–65.

Bohm, D., & Peat, F. D. (1989). *Science, order and creativity.* London, UK: Rutledge.

Bohr, N. (1957). *Atomfysik og menneskelig erkendelse (Atomic physics and human knowledge).* Copenhagen, Denmark: J.H. Schultz Forlag.

Bowles, S., & Gintis, H. (2011). *A cooperative species: Human reciprocity and its evolution.* Princeton, NJ, USA: Princeton University Press.

Broad, W., & Wade, N. (1985). *Betrayers of the truth: Fraud and deceit in science.* Oxford, UK: Oxford University Press.

Brydon-Miller, M., et al. (2003). Why action research? *Action Research, 1*(1), 9–28.

Bryman, A. (2001). *Social research methods.* Oxford, UK: Oxford University Press.

Campbell, T. C., & Campbell, T. M. (2007). *The China study.* Kent Town, Australia: Wakefield Press.

Chalmers, A. F. (1999). *What is this thing called science?* (3rd ed.). Buckingham, UK: Open University Press.

Charland-Verville, V., et al. (2014). Near-death experiences in non-life-threatening events and coma of different etiologies. *Frontiers in Human Neuroscience, 8*, 203. Online at http://journal.frontiersin.org/Journal/10.3389/fnhum.2014.00203/full

Committee on Science, Engineering and Public Policy. (2009). *On being a scientist: A guide to responsible conduct in research* (3rd edn). Washington D.C, USA: National Academies Press.

Cooper, D. R., & Schindler, P. S. (2003). *Business research methods* (8th ed.). New York, NY, USA.

Crowe, M. J. (1987). Ten misconceptions about mathematics and its history. In W. Asprag & P. Kitcher (Eds.), *History and philosophy of modern mathematics*. Minnesota studies in the philosophy of science (Vol. XI, pp. 260–277). Minneapolis, MN, USA: University of Minnesota Press.

D'Agostini, G. (1998). *Bayesian reasoning versus conventional statistics in high energy physics*. Invited talk at the XVIII International Workshop on Maximum Entropy and Bayesian Methods, München, Germany, July 27–31, 1998.

Darius, P., & Portier, K. (1999). Experimental design. In H. J. Adèr & G. J. Mellenbergh (Eds.), *Research methodology in the social, behavioural and life sciences* (pp. 67–95). London, UK: Sage Publications.

Das, G. (2009). *The difficulty of being good: On the subtle art of dharma*. New Delhi, India: Penguin Books.

Davies, E. B. (2010). *Why belief matters: Reflections on the nature of science*. Oxford, UK: Oxford University Press.

Depew, D. J., & Weber, B. H. (1996). *Darwinism evolving: Systems dynamics and the genealogy of natural selection*. Cambridge, MA, USA: MIT Press.

Editors of the Publications Division of the American Chemical Society. (2010). *Ethical guidelines to publication of chemical research*. Washington DC, USA: ACS Publications.

Efron, B. (2013). Bayes' theorem in the 21st century. *Science, 340*, 1177–1178.

Elliott, D., & Stern, J. E. (Eds.). (1997). *Research ethics: A reader*. Hanover, NH, USA: University Press of New England.

EVA. (2008). http://english.eva.dk

Feller, W. (1968). *Introduction to probability theory and its applications* (3rd ed.). New York, NY, USA: Wiley.

Feyerabend, P. (1975). *Against method: Outline of an anarchistic theory of knowledge*. London, UK: New Left Books.

Frank, P. G. (1954). The variety of reasons for the acceptance of scientific theories. *Scientific Monthly 79*, 139–145; Reprinted in Klemke, E. D., et al. (Eds.). (1998). *Introductory readings in the philosophy of science* (3rd ed., pp. 465–475). Amherst, MA, USA: Prometheus Books.

Franklin, A. (1989). The epistemology of experiment. In D. Gooding et al. (Eds.) *The uses of experiment: Studies in the natural sciences* (pp. 437–460). Cambridge, UK: Cambridge University Press.

Friedland, A. J., & Folt, C. L. (2009). *Writing successful science proposals* (2nd ed.). New Haven, CT, USA: Yale University Press.

Gauch, H. G, Jr. (2003). *Scientific method in practice*. Cambridge, UK: Cambridge University Press.

Gleick, J. (1987). *Chaos: Making a new science*. New York, NY, USA: Viking Penguin.

Gooding, D., et al. (Eds.). (1989). *The uses of experiment: Studies in the natural sciences*. Cambridge, UK: Cambridge University Press.

Goswami, A. (1993). *The self-aware universe: How consciousness creates the material world*. New York, NY, USA: Putnam.

Grassi, D., et al. (2005). Cocoa reduces blood pressure and insulin resistance and improves endothelium-dependent vasodilation in hypertensives. *Hypertension, 46*, 398.

Grassie, W. (2010). *The new sciences of religion: Exploring spirituality from the outside in and bottom up*. New York, NY, USA: Palgrave Macmillan.

Greenleaf, T. (1977). *The power of servant leadership: A journey into the nature of legitimate power and greatness*. New York, NY, USA: Paulist Press.

Gross, P. R., & Levitt, N. (1994). *Higher superstition: The academic left and its quarrels with science*. Baltimore, USA: The Johns Hopkins University Press.

Hacking, I. (1983). *Representing and intervening: Introductory topics in the philosophy of natural science*. Cambridge, UK: Cambridge University Press.

Hagen, S. (2003). *Buddhism is not what you think: Finding freedom beyond beliefs*. New York, NY, USA: HarperCollins.

Hanson, N. R. (1958). Observation. In N.R. Hanson (Ed.) *Patterns of discovery* (pp. 4–11, 15–19). New York, NY, USA: Cambridge University Press. Reprinted in Klemke, E. D., et al. (Eds.). (1998). *Introductory readings in the philosophy of science* (3rd ed., pp. 339–351). Amherst, MA, USA: Prometheus Books.

Hawking, S. (1988). *A brief history of time.* New York, NY, USA: Bantam Books.

Hawking, S., & Mlodinow, L. (2010). *The grand design: New answers to the ultimate questions of life.* New York, NY, USA: Bantam Books.

Hawking, S., & Penrose, R. (1996). *The nature of space and time.* Princeton, NJ, USA: Princeton University Press.

Hempel, C. G. (1948). Studies in the logic of explanation. In C.G. Hempel (Ed.) *Aspects of scientific explanation and other essays in the philosophy of science* (pp. 135–175). Baltimore, MD, USA: Williams and Wilkins Company. Reprinted in edited form in Klemke, E. D., et al. (eds.). (1998). *Introductory readings in the philosophy of science* (3rd ed., pp. 206–224). Amherst, MA, USA: Prometheus Books.

Hofstadter, D. R. (1989). *Gödel, Escher, Bach: An eternal golden braid.* New York, NY, USA: Vintage Books (Reprint of edition by Basic Books, 1979).

Hume, D. (1748/1955). *An enquiry concerning human understanding,* originally published in 1748, reprinted 1955 together with his autobiography and excerpts from his earlier *Treatise of human nature.* La Salle, IL, USA: Open Court Publishing Company.

Husted, J., & Lübcke, P. (2001). *Politikens filosofi håndbog (in Danish; English title: Politiken's handbook of philosophy).* Copenhagen, Denmark: Politikens Forlag.

Jeffreys, H. (1983). *Theory of probability* (3rd ed.). Oxford, UK: Oxford University Press.

Kalam, A. P. J. Abdul, & Tiwari, A. (1999). *Wings of fire.* New Delhi, India: Orient Longmann.

Kelly, E. F., et al. (2007). *Irreducible mind: Toward a psychology for the 21st century.* Lanham, Maryland, USA: Rowman & Littlefield.

Kelly, E. F., et al. (Eds.). (2015). *Beyond physicalism: Towards reconciliation of science and spirituality.* Lanham, Maryland, USA: Rowman & Littlefield.

Kitcher, P. (1982). Believing where we cannot prove. In P. Kitcher (Ed.) *Abusing science: The case against creationism* (pp. 30–54). Cambridge, MA, USA: The MIT Press. Reprinted in Klemke, E. D. et al., (eds.). (1998). *Introductory readings in the philosophy of science* (3rd ed., pp. 76–98). Amherst, MA, USA: Prometheus Books.

Klemke, E. D., et al. (Eds.). (1998). *Introductory readings in the philosophy of science* (3rd ed.). Amherst, MA, USA: Prometheus Books.

Koestler, A. (1989).*The act of creation.* London, UK: Arkana (first published by Hutchinson & Co., 1964).

Kuhn, T. S. (1970). *The structure of scientific revolutions* (2nd ed., enlarged). Chicago, IL, USA: Chicago University Press.

Lee, J. A. (2000). *The scientific endeavour: A primer on scientific principles and practices.* San Francisco, CA, USA: Addison Wesley Longman.

Levi, P. (1961). *Survival in Auschwitz,* translation of *Se questo è un unomo,* 1947. Springfield, OH, USA: Collier.

Lincoln, Y. S., & Guba, E. G. (1985). *Naturalistic inquiry.* Newbury Park, CA, USA: Sage Publications.

Locke, L., et al. (2007). *Proposals that work* (5th ed.). Thousand Oaks, CA, USA: Sage Publications.

Lovelock, J. (1979). *Gaia: A new look at life on earth.* Oxford, UK: Oxford University Press.

Lübcke, P. (1995). *Politikens filosofi leksikon (in Danish; English title: Politiken's encyclopaedia of philosophy).* Copenhagen, Denmark: Politikens Forlag.

Lyons, L. (2013) Bayes and frequentism: A particle physicist's perspective. *Contemporary Physics,54*(1), 1–16. Published on line January 18, 2013. Available at http://dx.doi.org/10.1080/00107514.2012.756312

Maxwell, J. A. (2005). *Qualitative research design: An interactive approach.* Thousand Oaks, CA, USA: Sage Publications.

Milgram, S. (1983). *Obedience to authority*. New York, NY, USA: Harper and Row.

Moitra, D. (2004). *Innovation, learning and firm flexibility in R&D outsourcing: A mixed-method study*. Ph.D. Dissertation proposal, Rotterdam School of Management Erasmus University, Rotterdam, The Netherlands.

Moody, R. A. (1975). *Life after life*. New York, NY, USA: Bantam Press.

Morgan, S. (2014). *Tutorial on the use of significant figures*. University of South Carolina, Columbia, South Carolina, USA. Available at http://www.chem.sc.edu/faculty/morgan/resources/sigfigs/sigfigs3.html

Morrison, M. A. (1990). *Understanding quantum physics*. Englewood Cliffs, NJ, USA: Prentice Hall.

Nagel, E., & Newman, J. R. (1956). *Gödel's proof*. New York, NY, USA: New York University Press.

Nathan, O. (1993). Science and spirituality. Essay in the Danish newspaper *Berlingske Tidende*, January 1, 1993.

National Academy of Sciences, National Academy of Engineering, and Institute of Medicine. (1992). *Responsible science, volume I: Ensuring the integrity of the research process*. Panel on Scientific Responsibility and the Conduct of Research, Committee on Science, Engineering, and Public Policy, Washington DC, USA: National Academies Press.

OECD. (1997). *The evaluation of scientific research: Selected experiences*. Paris, France: OECD Publication Service.

Pagels, H. (1982). *The cosmic code: Quantum physics as the language of nature*. London, UK: Michael Joseph.

Pais, A. (1982). *'Subtle is the lord...': The science and life of Albert Einstein*. Oxford, UK: Oxford University Press.

Pais, A. (1991). *Niels Bohr's times: In physics, philosophy and polity*. Oxford, UK: Oxford University Press.

Penrose, R. (1989). *The emperor's new mind: Concerning computers, minds, and the laws of physics*. New York, NY, USA: Oxford University Press (Paperback edition: Penguin Books, 1991).

Plato. (c. 380 BCE). *The republic* (G. M. A. Grube, Trans., Revised by C. D. C. Reeve). Indianapolis, IN, USA: Hackett Publishing Company.

Platt, C. (1998). What if cold fusion is real? *WiredMagazine6*(11), November, 1998. http://archive.wired.com/wired/archive/6.11/coldfusion.html

Popper, K. (1983). The open universe: An argument for indeterminism. In W. W. Bartley III (Ed.), *From the postscript to the logic of scientific discovery*. Rowman and Littlefield: Totowa, NJ, USA.

Popper, K. (1998). Science: Conjectures and refutations. Originally published in 1957 under the title Philosophy of science. A Personal report in Mace, C. E. (Ed.) *British philosophy in mid-century*. New York, NY, USA: MacMillan. Reprinted in edited form in Klemke, E. D., et al. (eds.). (1998). *Introductory readings in the philosophy of science* (3rd ed., pp. 38–47). Amherst, MA, USA: Prometheus Books.

Prigogine, I., & Stengers, I. (1984). *Order out of chaos: Man's new dialogue with nature*. New York, NY, USA: Bantam Books.

Pruzan, P. (1995). The ethical accounting statement. *World Business Academy Perspectives,9*(2), 35–46.

Pruzan, P. (2001). The question of organisational consciousness: Can organisations have values, virtues and visions? *Journal of Business Ethics,29*, 271–284.

Pruzan, P. (2015). The source of ethical competence: Eastern perspectives provided by a Westerner. In K. J. Ims, & L. J. Tynes Pedersen (Eds.) *Business and the greater good. Rethinking business ethics in an age of crisis* (pp. 117–148). Cheltenham, UK: Edward Elgar Publishing.

Pruzan, P., & Pruzan Mikkelsen, K. (2007). *Leading with wisdom*. Sheffield, UK: Greenleaf, and New Delhi, India: Sage Publications/Response Books (for South Asia only).

Pruzan, P., & Thyssen, O. (1990). Conflict and consensus: Ethics as a shared value horizon for strategic planning. *Human Systems Management 9*, 134–152.

Przeworski, A., & Salomon, F. (1998). *The art of writing proposals: Some candid suggestions for applicants to social science research council competitions.* Brooklyn, NY, USA: Social Science Research Council.

Radin, D. (1997). *The conscious universe: The scientific truth of psychic phenomena.* San Francisco, CA, USA: HarperCollins.

Randall, D. A., et al. (2007). Climate models and their evaluation (Chap. 8). In S. Solomon et al. (Eds.) *Climate change 2007: The physical science basis. Contribution of working group I to the fourth assessment report of the Intergovernmental Panel on Climate Change.* Cambridge, United Kingdom: Cambridge University Press.

Reichenbach, H. (1949). *The theory of probability.* Berkeley, CA, USA: University of California Press.

Reichenbach, H. (1951). *The rise of scientific philosophy.* Berkeley, CA, USA: University of California Press.

Revel, J. F., & Ricard, M. (1998). *The monk and the philosopher.* (translated from the French by J. Canti). New York, NY, USA: Schocken Books.

Russell, B. (1961). *Religion and science* (first published in 1935 by Home University Library). New York, NY, USA: Oxford University Press.

Russell, P. (2003). *From science to god: The mystery of consciousness and the meaning of light.* Novato, CA, USA: New World Library.

Salmon, W. C. (1967). *The foundations of scientific inference.* Pittsburgh, PA, USA: University of Pittsburgh Press.

Schrödinger, E. (1954). *Nature and the greeks.* Cambridge, UK: Cambridge University Press.

Searle, J. R. (1995). *The construction of social reality.* New York, NY, USA: The Free Press.

Sen, A. (2009). *The idea of justice.* London, UK: Allen Lane.

Sheldrake, R. (2012). *The science delusion: Freeing the spirit of enquiry.* London, UK: Hodder & Stoughton.

Siever, R. (1968). Science: Observational, experimental, historical. *American Scientist, 56,* 70–77. Reprinted in Miller, D. C., & Salkind, N. J. (2002). *Handbook of research design and social measurement* (6th edn, pp. 37–42). Thousand Oaks, CA, USA: Sage Publications.

Simon, H. A. (1983). *Reason in human affairs.* Stanford, CA, USA: Stanford University Press.

Snow, C. P. (1993). *The two cultures.* Cambridge, UK: Cambridge University Press.

Stace, W. T. (1967). Science and the physical world. In: W. T. Stace (Ed.) *Man against darkness.* Pittsburgh, PA, USA: University of Pittsburgh Press. Reprinted in Klemke, E. D., et al. (eds.). (1998). *Introductory readings in the philosophy of science* (3rd ed., pp. 352–357). Amherst, MA, USA: Prometheus Books.

Stern, J. E., & Lomax, K. (1997). Human experimentation. In D. Elliott & J. E. Stern (Eds.), *Research ethics: A reader* (pp. 286–316). Hanover, NH, USA: University Press of New England.

Stewart, I. (1996). The interrogator's fallacy. *Scientific American, 275*(3), 172–175.

Stirzaker, D. (1994). *Elementary probability.* Cambridge, UK: Cambridge University Press.

Summers, J. (1989). *Soho: A history of London's most colourful neighbourhood.* London, UK: Bloomsbury Publishing.

Swami Vivekananda. (1987). *Vedanta: Voice of freedom.* Calcutta: Advaita Ashrama.

Swazey, J. P., & Bird, S. J. (1997). Teaching and learning research ethics. In D. Elliott & J. E. Stern (Eds.), *Research ethics: A reader* (pp. 286–316). Hanover, NH, USA: University Press of New England.

Taylor, J. R. (1997). *An introduction to error analysis: The study of uncertainty in physical measurements* (2nd ed.). Sausalito, CA, USA: University Science Books.

Thagard, P. R. (1978). Why astrology is a pseudo-science. In P. D. Asquith & I. Hacking (Eds.) *Proceedings of philosophy of science association* (Vol. 1, pp. 223–224). East Lansing, MI, USA: Philosophy of Science Association. Reprinted in Klemke, E. D., et al. (eds.). (1998). *Introductory readings in the philosophy of science* (3rd ed., pp. 66–75). Amherst, MA, USA: Prometheus Books.

Truscott, F. W., & Emory, F. L. (1951). *A philosophical essay on probabilities* (translation of the 6th French edition of Laplace, J-P. *Essai philosphique sur les probabilities*). New York, NY, USA: Dover.

Varela, F. J. (1999). *Ethical know-how: Action, wisdom, and cognition.* Stanford, CA, USA: Stanford University Press.

Ward, P. D., & Brownlee, D. (2004). *Rare earth: Why complex life is uncommon in the universe.* New York, NY, USA: Copernicus Books.

Ware, M., & Mabe, M. (2009). *An overview of scientific and scholarly journal publishing Oxford.* UK: International Association of Scientific, Technical and Medical Publishers.

Warrell, D. A., et al. (Eds.). (2005). *Oxford textbook of medicine* (4th ed.). Oxford, UK: Oxford University Press.

Whitehead, J. (2004). Stopping clinical trials by design. *Nature Reviews Drug Discovery, 3,* 973–977. Available at http://www.nature.com/nrd/journal/v3/n11/abs/nrd1553.html

Wigner, E. (1960). The unreasonable effectiveness of mathematics in the natural sciences. *Communications in Pure and Applied Mathematics, 13*(1), 1–14. Also available free of charge at: www.dartmouth.edu/ ~ matc/MathDrama/reading/Wigner.html

Wills, E. (2009). Spirituality and subjective well-being: evidences for a new domain in the personal well-being index. *Journal of Happiness Studies, 10,* 49–65.

Zadek, S., Pruzan, P., & Evans, R. (1997). *Building corporate accountability: Emerging practices in social and ethical accounting and auditing.* London, UK: Earthscan.

Ziman, J. (1968). What is science? In *Public Knowledge* (pp. 5–27). Cambridge, UK: Cambridge University Press. Reprinted in Klemke, E. D., et al. (eds.). (1998). *Introductory readings in the philosophy of science* (3rd ed., pp. 48–53). Amherst, MA, USA: Prometheus Books.

Index

© Springer International Publishing Switzerland 2016
P. Pruzan, *Research Methodology*,
DOI 10.1007/978-3-319-27167-5

Printed in the United States
By Bookmasters